Climate Change in South Asia

This volume studies the challenges of climate change in South Asia and examines the role of the South Asian Association for Regional Cooperation (SAARC) in addressing them. It highlights the dangers posed by climate change in South Asia and underlines the need to strengthen and intensify regional cooperation to preserve, protect, and manage the diverse and fragile ecosystems of the region. The book examines policies and initiatives of the SAARC in tackling these issues and also analyzes their implementation by member countries.

Comprehensive and topical, this volume will be useful for scholars and researchers of South Asian Studies, environmental studies, climate change studies, public policy and governance, development studies, international relations, regional cooperation, and political studies. It will also be of importance to policymakers and NGOs working in this field.

Baniateilang Majaw is Assistant Professor at the Department of Political Science, Seng Khasi College, Shillong, India. He is the recipient of the Junior Research Fellowship for International and Area Studies by the University Grant Commission, New Delhi, India. He is the author of *Regional Politics in Meghalaya* (2015) and *Struggle for Hill State: An Unfulfilled Dream* (2018).

Climate Change in South Asia
Politics, Policies and the SAARC

Baniateilang Majaw

Routledge
Taylor & Francis Group

LONDON AND NEW YORK

First published 2020
by Routledge
2 Park Square, Milton Park, Abingdon, Oxon OX14 4RN

and by Routledge
52 Vanderbilt Avenue, New York, NY 10017

Routledge is an imprint of the Taylor & Francis Group, an informa business

© 2020 Baniateilang Majaw

British Library Cataloguing-in-Publication Data
A catalogue record for this book is available from the British Library

Library of Congress Cataloging-in-Publication Data
Names: Majaw, Baniateilang, author.
Title: Climate change in South Asia : politics, policies and the
    SAARC / Baniateilang Majaw.
Description: Abingdon, Oxon ; New York : Routledge, 2020. |
    Includes bibliographical references and index.
Identifiers: LCCN 2020008423 (print) | LCCN 2020008424 (ebook)
Subjects: LCSH: Climatic changes—South Asia. | South Asian
    Association for Regional Cooperation | Environmental change—
    South Asia.
Classification: LCC QC903.2.S64 M35 2020 (print) | LCC
    QC903.2.S64 (ebook) | DDC 363.738/740954—dc23
LC record available at https://lccn.loc.gov/2020008423
LC ebook record available at https://lccn.loc.gov/2020008424

ISBN: 978-1-138-36753-1 (hbk)
ISBN: 978-0-367-51416-7 (pbk)
ISBN: 978-1-003-05376-7 (ebk)

Typeset in Sabon
by Apex CoVantage, LLC

Dedicated to my late Mother, Mrs S. Majaw. Her inspiring words during my college days always made me feel safe and fulfilled throughout my difficult journey, till the completion of my higher studies.

This book is totally dedicated to my Mother.

Contents

Tables

Preface and acknowledgements

The challenges of climate change are highly demanding, interconnected, and increasingly viewed as the foremost problem of the 21st century. South Asia, the region where I dwell, is home to millions of the world's poorest and hungriest people. This region which consists of eight states of different sizes and capabilities has regularly been affected by climate change–related disasters for decades. Droughts, floods, tropical cyclones, storms, earthquakes, landslides, erosion of soil, heat waves are some of the common disasters which create misery for millions of people of South Asia. These disasters have adversely affected agricultural production, water and sanitation, infrastructure, and lives of the people. Much scientific research and evidence has projected that these climate change–related disasters will increase the risk of food security and human security, exacerbate water shortages, increase flood-related coastal damage, stunt economic growth as well as produce infrastructure damage, causing greater misery to the people of South Asia. As a result, this modest study *Climate Change in South Asia: Politics, Policy and the SAARC* developed in my mind, which I subsequently pursued in my doctoral research.

At the very outset, I wish to put on record my deep sense of gratitude to my supervisor, Dr Munmun Majumdar, for her invaluable and tireless guidance, encouragement, and constructive suggestions throughout the entire course of my work, without which I would not have completed this work. I offer a big "Thank you" to her for giving me an opportunity to work on climate change, which is one of the most contextual and relevant subjects in South Asia. I also express my thanks to North-Eastern Hill University (NEHU) Shillong, to all the teaching and non-teaching staff, my fellow research scholars and friends in the Department of Political Science. My gratitude goes to my fellow scholars for their help and support, especially Dr Baskhem Kshiar, Dr Morten Marbaniang, Dr Joplin Hek, and Dr Haokam Vaiphei (Thomas) who have lent their helping hands in the initial years of this project.

I wish to record my grateful appreciation and the scholarly advice, suggestions, encouragement, and motivation received from the professional researchers of the Institute for Defence Studies and Analyses (IDSA) New Delhi. I convey my gratitude to IDSA Library, New Delhi; the Maulana Abul

Kalam Azad Institute of Asia Pacific Studies (MAKAIAS) Library, Kolkata; Tata Energy and Research Institute (TERI), New Delhi; Jawaharlal Nehru University Library (JNU); North-Eastern Hill University (NEHU) Library; North Eastern Council (NEC) Library; and the Indian Council of Social Science Research – North-Eastern Region Centre (ICSSR-NERC) Shillong.

I am also grateful to the University Grant Commission (UGC) for awarding me the Junior Research Fellowship (JRF), through which I could obtain financial support which helped me throughout my research. I also thank the UGC for awarding me financial assistance under the study grant which enabled me to go to the SAARC Secretariat Library for the purposes of data collection or consulting archival materials in connection with its research work. I extend my thanks to the staff of the SAARC library as well as of the International Centre for Integrated Mountain Development (ICIMOD) Library for giving me valuable materials related to Himalaya glaciers which I have referred to in this manuscript. Further, I also extend my special thanks to Sir Ugyen Samdrup, the personal assistant at SAARC Secretariat (Environment Division), and Madam Kalpana Tamrakar, librarian at SAARC Library, who provided me with all the SAARC documents on climate change and other environmental disasters.

Further, it would be a betrayal if I don't acknowledge my gratitude towards Japfu Halls of Residence, a hostel where I spent crucial years of my life in writing this manuscript. The tears I shed and the memorable moments that I have shared in this place will accompany me forever.

This book will be useful for post-graduate-level students, researchers, climate policy scientists, policymakers (particularly from the South Asian region), research institutions, and international and UN organizations working at SAARC levels.

Baniateilang Majaw

Kutmadan, Sohra (Cherrapunjee)
Meghalaya, India

Acronyms

ACCSAP	Afghanistan's Climate Change Strategy and Action Plan
ADB	Asian Development Bank
ADC	Austrian Development Cooperation
ADMP	Agricultural Development Master Plan
ADPC	Asian Disaster Preparedness Center
AFSANA	Afghanistan Food Security and Nutrition Agenda
AMRUT	Atal Mission for Rejuvenation and Urban Transformation
ANDS	Afghanistan National Development Strategy
AOSIS	Alliance of Small Island States
AP	Agriculture Policy
AR4	Fourth Assessment Report
AR5	Fifth Assessment Report
BASIC	Brazil, South Africa, India and China
BCCPSP	Building Climate-Resilient Communities through Private Sector Participation
BCCRF	Bangladesh Climate Change Resilience Fund
BCCSAP	Bangladesh Climate Change Strategy and Action Plan
BCRWME	Building Climate Resilience of Watersheds in Mountain Eco-Regions
BINP	Bangladesh Integrated National Plan
BISP	Benazir Income Support Programme
BMZ	Federal Ministry for Economic Cooperation and Development
BNNC	Bangladesh National Nutrition Council
BRCH	Building Resilience to Climate-Related Hazards
CAPFs	Central Armed Police Forces
CCCs	Climate Change Cells
CCFF	Climate Change Financing Framework
CCPI	Climate Change Performance Index
CCS	Climate Change Secretariat
CCTF	Climate Change Trust Fund
CDM	Clean Development Mechanism
CDMP	Comprehensive Disaster Management Programme

CNG	Compressed Natural Gas
COPs	Conference of Parties
CPEIR	Climate Public Expenditure and Institutional Review
CRI	Climate Risk Index
DANIDA	Danish International Development Agency
DFID	Department for International Development
DMC	Disaster Management Centre
DNA	Designated National Authority
DRR-CCA	Disaster Risk Reduction – Climate Change Adaptation
ECRES	Enhancing the Climate Resilience of Endangered Species
EIA	Environmental Impact Assessment
EPA	Environmental Protection Agency
EPI	Environmental Performance Index
EPSL	Energy Policy of Sri Lanka
EU	European Union
FAO	Food and Agriculture Organization
FCB	Food Corporation of Bhutan
FFC	Federal Flood Commission
FFP	Food for Peace
FY	Fiscal Year
FYP	Five Year Plan
G8	Group of Eight
GCF	Green Climate Fund
GCRI	Global Climate Risk Index
GCISC	Global Change Impact Studies Centre
GDP	Gross Domestic Product
GEF/UNDP	Global Environment Facility/United Nations Development Fund
GEF	Global Environmental Facility
GHG	Greenhouse Gas
GHI	Global Hunger Index
GIM	Green India Mission
GLOFs	Glacial Lake Outburst Floods
GNH	Gross National Happiness
GNI	Gross National Income
GoB	Government of Bangladesh
GW	Giga Watts
HDI	Human Development Index
HFA	Hyogo Framework for Action
HKH	Hindu Kush-Himalayas
HPNSDP	Health, Population, Nutrition, Sector Development Programme
ICDS	Integrated Child Development Services
ICIMOD	International Centre for Integrated Mountain Development

IDMC	Internal Displacement Monitoring Centre
IFAD	International Fund for Agricultural Development
IFIs	International Financial Institutions
IFPRI	International Food Policy Research Institute
IGEG.CC	Inter-Governmental Expert Group on Climate Change
IHK	Indian Held Kashmir
INC	Initial National Communication
INDC	Intended Nationally Determined Contributions
IPCC	Inter-governmental Panel on Climate Change
IREDA	Indian Renewable Energy Development Agency
IUCN	International Union for Conservation of Nature
JICA	Japan International Cooperation Agency
KFW	Kreditanstalt für Wiederaufbau
LAPA	Local Adaptation Plans for Action
LDCF	Least Developed Countries Fund
LDCs	Least Developed Countries
LMDC	Like Minded Developing Countries
LULUCF	Land Use, Land Use Change and Forestry
M/DM&HR	Ministry for Disaster Managements and Human Rights
MAIL	Ministry of Agriculture, Irrigation and Livestock
MCCRMD	Mainstreaming Climate Change Risk Management in Development
MDGs	Millennium Development Goals
MDMS	Mid-Day Meal Scheme
MEAs	Multi-Lateral Environmental Agreements
MENR	Ministry of Environment and Natural Resources
MNFSR	Ministry of National Food Security and Research
MoAC	Ministry of Agriculture and Cooperatives
MoEc	Ministry of Economy
MoE	Ministry of Education
MoEF	Ministry of Environment and Forest
MoEST	Ministry of Environment, Science and Technology
MoFDM	Ministry of Food and Disaster Management
MoHFW	Ministry of Health and Family Welfare
MoPH	Ministry of Public Health
MsAPN	Multi-sector Action Plan for Nutrition
MSNP	National Multi-Sector Nutrition Plan
Mt	Million Tonnes
NACCC	National Advisory Committee on Climate Change
NAPA	National Adaptation Programme of Action
NAPCC	National Action Plan on Climate Change
NAPCCI	National Adaptation Plan for Climate Change Impacts
NATO	North Atlantic Treaty Organization
NCCAS	National Climate Change Adaptation Strategy for Sri Lanka
NCCC	National Climate Change Committee

NCCP	National Climate Change Policy
NCDM	National Council for Disaster Management
NCS	National Conservation Strategy
NCSA	National Capacity Needs Self Assessment
NDB	New Development Bank
NDMA	National Disaster Management Authority
NDRF	National Disaster Response Force
NDRRM	Natural Disaster Rapid Response Mechanism
NEA	National Environment Act
NEC	National Environment Commission
NEHAP	National Environmental Health Action Plan
NEPA	National Environmental Protection Act
NEPA	National Environmental Protection Agency
NES	National Environment Strategy
NFNP	National Food and Nutrition Policy
NFP	National Focal Point
NFP	National Food Policy
NFP	National Forestry Policy
NGOs	Non-Governmental Organizations
NGT	National Green Tribunal Act
NICRA	National Climate Resilient Agriculture Programme
NMEEE	National Mission on Enhanced Energy Efficiency
NMSH	National Mission on Sustainable Habitat
NNAF	National Nutrition Action Framework
NNFSC	National Nutrition and Food Security Coordination Committee
NNFSS	National Nutrition and Food Security Secretariat
NNP	National Nutrition Policy
NPAN	National Plan of Action for Nutrition
NPWC	National Policy for Wildlife Conservation
NREGA	National Rural Employment Guarantee Act
NSDS	National Sustainable Development Strategies
NSM	National Solar Mission
NSSD	National Strategy for Sustainable Development
NTP	National Transport Policy
NWM	National Water Mission
NWMP	National Water Management Plan
NWP	National Water Policy
ODF	Open Defecation Free
PARC	Pakistan Agricultural Research Council
PAT	Perform, Achieve and Trade
PBM	Pakistan Bait-ul-Mal
PCRWR	Pakistan Council of Research in Water Resources
PDS	Public Distribution System
PEPA	Pakistan Environmental Protection Act

PFM	Public Financial Management
PIL	Public Interest Litigation
PMCCC	Prime Minister's Council on Climate Change
PMD	Pakistan Meteorological Department
PNB	Punjab National Bank
PPP	Purchasing Power Parity
PRSP	Poverty Reduction Strategy Paper
REG	Regional Environmental Governance
RIPA	Regional Integrated Programme of Action
RVCC	Reducing Vulnerability to Climate Change
SAARC	South Asian Association for Regional Cooperation
SAC	SAARC Agricultural Centre
SACEP	South Asia Cooperative Environment Programme
SADKN	South Asia Disaster Knowledge Network
SAEO	South Asia Environment Outlook
SBI	State Bank of India
SCZMC	SAARC Coastal Zone Management Center
SDAN	Sustainable Development Agenda for Nepal
SDF	SAARC Development Fund
SDGs	Sustainable Development Goals
SDMC	SAARC Disaster Management Centre
SEA	Strategic Environmental Assessment
SEDMC	SAARC Environment and Disaster Management Centre
SFC	SAARC Forestry Center
SGRs	Strategic Grain Reserves
SIDs	Small Island Developing States
SMP	School Meals Programme
SMRC	SAARC Meteorological Research Centre
SNAP	Strategic National Action Plan
SNC	Second National Communication
SPCR	Strategic Programme for Climate Resilience
STORM	Severe Thunderstorm Observations and Regional Modelling
SUN	Scaling Up Nutrition
SUVs	Sport Utility Vehicles
SVPs	Sector Vulnerability Profiles
TAP	Tourism Adaptation Project
TOR	Terms of Reference
TPDS	Targeted Public Distribution System
TPES	Total Primary Energy Supply
UK	United Kingdom
UN	United Nations
UNCBD	UNFCCC, the Convention on Biological Diversity
UNCCD	UNFCCC, the Convention to Combat Desertification
UNCD	United Nations and Commonwealth Department

UNCED	United Nations Conference on Environment and Development
UNDP/NDMP	UNDP National Disaster Management Project
UNDP	United Nations Development Programme
UNEP	United Nations Environment Programme
UNFCCC	United Nations Framework Convention on Climate Change
UNGA	UN General Assembly
UNICEF	United Nations Children's Fund
UNISDR	United Nations International Strategy on Disaster Reduction
US	United States
USAID	United States Agency for International Development
USSR	Union of Soviet Socialist Republics
WB	World Bank
WCED	World Commission on Environment and Development
WFP	World Food Programme
WHO	World Health Organization
WSS	Water Sector Strategy
WTO	Warsaw Treaty Organization
WWF	World Wide Fund for Nature

1 Introduction

The challenges of climate change are very complex, interrelated, and increasingly viewed as the primary problem faced by the people in the 21st century. People all over the world are facing the effects of climate change, regardless of their geographical location and socio-economic status (Solomon *et al.*, 2007; Philander, 2008; Edenhofer *et al.*, 2014). Climate change is already having significant and costly effects on the people, their economy, and the weather conditions. From 1998 to 2017, more than 526,000 people died worldwide and losses of US$3.47 trillion (in PPP) were incurred as a direct result of more than 11,500 extreme weather related events (Eckstein, Hutfils and Winges, 2018: 5). The year 2016 had a record-topping occurrence of natural disasters like floods, earthquakes, and hurricanes which left US$175 billion in damage (Riley, 2017). After the 2016 drought across Southern Africa, 17 million people were expected to require food assistance before the 2017 harvest; Chinese floods caused US$14 billion in damage; flooding and landslides in Sri Lanka displaced hundreds of thousands; and weather-related events displaced 19.2 million people, twice as many as conflict and violence in 2015 (Lawrence-Samuel, Jackson and Thanki, 2017: 124; WMO, 2016). In 2016, Bolivia endured its worst drought in 25 years forcing this country to declare a state of emergency because of water security (Cohen, 2016; Jemia, 2016) and around 175,000 farmers in Morocco lost their jobs due to drought (Middle East Monitor, 2016). Climate change has had extreme impacts in India where intense floods destroyed 774 villages, affected over 5.4 million people, causing US$3 billion in damages (Krishnakumar, 2018; Sangomla, 2018; Faizi, 2018), with 600 people killed; over 25 million were affected by flooding in India, Bangladesh, Nepal, and Myanmar in the 2019 flood (*India Today*, 2019a). The United Nations Environment Programme's Adaptation Gap Report 2016 warns of increasing impacts and resulting increases in global adaptation costs by 2030 or 2050 that will likely be much higher than currently expected: "two-to-three times higher than current global estimates by 2030, and potentially four-to-five times higher by 2050" (UNEP, 2016: xii). Thus, it is one of the most complex issues the world is facing today. Scientists and policymakers alike have dispelled many uncertainties about climate change, but what is called for now is a response from all countries of the world to make sensible policy decisions to adapt as well as to mitigate it. Their goal is to reduce human susceptibility to its harmful effects.

I Understanding climate change

The scientific evidence confirms climate change will have widespread effects on the environment, socio-economic and other related sectors which include water accessibility, agricultural production and food security, human health, terrestrial ecosystems, biodiversity and the coastal zones (Solomon *et al.*, 2007; Philander, 2008; Edenhofer *et al.*, 2014; Barnett and Adger, 2007: 639–655; Busby, 2007: 2; UNEP, 2016: xii). Collectively, these effects will pose numerous challenges to human security and prosperity all over the world. What exactly does climate change mean? Some scholars have used the term *climate change* to refer to all forms of climatic inconstancy, regardless of the statistical nature or physical causes (Mitchell *et al.*, 1966: 79). One source of information says climate change is a long-term shift in the statistics of the weather and acknowledges it is a normal part of the earth's natural variability, which is related to interactions among the atmosphere, ocean, and land, as well as changes in the amount of solar radiation reaching the earth (NOAA, 2007: 1). The Inter-governmental Panel on Climate Change (IPCC) defines it as the change in the state of the climate that can be recognized over time whether due to natural variability or as a result of human activity which persists for an extended period (Bruce, Lee and Haites, 1996; Philander, 2008: 210, 544). This is in line with the official definition by the United Nations Framework Convention on Climate Change (UNFCCC) that climate change is the change that can be attributed "directly or indirectly to human activity that alters the composition of the global atmosphere and which is in addition to natural climate variability observed over comparable time periods" (UNFCCC, 1992: 7).

The change of climate is more visible in recent times. Average temperatures around the world have risen by 0.75°C (1.4°F) over the last 100 years and about two-thirds of this increase has occurred since 1975 (Hansen *et al.*, 2010). In the past, the increase of the temperature of the earth was the result of natural causes but presently it is being caused by the accumulation of greenhouse gas (GHG) in the atmosphere, which is a result of anthropogenic activities (Le Treut *et al.*, 2007). Scholars (Edenhofer *et al.*, 2014) have accepted that anthropogenic activities are now greatly influencing global climate change processes. Anthropogenic GHG emissions since the pre-industrial era have driven large increases in the atmospheric concentrations of carbon dioxide (CO_2), methane (CH_4), and nitrous oxide (N_2O), thereby drastically enhancing the greenhouse effect, causing the earth's average temperature to rise (Pachauri and Meyer, 2014: 4). Total anthropogenic GHG emissions have continued to increase between 2000 and 2010, despite a growing number of climate change mitigation policies (Pachauri and Meyer,

2014: 4). This perpetual increase of GHG emissions is increasing the earth's temperature, causing the crisis that the world is facing today (Hansen and Sato, 2012; Shakun and Carlson, 2010: 1801–1816). In fact, the earth has warmed significantly. Fourteen out of the 15 warmest years in the modern meteorological record have occurred from 2000 to 2015 (Carrington, 2015). The record of 2014 as the hottest year ever recorded was broken by the year 2015, which in turn was broken by 2016, which was about 1.1°C above pre-industrial levels (Riley, 2017; Harris, Roach and Codur, 2017: 5). Global mean temperature in 2018 was colder than 2015, 2016, and 2017, but warmer than every previously observed year prior to 2015. But June 2019 was the warmest June since records began in 1850, and July 2019 was likely be both the warmest July on record and the single warmest month ever recorded (Hausfather, 2019). Experts agree the incessant rise of global heating to 2100 is a real possibility (Vince, 2019). A long series of scientific research and international studies has agreed, with more than 90 per cent certainty that climate change is due to the GHG produced by anthropogenic activities or human-induced interference with the climate system such as deforestation, burning of fossil fuels etc. (Houghton, 2010: 337–351; Denman *et al.*, 2007; Bousquet *et al.*, 2006: 439–443; Ramanathan and Feng, 2008: 14245–14250; Anderegg *et al.*, 2010: 12107–12109). These findings are recognized by the national science academies of all the major industrialized countries (IPCC, 2007; The Royal Society, 2014). Thus, scholars agree that the global climate change of the last few decades has been primarily attributable to the fossil-based industrial growth of the advanced countries (Magness, Lovecraft and Morton, 2011: 459; Vitousek *et al.*, 1997: 494–499). There is strong evidence from earth scientists who claim that where major climate change is occurring global warming has not always been as intense as it is today (Pickering and Lewis, 1997: 39–103; Hausfather, 2019; Vince, 2019).

There are other scientists who have argued changing climate is just a phenomenon which has been taking place accurately from before, it is continuing now, and it will happen in the future too (Cox, 2007; Calvin, 2002). So, it is a global phenomenon. It is periodic or quasi-periodic and occurs on all temporal scales through millions of years (Pickering and Lewis, 1997: 39–103). Due to the conflicting views on the causes, it leads to a view that man-made causes may coincide with a cyclical period of climate change, nevertheless leading to the worsening of problems being faced by people in most parts of the world today, especially those in the poor countries (Alley, 2000). So, it is generally stated such change must be due to natural changeability in addition to the outcome of human activities ever since humans sometimes endeavour to control climate deliberately for their own personal benefit by disturbing the environment.

Rising temperatures may not sound like much; after all, it is less than a typical temperature change between night and day. But it is assumed climate change–related effects are threatening to bring huge repercussions to the earth (Stern, 2006; Fairclough, 1991). The rise in global temperature is harming the earth in several ways, which include (i) the likelihood to affect the global water cycle by making the arid and semi-arid areas drier, causing water shortages and an intense amount of distress to millions of people in dry regions which are degrading into deserts (DARA, 2010; Allison, 2009); (ii) perennial ice cover in the Arctic, an area equivalent to the size of Norway, Denmark, and Sweden combined, has vanished in the last 30 years; (iii) Antarctica has been losing 100 km^3 (24 mi^3) of ice per year since 2002, leading to sea level rise (Comiso and Hall, 2014: 389–409; McMillan *et al.*, 2014: 3899–3905; Mengel *et al.*, 2016: 2597–2602; Hansen and Sato, 2012; Shakun and Carlson, 2010: 1801–1816); (iv) global sea level has risen by 21 cm (8 in) since 1880 and the rate of rise is accelerating which is now at a pace that has not been seen before – as of 2017, global mean sea level was 77 mm (3 in) above the 1993 average, which is the highest annual average in the satellite record since 1993 to date (Lindsey, 2018; Church and White, 2011: 585–602; Hunter, 2010: 331–350); and (v) it increases the rate of strong cyclones, storm surges, etc. which is subsequently leading to coastal erosion and sea water intrusion (Yu and Chiu, 2011; Emanuel, 2007; Elsner, Kossin and Jagger, 2008). Further, according to the conservative scientific consensus, a 1.5°C increase in global temperatures due to anthropogenic influence will generate a global sea level rise of between 1.7 and 3.2 ft by 2100, even when all the countries of the world will collectively manage to keep global temperatures from rising to 2°C, by 2050 (Muggah, 2019). In 2012, the National Oceanic and Atmospheric Administration estimated that sea level rise would increase by 2 m (6.6 ft) depending on anthropogenic activities and the impacts of warming air and the melting of ice bodies (Pfeffer, Harper and O'Neel, 2008; NOAA, 2012). If any of these hypotheses are correct, the impacts of climate change are likely (Bruce, Lee and Haites, 1996: 6) to impose costs on future generations, particularly in regions where damages occur, including those areas with low GHG emissions. These impacts will be tremendous and may necessitate a drastic shift in the way we live.

II Drastic impacts of climate change

The impacts of climate change will not be uniformly distributed across the globe, with developing countries more likely to experience its negative effects. About 80 per cent of the damages cause by it may be felt in the developing countries (Mendelsohn, Munasinghe and Niggol, 2005: 553–569). By 2050, between 350 million and 600 million people living in the equatorial regions, particularly in Africa, are projected to experience water stress (Niang *et al.*, 2014: 1199–1265). So, food production will be affected and becomes a major challenge due to droughts and shifts in rainfall (NRC, 2010; Niang *et al.*, 2014: 1199–1265). Loss of snowmelt from the Himalayas will reduce the flow of water into the Indus, Ganges, and Brahmaputra basins leading to water shortages in Pakistan and India, threatening agricultural production (Adams *et al.*, 2013; MoE, 2009: 2). South Asian cities like Kolkata and Mumbai will face increased flooding (Dasgupta *et al.*, 2009: 22–24), warming temperatures, and intense cyclones. Meanwhile in South East Asia, Vietnam's Mekong Delta, which produces most of the country's rice, is susceptible to sea level rise (Hijioka *et al.*, 2014: 1327–1370). By the mid-21st century, climate change can increase crop yield up to 20 per cent in East and South East Asia, while decreasing yield up to 30 per cent in Central and South Asia (Hijioka *et al.*, 2014: 1327–1370). Sea level rise and more severe storms and coastal flooding will continue to affect coastal areas (Wong *et al.*, 2014: 361–409). Coastal development and population growth in areas such as Cairns and Southeast Queensland (Australia) and Northland to the Bay of Plenty (New Zealand) will place more people at risk and put infrastructure at risk (Reisinger *et al.*, 2014: 35–94). On the contrary, areas that are already affected by drought, such as Australia and eastern parts of New Zealand, are likely to experience reductions in water available for irrigation and forestry production (USGCRP, 2016: 312; Reisinger *et al.*, 2014: 35–94; NRC, 2010).

Being a higher-latitude country, the United Kingdom (UK) may fare better than many nations as global warming progresses. But that is not to say the UK will escape the costs of climate change, given its large coastline where rising sea levels can be an obvious threat (Stern, 2006). Wide-ranging impacts of climate change have been documented in Europe such as in glaciers melting, sea level rise, longer growing seasons, species range shifts, and heat wave–related health impacts (Kovats *et al.*, 2014: 1267–1326). In Southern Europe, higher temperatures and drought may reduce water availability, hydropower potential, summer tourism, and crop productivity, hampering economic activity more than other European regions (Kovats *et al.*, 2014: 1267–1326). In Central and Eastern Europe, summer precipitation is projected to decrease, causing higher water stress, forest productivity reduction, and increasing incidence of wild fires (Kovats *et al.*, 2014: 1267–1326). In Northern Europe, it is expected to bring mixed effects, from the increase of crop yields and increased forest growth to frequent winter floods, endangered ecosystems, where lost is likely to outweigh benefits (Kovats *et al.*, 2014: 1267–1326).

In the drier areas of Central and South America climate change will likely worsen drought, leading to desertification of agricultural land. Dearth of water will have adverse consequences on food supply (Magrin *et al.*, 2014: 1499–1566). In temperate zones, soybean yields are projected to increase (Magrin *et al.*, 2014: 1499–1566). Sea level rise is projected to increase risk of flooding, displacement of people, salinization of drinking water resources, and coastal erosion in low-lying areas. These risks threaten fish stocks, recreation, and tourism (Magrin *et al.*, 2014: 1499–1566). In North America warming in the western mountains will reduce snowpack, boost winter flooding and diminish summer flows, exacerbating competition for over-allocated water resources (Romero-Lankao *et al.*, 2014: 1439–1498).

In the Polar region, climate changes may reduce the thickness and extent of glaciers and ice sheets (Larsen *et al.*, 2014: 1567–1612). Small island nations, wherever they are located (in the tropics or higher latitudes), are highly vulnerable to changes in sea level (Mimura *et al.*, 2007: 687–716). Sea level rise is projected to worsen inundation, storm surge, erosion, and other coastal hazards. These impacts would threaten vital infrastructure, settlements, and facilities that support the livelihood of many island communities (Larsen *et al.*, 2014: 1567–1612; Mimura *et al.*, 2007: 687–716). By 2050, between 665,000 and 1.7 million people in the Pacific region are expected to be forced to migrate due to rising sea levels, including the entire populations of islands such as Fiji, the Marshall Islands, and Tuvalu (Milman, 2015; Wyett, 2013). By the end of this century, the Maldives is projected to lose around 77 per cent of its land area (Tol, 2007: 741–753; Woodworth, 2005: 1–18).

Climate change will have different effects in different parts of the world. Some places will experience more warm temperatures than others. There are some regions which will get more rainfall, while others will experience more frequent droughts (Clark, 2011). Some low-lying coastal regions will be subject to more frequent flooding or even permanent inundation (de Almeida and Mostafavi, 2016: 6–8; Nicholls *et al.*, 2007: 315–356). Large areas of Bangladesh, the Netherlands, along with the US state of Florida and the city of New Orleans, are at great risk of even slight increases in sea level (Dasgupta *et al.*, 2009: 22–24; Nicholls *et al.*, 2007: 315–356). By 2100, humans will witness the disappearance of most low-latitude glaciers, including two-thirds of the so-called third pole of the Hindu Kush-Karakoram-Himalayan Mountains and Tibetan plateau, which feeds many of Asia's important rivers (Shannon *et al.*, 2019: 325–350). So, many rivers in South Asia will face floods more often. A wide equatorial belt of high humidity will cause intolerable heat stress across tropical Asia, Africa, Australia, and the Americas (Vince, 2019). All these consequences will have social, economic, and environmental effects, such as risk of disasters, water stress, food security, health risk, exploitation of natural resources, social and economic conflicts and migrations (Raleigh, Jordan and Salehyan, 2008; Beniston, 2010: 557–568). However, the poor who are economically reliant on climate-sensitive sectors such as agriculture and fisheries are particularly vulnerable to these changes in climate conditions. Countries with limited human, institutional, and financial capacity will be in danger as they are the least able to adapt to more frequent and more severe storms, droughts, and floods (Neumayer and Pluemper, 2007: 551–566: Pettengell, 2010: 10).

Climate change may not directly result in violence. Rather, there is growing consensus that it will instead act as a "threat multiplier," aggravating existing challenges and sources of tension such as weak governance, poverty, historical grievances, and ethnic differences (Ahmed, 2011: 335–355; Keil, 2014: 162). With climate change making many fragile parts of the world hotter, drier, and less predictable, its allied impacts will likely have severe consequences like mass migrations of the people, particularly in tropical regions, putting at risk the infrastructure, agriculture, human health; reduce daily labour supply; strain water resources; create rising competition for scarce natural resources, overwhelming state institutions by placing additional stress on social, economic, and natural systems and the environment (Brown, 2008; UNEP, 2009: 13; McLeman and Smit, 2004,2006; Laczko and Collett, 2005; Hazarika, 1993). First, sea level rise and an increase in the frequency and intensity of extreme weather events pose several challenges to physical infrastructure especially in coastal areas, from the United States (Burkett and Davidson, 2012) to Bangladesh. Second, due to rising sea water temperatures the oceans are turning acidic at what may be the fastest pace in 300 million years, with potential severe consequences for marine ecosystems, asserts a report in *Science* magazine (Hönisch *et al.*, 2012: 1058–1062; Zabarenko, 2012). Among the first victims of ocean warming and acidification are coral reefs and coral degradation, which will negatively impact island communities and livelihoods, including the tourism industry (Mia, 2019). Third, climate change and climate variability will increase the existing poverty and exacerbate inequalities (Hallegatte *et al.*, 2016; UNDP, 2011). Fourth, water shortage as a result of heat waves and more intense droughts may generate tensions between the states of water-scarce regions like the Middle East, North Africa, and Southern Europe (Hoerling *et al.*, 2012: 2146–2161). Fifth, climate change is expected to cause significant health threats such as malnutrition and contribution to the outbreaks of infectious diseases like malaria and diarrhoea, because of poor water and food quality and extreme weather events (Hallegatte *et al.*, 2016; Kellet, 2014). The population of the low-income countries, particularly the poor and weaker groups, is facing these health risks (Smith *et al.*, 2014; Kellet, 2014). In Africa, children and pregnant women are particularly vulnerable to climate-related health risks. So, it will have an effect on the health of the poor workers (Niang *et al.*, 2014: 1199–1265). Sixth, there is growing evidence climate change can have a major influence on food security – where it may lead to 170 million more undernourished people by 2080 (Sternberg, 2012: 519–524; Otto *et al.*, 2012; Rahmstorf and Coumou, 2011; Niang *et al.*, 2014: 1199–1265). Thus, climate change will deteriorate the living conditions of farmers, fishermen, and forest-dependent people who were already in need of food (Edame *et al.*, 2011: 205–223). Seventh, it can drastically alter plant phenology since temperature is a major determinant of the rate of plant development (Cleland *et al.*, 2007: 357–365). Eighth, sudden-onset natural disasters such as flooding, tsunamis, and hurricanes, and slow-onset changes such as water scarcity or soil degradation, to an environment will result in the permanent displacement of people and, consequently, increase internal and international migration (Reuveny, 2007: 656–673; Black *et al.*, 2013). Finally, climate change is suspected as a prime cause of the increase in wildfires (Harris, Roach and Codur, 2017: 12; Richtel and Santos, 2016; Austen, 2016).

Climate change–related natural disasters have become primary causes of forced migration with an average of nearly 27 million people having been displaced annually by its impacts (IDMC, 2014). Many, if not most, of these impacts are likely to be significantly affecting poor rural communities who are the least responsible for causing them as they have limited capacity to cope with the consequences due to the lack of adequate infrastructure, public services, and social protection systems (UNRISD, 2012, 2016). The massive damage associated with climate change–related hazards is unprecedented – massive loss of human life, adequate standard of living, education, food and water availability, health, infrastructure damage, total disruption of daily activities, human displacement, human rights etc. (Humphreys, 2008; Jodoin and Lofts, 2013). For various reasons, this issue is a basic factor which needs to be recognized in understanding how it fits into the wider picture (Burroughs, 2007: 2; Giddens, 1994). Eventually, it has rapidly moved up to the political agenda as the scientific evidence has become better known in recent times (Stafford-Smith *et al.*, 2016: 1–9; Adams *et al.*, 2016: 11; FitzRoy and Papyrakis, 2010: 1). Without policy intervention, perpetual anthropogenic activities in a business-as-usual scenario will be disastrous, as global temperatures are expected to continue to rise. Therefore, rather than ignoring its multifaceted impacts, it is now recognized as an international environmental issue or a threat multiplier because its disasters may hamper countries in the achievement of sustainable development goals (SDGs). That is why climate change has intensely worried both the developed and the developing countries, particularly when they realized the changes are likely to boost the power of the strongest tropical cyclones, heat waves, and other climate change–related disasters (Elsner, Kossin and Jagger, 2008; Lee, 2009: 10). As it is a global problem encompassing an incredibly diverse mix of human society, the United Nations (UN) is looking at climate change as an issue posing serious challenges to the world community to achieve its ambitious targets set by the UN General Assembly in 2015 known as the SDGs –a target committed to by all the member states of the UN.

Known as Agenda 2030 for Sustainable Development, the SDGs carry on the work begun by the Millennium Development Goals (MDGs), which initiated a global campaign from 2000–2015 to end poverty in its various dimensions (Woodbridge, 2015: 1). In September 2000, the world leaders adopted the UN Millennium Declaration which provided the basis for the pursuit of the MDGs. A global consensus was successfully forged around the importance of poverty reduction and human development. Thenceforth, the world community tried to uplift a large segment of the poor and vulnerable. The world reached the poverty target five years before the 2015 deadline. In developing regions, the amount of people living on less than $1.25 a day fell from 47 per cent in 1990 to 22 per cent in 2010, and about 700 million fewer people lived in conditions of extreme poverty in 2010 compared with 1990. However, results fell short of international expectations and of the global targets set to be reached by the 2015 deadline (UN, 2013: v–vi). The progress has been uneven, particularly in Africa, least developed countries (LDCs), landlocked developing countries, and Small Island developing States, and some of the MDGs remain off-track, in particular those related to maternal, newborn, and child health and to reproductive health (Kutesa, 2015: 5).

In September 2015, world Heads of State or Government unanimously adopted the SDGs, the successor framework to the MDGs which were agreed by governments in 2000 and which came to an end in 2015 (Stafford-Smith *et al.*, 2016: 1–9). The Agenda 2030 for SDGs adopted by all UN Member States in 2015 provides a shared blueprint for peace and prosperity for people and the planet, now and into the future. It consists of 17 goals and 169 targets, spanning the three dimensions of economic, social, and environmental development (Nilsson, Griggs and Visbeck, 2016: 320–322). The Agenda 2030 is universal, not just because the SDGs are global in scope, but because all countries have to do something to achieve them as no country can consider itself to be sustainably developed and having already done its part to meet the SDGs (Adams *et al.*, 2016: 11). Hence, all national governments as well as other stakeholders, including local governments, business, and the civil society, are expected to identify, implement, and report on specific actions that lead to their achievement. The governments have to translate these goals and targets into the national policies, to resource and implement these policies and to measure their implementation. The civil society organizations are expected to play an important role in popularizing SDGs as well as take on the role for monitoring the implementation of the SDGs (Abhiyan, 2017: v).

The Agenda 2030 offers the opportunities and challenges especially for developing countries as the cost of its implementation will be high though it can bring long-term benefits (Adams *et al.*, 2016: 11). Achieving success in a framework as huge as the SDGs requires a massive amount of financial investments which will be difficult for the poor countries. But, if national governments do not take urgent action, the climate crisis will carry on its devastation around the globe and affect millions of people, particularly those living in the lowest-income countries; will widen the inequality gap; exacerbate poverty for people and countries that have done next to nothing to cause the climate crisis (Lawrence-Samuel, Jackson and Thanki, 2017: 124). Therefore, amidst the 17 goals which form the centrepiece of the Agenda 2030, there is SDG 13 which stresses on tackling an urgent action to combat climate change and its impacts. The SDG 13 makes reference to the lead role of the UNFCCC in the domain of negotiating international climate policy. As climate change is a global problem encompassing an incredibly diverse mix of human society, it is taken as a significant subject for the study.

III Climate change in the South Asian region

The impacts of climate change will be felt worldwide. Some parts of the earth will be more vulnerable to it than the others. It is stated earlier the developing countries are more likely to experience the impacts of climate change. So, a brief discussion of the vulnerability of South Asia to the impacts of climate change is necessary. Climate change–related disasters will be relentless in South Asia – defined as Afghanistan, Bangladesh, Bhutan, India. the Maldives, Nepal, Pakistan, and Sri Lanka because the ecosystems of this region vary from desert to the Himalayan mountainous range and its glaciers to the plains, delta, and the tropical forest as well as the low-lying coastal areas (Adams *et al.*, 2013: 123). With around 1.8 billion people, this region covers almost 5.1 million km^2 (1.9 million mi^2) of its total land area and about 11.51 per cent of the Asian continent. It is a home to millions of the world's poorest and hungriest people who have for decades been regularly affected by ecological disasters (Sawe, 2018; Pandey, 2015; Islam and von Braun, 2008: 9).

In this region climate change is the mother of all ecological hazards and extreme weather events (WHO, 2016: 6), whose effects have been profound, regular, and enormous. Droughts, floods, tropical cyclones, storm surges, earthquakes, landslides, erosion of soil, heat waves, erratic rainfall are some of the common disasters which put miseries to millions of people of South Asia (Hasemann *et al.*, 2014: 7; Islam, Hove and Parry, 2011). These disasters have adversely affected agricultural production, water and sanitation, infrastructure, physical security, and lives of the people (Islam, Hove and Parry, 2011). Scientific evidence has predicted climate change–related disasters will increase the risk of food security, human security; exacerbate water shortages, sea water intrusion; increase flood-related coastal damage; stifle economic growth as well as cause infrastructure damage etc. (Edenhofer *et al.*, 2014; Ortiz *et al.*, 2008; Vinke *et al.*, 2017; Philander, 2008). Together these disasters will cause miseries to the people of South Asia for the reason that the majority of the people here are low- or lower-middle-income communities who are now struggling to support their daily needs. The Fourth Assessment Report (AR4) and Fifth Assessment Report (AR5) of the IPCC published in 2007 and 2014 respectively (Solomon *et al.*, 2007; Denman *et al.*, 2007; IPCC, 2007; Le Treut *et al.*, 2007; Philander, 2008; Edenhofer *et al.*, 2014) have highlighted that climate change would bring about some of the challenges to the countries of South Asia where the poor people will be hard hit by its consequences.

The Himalayan glaciers which hold water reserves that flow into rivers such as the Indus, Ganga, Brahmaputra, and Meghna are critical to millions in Pakistan, Nepal, Bhutan, India, and Bangladesh (Jimenez Cisneros *et al.*, 2014: 229–269). Climate change can have devastating effects on these Himalayan glaciers. Rapid melting of these glaciers due to rising temperatures can trigger havoc and have long-lasting effects on the region's water supply (Vinke *et al.*, 2017). Initially, as the mountain glaciers melt, lakes of the region will have more water and natural dams will be formed. Lakes caused by melting glaciers of the Himalayas which are primarily attributed to climate change can burst their banks, leading to glacial lake outburst floods (GLOFs). The GLOFs are posing catastrophic threats particularly to the mountainous communities in Pakistan, Nepal, Bhutan, and India who are facing the danger of flash floods (Chugh, 2019; Cruz *et al.*, 2007: 493; ICIMOD, 2011). But over time, as the water level would dwindle, water supply in the Indus, Ganga, and Brahmaputra could decrease by 8.4 per cent, 17.6 per cent, and 19.6 per cent, respectively (Hasemann *et al.*, 2014: 5; Chugh, 2019). As the source water decreases, the downstream populations will see water shortages or even drought (Chugh, 2019). The reduction of water flow will later increase the tensions over water supply that already exist in the region, mainly between India and Bangladesh, as well as between India, Pakistan, and Nepal (Nabeel, 2019; Nishat *et al.*, 2013; Renner, 2011; Hasemann *et al.*, 2014: 5).

Glacial melting is not only leading to periodic floods and water shortages, it will also result in rising sea levels where the effects are already felt in South Asia (Huq and Ayers, 2008; Pachauri and Reisinger, 2007b). Sea level rise in this region is expected to continue at a rate between 1 and 3 mm per year, which is slightly more than the global average (Shivakumar and Stefanski, 2011: 27). The low-lying coastal areas of Bangladesh in particular are already threatened by a 1 m sea level rise which could inundate its flat vicinity (World Bank, 2012a; Nishat *et al.*, 2013; Dasgupta *et al.*, 2007). Almost 25 per cent of Bangladesh's coastal population are under threat of tropical cyclones. Presently, saline intrusion into the agricultural lands and into the drinking water (Khan *et al.*, 2011: 18–33) reaches at 100 km upland from the Bay of Bengal. Large coastlines of India, Sri Lanka, Pakistan, and the Maldives are threatened by sea level rise (Adams *et al.*, 2013; Frykman and Seiron, 2009; Chaudhry, 2017: 34; Butt, 2015; Mohan, 2014; EPW, 2009). Worse is that by the 2070s, the population and assets of port cities like Kolkata, Mumbai, Chennai, Surat, and Dhaka will be exposed to coastal flooding, predicts the AR5 of the IPCC (Jimenez Cisneros *et al.*, 2014; Mohan, 2019).

Monsoons have become less predictable and more intense in this region (Sheppard, 2019; Tesfaye *et al.*, 2017). This increasing frequency of rainfall means a higher risk of rivers and dams bursting, which often results in severe floods, landslides, and mudflows bringing miseries to the people in the low-lying areas (Sheppard, 2019; Shrestha *et al.*, 2000: 317–327; Mirza, 2002: 127–138). Supplemented by rising temperatures, extreme rainfall and intense flooding are expected to increase the incidence of disease to those living in the affected areas particularly in the slums (Tesfaye *et al.*, 2017; Shaw, Mallick and Islam, 2013: 59–60). Contaminated urban floodwaters have already caused exposure to disease and toxic compounds in India and Pakistan (Jimenez Cisneros *et al.*, 2014). As temperature is projected to increase up to 4°C and the expected rise of precipitation during the rainy season by 20 to 30 per cent (Vince, 2019; World Bank, 2012a; UNDP, 2004), South Asian countries will therefore be susceptible to extreme flooding.

Heat stress has negatively impacted wheat yields in the region (Chugh, 2019). While annual temperature is projected to increase by 1.4–1.8°C in

2030 and 2.1–2.6°C in 2050, heat-stressed areas in the region can increase by 12 per cent in 2030 and 21 per cent in 2050 (Tesfaye *et al.*, 2017: 959–970). Owing to heat stress, almost half of the Indo-Gangetic Plains, which is the major food basket of the region, may become inappropriate for wheat production by 2050 (Aryal *et al.*, 2019: 1; Ortiz *et al.*, 2008: 46–58). Even a somewhat modest warming of 1.5–2°C can relentlessly impact the accessibility and stability of water resources owing to increased monsoon variability and glacial melt water, thereby threatening the future agricultural production, and fluctuate food supplies, inflate prices, and aggravate the situation of food insecurity and poverty in South Asian countries, which will adversely affect the livelihoods of millions of poor people in the region (Vinke *et al.*, 2017: 1569–1583; Bandara and Cai, 2014: 451–465; Shankar *et al.*, 2015:1657–1671; Wang, Lee and Son, 2017:517–534). When agriculture provides livelihood to over 70 per cent of its people, employs almost 60 per cent of the labour force, and contributes 22 per cent of the regional gross domestic product (GDP) (Wang, Lee and Son, 2017:517–534), the losses of GDP will have huge impacts on the region, particularly on those communities whose livelihood depends on agriculture (Ahmed and Suphachalasai, 2014).

In this region, increasing temperatures, sea level rise, cyclones, droughts, flooding of river systems fed by melting glaciers, and other extreme weather events have intensified internal and international migration patterns over the years (Podesta, 2019: 3). Between 2008 and 2013, (IDMC, 2014) more than 46 million people were displaced by sudden-onset disasters – e.g. cyclones, floods etc. Within India, some 26 million people got displaced in the same period (The Nansen Initiative, 2015: 7). In this densely populated region, a single event often displaced millions of people. Around 11 million people were displaced in 2010 when Pakistan's Indus River flooded (IDMC, 2012), while in October 2013 the Indian government evacuated over 1 million people in anticipation of Tropical Cyclone Phailin reaching its shores, with 13.2 million people ultimately affected by the disaster (The World Bank, 2014). While the vast majority of displacement has been internal, a few instances of cross-border displacement have been reported in the wake of sudden-onset disasters, such as between India and Bangladesh when Cyclone Aila struck the latter in 2009 (Leighton, Shen and Warner, 2011; Shamsuddoha *et al.*, 2011; McAdam and Saul, 2010). When Cyclone Fani lashed eastern India in 2019, only three people were killed whereas millions of them were displaced (Sang, 2019).

In 2015 alone, the world witnessed over 19 million people (IDMC, 2015) displaced internally as a result of sudden-onset disasters. In South Asia, 7.9 million (41 per cent) people were displaced in the same year. Slow-onset disasters like desertification in remote Nepalese villages in the Himalayas (Khatri, 2013) and drought in Kuchi nomadic communities in Afghanistan (Consulting, 2014) have driven people to move internally and abroad in search of alternative livelihoods. Sea level rise coupled with saline intrusion and erosion pose unique challenges in the coastal areas forcing people to flee (Kothari, 2014: 130–140). Although there are people who have been displaced and moved beyond their national border, the overall number of these migrants is not known (The Nansen Initiative, 2015: 7). In Pakistan and Afghanistan there have been increasing numbers of migrants who moved from rural to urban areas (Abid *et al.*, 2016: 447–460; Saba, 2001: 279–289). There are also those who have migrated across the international border. Besides the prolonged conflict, food insecurity, environmental disasters have pushed huge numbers of Afghans to Pakistan and Iran (Saidi, 2017; Garrote-Sanchez, 2017: 3). These Afghan migrants are now facing a mass deportation threat (Kajjo and Jedinia, 2019; Lazarus, 2018). Natural disasters are responsible for rural-urban migration in India and Bangladesh (McDonnell, 2019). India has 139 million internal migrants (Sharma, 2017). In Bangladesh, the rural-urban migration has swelled over the years, i.e. from 5 million in 1961 to 56 million in 2014 (Ali *et al.*, 2015). There are large numbers of migrants from Bangladesh which have illegally moved to India (Sood, 2017: 35–36; Lal, 2016). This illegal migration from Bangladesh to North East India particularly to Assam, Tripura, and Meghalaya has become a tense political issue between the two countries with India claiming national security threats due to the increasing presence of Bangladeshis (Sood, 2017: 37; Werz and Hoffman, 2015: 99–108).

Severe climate change consequences in Nepal have displaced thousands of Nepalese internally and the open border with India has encouraged them to move externally but it is hard to say how many of them have migrated (Regmi and Paudyal, 2009: 1–13; Butler, 2009; Sharma and Thapa, 2013). The Maldivians have often moved from one island to the other in search of better livelihoods or because of environmental reasons like severe sea level rise (Stojanov *et al.*, 2016: 1–15; Kelman *et al.*, 2019: 285–299; Sood, 2017: 39; Rabbani, Shafeeqa and Sharma, 2016: 132). The Maldives is expected to become a microcosm of adverse climate change consequences and a producer of climate migrants for the bleak future predictions placed on the ability of humans to continue living on the sinking archipelago (Boyle, 2012). Rural-urban migration has been a common phenomenon over more than three decades in Sri Lanka (Manel, Punpuing and Perera, 2017: 127). Displacement of coastal dwellers is increasing as economic losses are substantial because of the hazard of coastal erosion (Martin, 2017: 20). Monsoon rains, droughts, floods, and landslides are the other disasters displacing people in Sri Lanka (Perera, 2019; Martin, 2017: 22–23; Lingam, 2017). Therefore, in the absence of adaptation measures climate change will have multifaceted impacts in South Asia.

South Asia is likely one of the regions of the world that is facing major casualties from climate change–related extremes such as floods, droughts, sea level rise, extreme temperatures, food production, and health of the people. These disasters have threatened human settlements and pushed millions to move internally and externally. Displacement of environmental disaster victims has been responsible for population growth, increased poverty, and also increases in the risk of social tension and conflict in receiving areas, contributing to xenophobia, persecution, etc. (The Nansen Initiative, 2015: 7; Ghimire, Ferreira and Dorfman, 2015: 614–628). With the region's population expected to rise to 2.00 billion if fertility rates follow the low scenario, to 2.32 billion in the medium scenario, and to 2.68 billion in the high scenario (Bloom and Rosenberg, 2011: 2) by 2050, these problems identified by the IPCC's reports and other research will be compounded sooner rather than later. It is also anticipated that population movements in the context of disasters and climate change in this region are likely to increase over the next decades (The Nansen Initiative, 2015: 7). Thus, rather than pushing the region towards greater instability, climate change–related disasters thrust and provide the impetus for serious regional cooperation because concerted action is needed to enable SAARC countries to adapt to its effects. Therefore, several initiatives have been taken by SAARC on climate change ever since the third summit held in Kathmandu 1987. As early as 1987 the Heads of State or Government of SAARC have been addressing in a systematic way the necessity to intensify regional cooperation for preserving, protecting, and managing the ecosystem of the region which is known for its diversity as well as for its fragility. Successive summits have reiterated the need to strengthen and intensify regional cooperation to preserve, protect, and manage the diverse and fragile ecosystems of the region, including the need to address the challenges posed by climate change and natural disasters. It is within the framework of functionalism that seeks cooperative ventures based on mutual advantage and in which participating states have common problems to address as well as provides for re-evaluating their interests in the context of mutual benefit that this manuscript, titled *Climate Change in South Asia: Politics, Policies and the SAARC*, is carried out.

IV Literature review

In recent years, the number of studies on climate change has increased, contributed by different scholars. Some of the scholarly works (Solomon *et al.*, 2007; Parry *et al.*, 2007; Edenhofer *et al.*, 2014; Baghel, Stepan and Hill, 2016; Karim, 2013; Douglas, 2009) have provided detailed information on the effects of the anthropogenic activities on the atmosphere, containing an evaluation of observed climatic changes using the latest measurement techniques. Other information which these works described include a detailed review of climate change observations and modelling for each continent as well as the first probabilistic evaluation of climate model simulations. They also highlight that climate change will bring about several challenges to South Asia: (i) glacial melting of the Himalayas will increase flooding and affect long-term water resource availability; (ii) pressurize natural resources and the environment owing to rapid urbanization, industrialization, and economic development; (iii) crop yields will likely decrease up to 30 per cent by the middle of the 21st century; (iv) periodic floods and droughts will impact on the health of the population; and (v) sea level rise will exacerbate inundation, storm surge, soil erosion, and other coastal hazards. With regard to the future impacts of climate change on South Asia, Nicholas Stern (2006) predicts that this region, which consists of eight states of different sizes and capabilities, will be extremely vulnerable to climate change. Though the impacts are likely to be uneven, it is possible climate change will be most felt in this region. Mirza and Ahmad (2005b) speak about the pressing issue of water resources in South Asia – a region characterized by abundant water, with devastating floods during the monsoon, that faces water scarcity and droughts during the dry period. Erratic water flow often causes extensive damage to infrastructure, property, livestock, and human populations as well. Agriculture and other key economic sectors suffer greatly because of disasters. These events will increase unless adequate adaptation measures are designed and implemented with the active participation of the stake-holders. Another report (UNEP, 2009) provides an overview of the state of the natural environment in South Asia, including emerging trends, taking into account socio-economic factors. This report concludes that South Asia is very vulnerable to climate change where its impacts have been observed in the form of glacier retreat in the Himalayan region, where thousands of glaciers will likely shrink, which will affect the water volume of the Indus, Ganges, and Brahmaputra, the lifeline of millions of people in South Asian countries. This will exacerbate the challenges of poverty reduction, agriculture production, and improving access to safe drinking water. The current climate warming has the power to melt the Hindu Kush-Himalayas (HKH) ice sheets. Some studies therefore (Mool *et al.*, 2001; Ives *et al.*, 2010; Bajracharya and Shrestha, 2011; Price *et al.*, 2014) argue that glaciers of the HKH, which are nature's renewable storehouse of fresh water, the perennial sources of rivers that are used as renewable resources for irrigation, drinking water, energy, and industry, are retreating. The sudden breaching of these ice sheets can result in the discharge of huge amounts of water and debris with catastrophic effects to the downstream settlements and agricultural areas. Another study (Shrestha, Mool and Bajracharya, 2007) demonstrates the potential threats of GLOF hazards to Nepal and Bhutan in particular, as well as raising the awareness of the public at large to the potential impacts of climate change in the Himalayas in general. Other works (Lal *et al.*, 2011; Alagh, 2001) address the important topic of food shortage in South Asia, a region where 30 per cent of the people are food insecure. Owing to water scarcity the problem of food shortage may be aggravated by the projected climate change, especially by rapid melting of the glaciers of the Himalayas and the increase in variability in monsoonal rains and frequency of extreme events. The melting of glaciers will also affect the coastal population where they may be displaced when the sea level rises.

Umesh Kulshrestha's edited book (2017) addresses the problem of air pollution responsible for climate change with a critical analysis of various aspects. This book illustrates the sources of air pollution, drivers of climate change, impact of air pollution, regional resilience, deposition of air pollutants to various ecosystems, demographic shift, and mitigation options for air pollution and climate change. According to Muthukumara *et al.* (2018) higher temperatures and erratic rainfall patterns will reduce living standards of the communities across South Asia. To avoid the decline of living standards of the people the book stresses the need for the change of employment from agricultural to non-agricultural sectors. The work of Prakash, Saravanan and Chourey (2011) highlights the need for an interdisciplinary approach to water and health in South Asia. Sinha, Bauer and Bullen (2015) stress the influence of climate change–related disasters such as floods, landslides, droughts, degradation, and urban pollution which can have great impact on the living standards of the people in the dry and marshy regions of India, the hilly region of Nepal, and the coastal areas of Sri Lanka and Bangladesh. Barua, Narain and Vij (2018) see climate change in South Asia not only as an environmental problem but as a common problem and as such needs governance from an interdisciplinary approach and across different levels: local, state, and national, the need of financing mechanisms, and a top-down approach in adaptation and mitigation in order to achieve desired outcomes. Mukhopadhyay, Karisiddaiah and Mukhopadhyay (2018) discusses the necessity to deal with climate change in South Asia from a regionally focused standpoint. They stress the need not to merely depend on foreign help but to also act independently. Another work (Damodaran, 2010) discusses the complexion of global environmental negotiations in searching for a solution to the climate change problem, and the inconsistent implementation of national environmental policies by emerging economies including India as they are concentrating on development and eradication of national poverty. According to Paton and Johnston (2017) the Indian Ocean Tsunami and other climate change–related disasters have warned about the susceptibility of different communities to their devastating effects. This volume provides valuable insights into how societal resilience can be developed and sustained, the resources and strategies required at each level to facilitate resilience and how they can be integrated to develop a sustained capacity to adapt to the consequences of natural hazards.

A review of some of the existing literature reveals that although there is ample literature on climate change, there is, however, very little writing on regional cooperation and climate change in South Asia, a region which is facing the risk of several climate change effects. All these works focus on some levels but do not study mitigation and other adaptation measures at the regional level. Thus, a significant gap still exists in the current literature of climate change in this region. Hence, this manuscript may fill up some of the gaps as it focuses not only on climate change and its effects such as (i) floods, (ii) droughts, (iii) food shortages, and (iv) the rise in temperature, but it also looks into the adaptation measures at the national and regional level.

V Milestone in environmental protection

Early interest in the study of environmental devastation stems from an appreciation that human societies are constrained by environmental limits, the infringing of which can lead to socio-economic catastrophes. Some have imagined the global climate change challenge as akin to a "tragedy of the commons" (Hardin, 1968: 1243–1248; Catton, 1982). According to Sharon Beder (2007: 13), man's overexploitation of the earth's resources followed by continental economic growth is the main cause of environmental decline. This persistent mistreatment of the earth may not continue forever, if the crisis is not addressed in time with appropriate binding agreements. A number of publications have predicted the future doom of the earth is waiting. These studies have conceptualized the environmental problem known as climate change as a security threat to humanity (Vivekananda, 2017; Redclift and Grasso, 2015: 382–401; O'Brien, Asuncion and Berit, 2010: 3–22; Barnett, 2003: 7–17; Schubert *et al.*, 2008). Similarly, ecological problems such as ozone depletion become salient in international affairs when several thinkers begin to discuss multilateral environmental agreements (Haas, Keohane and Levy, 1994; Susskind, 1994; Russett and Oneal, 2001). Likewise, a number of scholarly monographs have started to provide a multi-theoretical treatment of climate change (global warming) in the field of International Relations (Eckersley, 2004; O'Neill, 2009). The reason is climate change represents one of the most serious and far-reaching environmental challenges facing humankind today (Luterbacher and Sprinz, 2001: 1–18). The international consensus of scientific opinion piloted by the successive IPCC reports has agreed global temperature is increasing and its main causes are the accumulation of carbon dioxide and other GHG in the atmosphere (Solomon *et al.*, 2007; Watson and the Core team, 2001). Scientific opinions (Houghton *et al.*, 2001) have also predicted the threat carried by it will become more severe over the coming decades. Such opinions have started as early as 1988, when scientists cautioned the human tinkering with the earth's climate would amount to an unintended, uncontrolled globally pervasive experiment whose ultimate consequences could be second only to a global nuclear war (Orville, 1958: 54). As a result, the vulnerability and fragility of the earth with regard to climate change has been the focus of the field of International Relations. However, the discussion needs to focus on the milestone in environmental protection; thereby it is illustrated in the subsequent paragraphs.

An American environmentalist movement under the name of the Sierra Club which was founded in 1892 launched the first grassroots political campaign on behalf of wilderness preservation (Hillstrom, 2010: 214). Under the influence of the ideas of Ralph Waldo Emerson and Henry David Thoreau, the Club owes a great deal to George Perkins Marsh who published *Man and Nature or, Physical Geography as Modified by Human Action* in 1864 (Lowenthal, 2003). This publication not only analyzed the historical

accident of civilization, particularly deforestation and watershed degradation in various regions of the world, it also inspired generations of environmentalists around the globe. This Club is widely recognized (Giddens, 2011:
48–75) as the world's first significant environmental organization which is
devoted to the protection of wilderness areas.

Robyn Eckersley's view (2010: 258) is that the problems of the environment have never been a vital concern in the discipline of International Relations, which traditionally focuses on the questions of security and inter-state
conflicts. Ever since the 1960s, however, events like chemical pollution and
global warming have resulted in a widespread increase in environmental interest. There are a number of seminal books on environmental issues that were
published in the 1960s and 1970s which most likely thrust environmental
concerns much higher on the popular political agenda. *Only One Earth: The
Care and Maintenance of a Small Plane* (Ward and Dubos, 1972) and *Silent
Spring* (Carson, 1965; Smil, 1993: xi) have portrayed the link between environment and health damage. In fact, one of the famous studies for this subject
done during the late 1960s and early 1970s was commissioned by the Club
of Rome. The scientists, educators, economists, humanists, industrialists, and
civil servants under the leadership of Italian businessman Aurelio Peccei and
Alexander King of Scotland formed this Club of Rome in 1968. The study
published *The Limits to Growth* in which the Club of Rome predicted the
running out of key resources and the earth's ability to absorb waste product
by 2010 (Meadows *et al.*, 1972). Paul Ehrlich's (1968) *The Population Bomb*
focuses on the need to restrict population growth if the world is not to reach
the limits to growth. The global oil crisis of the 1970s (Scholte, 2005: 287)
also highlighted the impending exhaustion of many vital natural resources.
The Limits to Growth, however, has shown the existing exponential growth
rates of population and economic activity. The study states:

> If the present growth trends in world population, industrialization, pol-
> lution, food production and resource depletion continued unchanged,
> the limits to growth on this planet will be reached sometime within
> the next one hundred years. Moreover, the most probable result will
> be a rather sudden and uncontrollable decline in both population and
> industrial capacity.
>
> (Meadows *et al.*, 1972: 23–24)

Due to massive population growth and resource depletion, Paul Ehrlich's
work (1968) foresees a gloomy prospect for the world. The Club of Rome
paints an apocalyptic future for the earth if the environmental effects of
the economy are not addressed in appropriate time. There are those who
argued the world has reached a point where the explosion of economic
and pollution growth will exceed the carrying capacity of the earth (Ravell,
2008: 56–75). Therefore, *The Limits to Growth* makes headlines around the
world and begins a debate about the limits of the earth's capacity to support

human economic expansion (Atkinson and Davis, 2001: 165–166). While the idea of *The Limits to Growth* has appealed to the ordinary man's general sagacity, it also worried Western intellectuals and angered the economists, conservatives, and politicians alike who viewed any criticism of economic growth as a direct attack on capitalism (Ekins, 1993: 269–288; Suter, 1999: 1–5; Norgaard, 2001: 167–169), a system in which economic profit is structured according to market principles.

Under the influence of some prominent publications such as *The Limits to Growth* (King and Schneider, 1972; Meadows *et al.*, 1972), the information has raised considerable public attention, highlighting the environmental dangers of the relentless pursuit of material growth. *The Ecologist* not only discusses the need to stabilize the economy, it also argues that economic growth cannot continue into the future without the appended outrageous disasters (Dann, 1999: 343). *A Blueprint for Survival* (Goldsmith, 1972) is another influential environmentalist text to which an entire issue about the environment has been dedicated. It offers dire predictions of impending ecological catastrophe unless exponential economic growth is replaced with steady-state economic development (Eckersley, 2010: 260). The *Blueprint for Survival* states:

> The principal defect of the industrial way of life with its ethos of expansion is that it is not sustainable. . . . By now it should be clear now that the main problems of the environment do not arise from temporary and accidental malfunctions of existing economic and social systems. On the contrary, they are the warning signs of a profound incompatibility between deeply rooted beliefs in continuous growth and the damning recognition of the earth as a space ship, limited in its resources and vulnerable to thoughtless mishandling.
>
> (Goldsmith, 1972)

The escalation of transboundary environmental problems from the 1970s onwards sees the emergence of a subfield of International Relations (Johnson, 1973: 87–88; McLin, 1972a: 1–12, 1972b: 1–7; Caldwell, 1974; Report of the UNCHE, 1973) which is concerned with international environmental cooperation that focuses primarily on the management of ecosystems. These publications coincided with the first UN Stockholm Conference on the Human Environment which formalized the emergence of the environmental crisis as a global issue. This UN Conference which took place at the Royal Opera House in Stockholm on June 5, 1972, in its section I, contained the Declaration on the Human Environment (Report of the UNCHE, 1973). While comparing this declaration on the environment with the Universal Declaration of Human Rights, 1948, it has been observed that it is essentially a manifesto expressed in the form of an ethical code intended to govern and influence future action and programmes, both at the national and international levels (Starke, 1989: 406).

Several climate conferences were held in the late 1960s and the early 1970s. The year 1979 saw the world's first World Climate Conference. Further, 1985 is the year when the Vienna Convention for the Protection of the Ozone Layer was created and two years later in 1987 saw the signing of the Montreal Protocol under the Vienna Convention. Again in 1987, the Brundtland Commission's report, *Our Common Future*, which highlighted environmental problems, was released. This report (WCED, 1987) emphasized the close relationship between economic development and environmental concern. A year later in 1988, the UN set up the IPCC and considered the issue forceful enough to expand into the UNFCCC, which was duly adopted at the Rio de Janeiro Earth Summit in 1992 (Brown and Kutting, 2008: 164). The framework entered into force on March 21, 1994. The conference established a yearly meeting, a conference of the parties or COP meeting to be held to continue work on Protocols which would be enforceable treaties. The year 1995 sees the creation of the phrase "preventing dangerous anthropogenic interference with the climate system" (also called "avoiding dangerous climate change") which first appeared in a policy document of a governmental organization, the IPCC's Second Assessment Report: Climate Change 1995 (Houghton *et al.*, 1996). In 1996, the European Union (EU) adopted a goal of limiting temperature rises to a maximum 2°C rise in average global temperature. The year 1997 sees the creation of the Kyoto Protocol under the UNFCCC. Henceforth, climate change becomes the focus of the scholars of International Relations and increasingly observed as the most pressing issue.

In addition to these events, political momentum for the anxiety of climate change was building when political debate on the problem of climate change occurred in 1989. The first international political agreement dealing explicitly with climate change was a resolution adopted at the 43rd session of the UN General Assembly on January 27, 1989 (IPCC, 1990: 122). This resolution was formulated as an unswerving response to a proposal put forward by the government of Malta as an indirect response to the growing international consensus reflected in meetings beginning with the first World Climate Conference in 1979 (Müller, 2002: 5; Zillman, 1997), including the conference at Villach, Austria, in October 1985 to which the threat posed by the accumulation of GHG in the atmosphere was acknowledged.

In March 1989, at the invitation of France, Norway, the Netherlands, representatives of 24 countries, including 17 Heads of State or Government, were invited to a summit meeting at The Hague to consider the climatic issue (Hague Declaration on The Environment, 1989a: 1308–1310; Climate Change Declaration, 1989). At this summit (The Hague Declaration) the participating countries declared their vow to aptly address the problem of climate change. The participating countries came up with a goal of developing a new regime (along the lines of the UN Security Council) on global environmental issues (Hague Declaration on the Environment, 1989b: 1308–1310). The way the heads of the state have discussed the crisis at that time had certainly reflected the increasing importance of climate change.

Another political meeting took place in Dakar (Synthesis Report, 2009: 1–39), Senegal, where the heads of French-speaking countries met in May 1989 to endorse The Hague Declaration and to set up an agenda for heartening action. Numerous other political initiatives were taken in the same year, which paved the way for a focused debate on various aspects of climate change in Noordwijk in the Netherlands. This resulted in the adoption of the Noordwijk Declaration on Climate Change by 67 countries in November 1989; the declaration assumed:

> For the long term safeguarding of our planet and maintaining its ecological balance, joint effort and action should aim at limiting or reducing emissions and increasing sinks for greenhouse gases to a level consistent with the natural capacity of the planet. Such a level should be reached within a time frame sufficient to allow ecosystems to adapt naturally to climate change, to ensure that food production is not threatened and permits economic activity to develop in a sustainable and environmentally sound manner. Stabilizing the atmospheric concentrations for greenhouse gases is an imperative goal. The IPCC will need to report on the best scientific knowledge as to the options for containing climate change within tolerable limits. Some currently available estimates indicate that this could require a reduction of global anthropogenic greenhouse gas emissions by more than 50 per cent.
>
> (Noordwijk Declaration, 1989)

Due to the growing emission of greenhouse gases, at Noordwijk they felt certain things need to be reversed in order to stabilize the atmospheric concentration of GHG. Hence, it was agreed that sink (forest preservation and replanting) management should be improved. Thus, the target was formulated at Noordwijk:

> Carbon dioxide target: In the view of many industrialized nations, such stabilization of carbon dioxide emissions (at 1990 levels) should be achieved at the first step at the latest by the year 2000. Forest target: Agrees to pursue a global balance between deforestation on the one hand and sound forest management and afforestation on the other. A world net forest growth of 12 million hectares a year at the beginning of the century should be considered as a provisional aim.
>
> (Beukering and Vellinga, 1996: 200)

Other reports have depicted climate change as "a threat multiplier which exacerbates existing trends, tensions and instability," posing "political and security risks that directly affect European interests" (Ahmed, 2011: 335–355; High Representative for Common Foreign and Security Policy,

2009: 2). That is why climate change stands out as the quintessential global-scale collective action problem that requires a cautious policy coordination and multilevel governance (Esty and Moffa, 2012: 777). Though national policies and local measures are required to advance climate action at the national level, international cooperation remains crucial for climate protection. Absent a collective commitment to mitigate climate change within state territories, some states may behave as free riders or jeopardize the efforts undertaken by other nations by continuing to emit large amounts of GHG (Ostrom, 2010: 550–557; Peel, Godden and Keenan, 2012: 249). So, the UNFCCC which was adopted at the 1992 Earth Summit (stated earlier) set a non-binding goal to limit GHG emission to 1990 levels by the year 2000. Owing to the US objections for commitment (Goldstein and Pevehouse, 2008: 389), the goal was not met. Another breakthrough to lessen emissions was observed with the adoption of the Kyoto Protocol in 1997. The Kyoto Protocol, entered into force in 2005, sets binding targets for countries to reduce GHG emissions by 2012 (McGovern, 2006). Beyond this commitment period, parties at the Group of Eight (G8) Summit have agreed to at least halve the global emissions levels by 2050 (Breidenich *et al.*, 1998: 315–331). The Kyoto Protocol approves a complex formula for reducing emissions to 1990 levels in the Global North over about a decade (Victor, 2001; Grubb, Vrolijk and Brack, 1999). But it meets serious differences among the developed countries on the one hand and the developing countries on the other. Their objections reflect their record of development. The refusal of the Chinese and the US governments to ratify the 1997 Kyoto Protocol has severely demeaned the key attempt to set legally binding limits on GHG emission (Redclift and Sage, 1999: 140–147). To reach a secure climate future both developed and developing nations agreed to step up their efforts through the "Bali Road Map" (Bali in December 2007). The Bali Road Map including The Bali Action Plan (2008: 3) tabled the course for a new negotiating process under the UNFCCC. It also includes the current negotiations under the Kyoto Protocol, and their 2009 deadline which focuses on further quantified emissions cut commitments for industrialized countries, as well as negotiations on the ongoing work pertaining to key issues including technology, adaptation, and reducing emissions from deforestation (He, 2010: 5–33).

The conference which was supposed to find a replacement for the Kyoto Protocol, Copenhagen in 2009, does not succeed. Another major international conference on which hopes now rest is the Paris Agreement of 2015. Both the Agenda 2030 and the Paris Agreement of 2015 are negotiated in parallel and adopted in 2015. While SDG 13 is supported by existing international law, its successful implementation still faces a number of obstacles. The success of SDG 13 in particular, as well as that of the other SDGs in general, hinges to a great extent on the efficacious implementation of the Paris Agreement (von Stechow *et al.*, 2016). So, it is a voluntary agreement endorsed by the world leaders in 2015. However, all the State Parties that ratify the Paris Agreement have made bottom-up commitments and will submit updated climate plans called Intended Nationally Determined Contributions (INDCs) every five years. The INDCs are at the heart of the Paris Agreement in order to achieve its long-term goals. This must lead to a GHG emissions reduction of 70 per cent to 80 per cent by the second half of the century. The success of this Agreement depends on negative emission, a condition which doesn't go down well with politicians and industry bosses of some countries like those in the US (Hickel, 2017). Other countries are doing very little to avert an impending catastrophe of climate change (Manne, 2013). It can be stated the Paris Agreement is already facing its litmus test.

Climate change is accepted as a real and collective problem, but it has also become a complex and complicated issue to deal with. Despite the existence of climate agreements, governments all over the world (Vidal, 2012) do not see its multifaceted impacts as urgent or of high priority. Thus, climate negotiators faced a dilemma when they realized the movement lacks significant commitment (Rootes, 1999: 1–12). The *raisond'être* is that most of the populous developing states always look to the industrialized countries to take on greater responsibility for meeting this challenge based on their historical emission. Such a stand by these countries has been potentially undermining the level of commitment in the global effort against climate change.

2 Theories of regional cooperation

Climate change is a global problem, but effective action requires that governments decide to act. So, ever since the early 1990s, it is witnessed the growth of consciousness on regulating climate change. Adaptation to climate change has become one of the major global problems today which can only be achieved by substantial collaboration across countries. Adaptation means taking actions to trim down liability to climate change (Javeline, 2014: 420–434). Such measures may include protection of coastal cities, adaptation to a warmer climate, to bring water to highly populated and arid regions and most importantly reduction of GHG. For this reason, climate experts and scholars of International Relations are taking keen interest in mitigating climate change. To do this, this chapter turns to the theories of Regional Cooperation, to see how climate change and adaptation are viewed by these theories. This chapter will start with a dialogue on Regional Cooperation and Regionalism which will be followed by the discussion on theories of Regional Cooperation and determines a particular theory that will be adopted as a framework of the study.

I Regional cooperation

According to Aristotle (Ebenstine, 1960: 67), man cannot live alone. He is a political and social animal destined by nature for political life and associates himself with his fellow men to form associations to fulfil his common needs. To him all associations are political (Ebenstine, 1960: 67) in as much as they aim at a common good through joint action ever since the will to live together among men is a feeling of closeness. Likewise, at the international level, nation-states organize themselves into some form of regional cooperations to achieve their common interests and common needs. Existing forms of regional cooperation came into being mostly in the nineteenth century and further developed in the twentieth century. By the start of the 21st century, they play a vital role (Baylis and Smith, 2011: 591; Padelford, 1954: 204) in the contemporary International Relations for the maintenance of peace and security as well as facilitating social and economic cooperation among states, regulate inter-states relations and control state activities. The endeavour is to discourage (Gopal, 1990: 2–3) war and encourage peace and order. As such, regional cooperation seems to be an appealing scheme especially for the weak and poor countries. Since they have achieved considerable importance in the field of International Relations, understanding regional cooperation is vital as it can offer a scale for observing the peaceful cooperation of humankind. Presently, regional cooperation (Feil, Klein and Westerkamp, 2009: 7) has been part and parcel of the global economic order. With an objective to build harmony between neighbours, regional cooperation is also a general way for supporting and improving relations between parties or for achieving transboundary goals. Most importantly, the logic is that regional cooperation over issues of mutual interest, such as the environmental disaster will build trust involving even the conflicting parties.

Before the start of a discussion on regional cooperation, perhaps it is necessary to define the terms regional and cooperation concisely. These words may provide the essential bases for the emergence of the phenomenon of regional loyalty which ultimately finds expression in the existing regional cooperation. To the *International Encyclopaedia of the Social Sciences* (Sills, 1972: 382) the word regional means the systematic approach to space in the sense of the human habitat. Hence, regional relate to a small geographical area. Whereas, cooperation is a joint or collaborative actions which occur when actors regulate their actions through the process of organization making that is directed towards some goals in which there is common interest or hope of reward for the parties involved. It may be a voluntary or involuntary, (Sills, 1972: 384–390) direct or indirect, formal or informal but always there is a combination of efforts towards a specific end in which all the participants have a stake, real or imagined. It is a result of shared values and norms (Smelser and Baltes, 2001: 2751) and norm-conforming behaviour. Cooperation also implies reciprocity, a willingness to give as well as take the voluntary acceptance of limits on all sides rather than one (Edwards, 2004: 3–4). It involves going a step further by stating that cooperation (Welsh and Willerton, 1997: 33–36) requires at the most basic level, the presence of common problems and tasks, which lead to a commonality of expectations and the overlapping of interests on the part of the nation-states. For other scholars (Laszlo, Kurtzman and Bhattacharya, 1981: 12) cooperation refers to something that aspire a motivation for the achievement of necessary levels of collective self reliance. It means such a close degree of economic intertwining that (Rourke, 2005: 466; Rourke and Boyer, 2004: 355–365), by formal agreement or informal circumstance, the countries who are involved begin to surrender some degree of sovereignty and act together as a unit. It is supposed to create the necessary atmosphere for converting foes into friends (Sridharan, 2008: 2). To Robert O. Keohane (1994: 102–107), the term cooperation is highly political and therefore does not imply an absence of conflict. It is typically mixed with conflict and takes place only in situations in which actors perceive that their policies are actually or potentially in conflict. So, cooperation (Keohane,

1994: 102–107) must not be viewed as the absence of conflict, but as a reaction to a conflict or potential conflict.

In International Relations, regional cooperation (Weigal, 2002: 191) implies that states position in the same geographical area cooperate with each other to resolve common problems or to achieve certain objectives beyond the capacity of individual national attainment. It also refers (Vasilyan, 2006: 102) to the realization of states within a region whereby autonomous national stances are brought together for the achievement of a common goal on a bilateral, trilateral or multilateral platform. It envisions that states cooperate (Syed, 1999: 5–10) only to the extent whereby the accomplishment of one state should not be at the cost of the other, as it visualizes (Chhibber, 2004: 2) on developing a sense of common interest required to successfully thwart any external interventions. It originates from the need to explore different ways in which (Gill, 2005: 95) the countries of the region can cooperate in a meaningful manner not only at the socio-economic level but also at the political level. The motive of having regional cooperation is to evolve a coordinated approach on certain issues which are of common concern between the neighbours. Regional cooperation incorporates a diverse set of actions for assisting each other such as controlling tropical diseases, disaster assistance, military alliances, nuclear programme and environmental management. Since it comes in many forms and shapes in the international community (Stubbs, 2000: 297–318), its fundamental perception has been analyzed through different approaches (Coleman and Underhill, 1998; Mattli, 1999). To Niklas Swanstrom (2002, 11–12) therefore, the concept of regional cooperation can be a useful analytical tool but it may also be dangerously applicable in a wide spectrum of areas depending on the countries involved.

It is apparent that without the spectre of purpose, difficulty or conflict, there is no need for regional cooperation. From the discussion it is very clear regional cooperation is a collective technique meant to signify a cooperative process which transcends the boundary of any one sovereign state in a particular region. Thus, regional cooperation is a process of voluntary adjustment among neighbouring states. Even when they are conflicting to each other, countries merge together for the purpose of mutual benefits based on national importance. For the study of International Relations, it is clear that regional cooperation implies states which position in the same geographical area cooperate with each other to resolve common problems or achieve certain objectives that are beyond the capacity of individual national attainment.

South Asia has been characterized by a scholar as a region which lack contacts and cooperations (Palmer, 1975: 889). There are scholars who have dismissed (Brecher, 1966; Cheema, 1986) the possibility of regional cooperation in this region due to its history of internal political instability. However, the growing popularity of the concept of regional cooperation made the former president of Bangladesh, Zia-ur Rahman, to put his earliest proposal for establishing a framework for regional cooperation in the region way back in May 2, 1980 (Umar, 1992: 14; Gonselves, 1995). The South Asian

Association for Regional Cooperation (SAARC) is an organization of South Asian nations which was established on December 8, 1985 despite divisiveness and initial opposition from India and Pakistan. Bangladesh, Bhutan, India, the Maldives, Nepal, Pakistan and Sri Lanka are its founding members. In April 3, 2007, Afghanistan joined as a full member of SAARC (Dawn, April 04, 2007). One of the objectives of SAARC is 'to contribute to mutual trust, understanding and appreciation of one another's problems' (SAARC Charter, 1985: 1). When regional cooperation is increasingly important as a means to ease cooperation in realizing common purposes and goals (Wright, 1970: 199) in a given region, a discussion of theories of international relations may help understand the context of regional cooperation better. But, a brief look on regionalism is considered necessary.

II Defining regionalism

There is a growing body of literature (Legrenzi and Harders, 2008: 1) which centres on the notion that regional cooperations are becoming essential actors in world politics. The early regional organizations may also be considered as cases of regionalism. Martin Griffiths (2005: 723) defines regionalism as the growing political or economic process of cooperation among states in particular geographic region. It is so since a region represents a dynamic, spatial concept that is defined by the blending of geographical proximity. Graham Evans and Jeffrey Newnham (1998: 474) describe regionalism rather abstractly as a complex of attitudes, loyalties and ideas which concentrates the individual and collective minds of people upon what they perceive as their region. This can be due to the feeling of loyalty to one's own region. To David L. Sills (1972: 3778) the term represents the regional idea in action as an ideology, as a social movement, or as the theoretical basis for regional planning. It is also applicable to the scientific task of delimiting and analyzing regions as entities lacking formal boundaries. Whereas, to other scholar (Lawson, 2008: 16) it refers to the general phenomenon as well as the ideology urging for a regionalist order, either in a particular geographical area or as a type of world order. This explains regionalism as an understanding (Fawcett, 2005b: 22) whereby states cooperate and organize strategy within a given region. Its endeavour is to pursue and promote common goals in one or more concern areas. While it may be manifest in different forms in different regions, it does convey an idea that nations and peoples in a specific international region expresses a common sense of identity and pursue a common objective of "greater coherence" through "structures, processes and arrangements . . . in terms of economic, political, security, socio-cultural and other kinds of linkages" (Dent, 2009: 7). It is something where political leaders often conduct foreign policy and approach their external environment through a regional community perspective. So, the term regionalism refers to not only the growth of economic cooperation but also to the growth of regional identity and consciousness among neighbouring countries altogether.

Regionalism is a very prominent phenomenon in the contemporary International Relations. The attempts to explain regionalism in conceptual terms in the post-war years started with the emergence of regional integration (Cai, 2010: 7) in Western Europe. In the early stages, regionalism derives its inspiration from liberal theories of International Relations which stressed on the mutual gains to be derived from multilateral cooperation. This is due to the belief regionalism will be one of the best ways of addressing tensions in International Relations (Doyle, 1986: 1151–1169; Aranzadi, 2006). Within the realist tradition, regionalism offers the promise of cooperation among like-minded states to prevent the challenges coming from the assumed model of anarchy (Donnelly, 2000; Hoffmann, 1965). From the 1990s onward, constructivism inspires the study of International Relations with its focus on ideas and interests when it stresses on regional awareness and regional identity, on the shared sense of belonging to a particular regional community (Ruggie, 1998). According to the constructivists, regional cohesion depends on a sustained and durable sense of community based on mutual responsiveness, trust and high levels of what may be called "cognitive interdependence" (Hurrell, 1995: 64; Carlsnaes, Risse and Simmons, 2002: 143). For constructivism, the International Relations pattern at the regional level is informed by elite and cultural considerations of identity and community (Ganesan, 2010: 1). Another body of literature that stand under the rubric of regionalism theory is functionalism. This theory is an expression of man's thoughts based on the trans-national teamwork written in a period when Europe was confronting a deep crisis (Popoviciu, 2010: 162–172). Its advocates have maintained that transnational programmes are easier to start when they are non-controversial in nature and, once begun, would lead automatically to new projects and relationships (Rothchild, 1967: 143–144). This logic rests upon the idea of David Mitrany (1966) that nation-states, as now composed, are unable to solve increasingly complex economic and social problems of a border-crossing nature.

Despite the differences in the core assumptions regarding the motivation of states for cooperation between the various schools of thought, regionalism echoes in the study of International Relations. The stimulus may differ, states often engage in joint efforts at the regional level. As the main purpose of this Chapter is to discuss the theories of regional cooperation, the succeeding paragraph will therefore examine four theories and will take functionalism as a suitable approach since it originally dominated the integrated thinking of the Europeans.

III Theories of regional cooperation and climate change

(i) Realism

The study of International Relations has normally been dominated by several schools of thought. The most dominant of these is realism which came into the trend after World War II. It is also dominated the discourse during the Cold War. Cooperation is very much part of realism, but mostly as an alliance against the enemy. For realism, survival is the key determinant of states actions (Laferrière and Stoett, 1999: 83–84) as they are inherently death-fearing and power-seeking. States are rational actors who take insightful steps to assure they will survive whatever collisions may await them. There are those who argued (Bull, 1997; Alderson and Hurrel, 2003) that international society within the realm of realism will be anarchical being dominated by individual states where each one of them strives to maximize their own power and security. To realists, ethics, moral values and justice have no place in international politics and are instead viewed as 'oxymoronic expression[s]' (Franceschet, 2002: 347–357; Okereke, 2010: 462–474). Any cooperation that might occur would most likely take the form of transitory alliances, which would balance the power among opposing blocs of states (Morgenthau, 1978).

From the perspective of realism, insecurity pervades regional cooperation and breeds an ongoing struggle between states for power and survival (Frankel, 1996: ix–xx). There is no institution holding power over them as the states endure to pursue aggressive tactics in order to maximize their power (Waltz, 2001). States are always being on guard against their neighbours since they are for all time in potential danger of invasion. When the states are inclined towards conflict and competition, regional cooperation will usually prove elusive. Within realism therefore regional cooperative venture will enhance the power of individual states.

In the field of economic cooperation few scholars have argued cooperation might be feasible if a single actor with a preponderance of power exists and is willing to use its power resources (Gilpin, 1975; Kindleberger, 1973). This actor is identified as a "hegemon" (Keohane, 1980: 1967–1977). As such, this forecast that the degree of cooperation will be directly relative to the scale to which one actor dominates international politics. Acting either benevolently or malevolently, the "hegemon" has the resources to transform international structures so that coordinated policies to address perceived collective action problems result. So, if regional cooperation is possible at all, it will be only under conditions of hegemony, where a dominant state is able to use its influence and military power to create and enforce the institutional rules necessary to sustain cooperation between states (Keohane, 1984: 57). For instance, the United States (US)-led North Atlantic Treaty Organization (NATO) and the Soviet-led Warsaw Treaty Organization (WTO) offer a very clear picture where the US and the Union of Soviet Socialist Republics (USSR) respectively, dominated the two organizations. Realism therefore, falls short of providing any real ground for purposive or meaningful action in the regional cooperation (Carr, 2001: 86). It often portrays as the advocate of an aggressive policy (Dunne and Schmidt, 2005: 178). Applied to the climate change issue, one actor may be able to issue threats and coax another into changing its activities that contribute to climatic change. War has often been used as a means to achieve foreign policy goals related to natural resource issues (Westing, 1980). One major actor possibly will threaten to use trade sanctions against a climate violator, and, if implemented, deprive the target country of welfare (Luterbacher and Sprinz, 2001: 34). Indeed, the global community has a hegemon (example US) within the international political system who has the power and resource to influence others to cooperate and mitigate climate change. This, as a result, suggests that the chances for international cooperation on climate change are high. However, the hegemon itself is not in favor of international cooperation on climate change. Ever since the 1980s, U.S. political leaders have been resistant-symbolically and operationally-to domestic action and international cooperation on climate change (Jamieson, 2014). In a 2010 Gallup survey of 1,014 adults in the U.S., 74 per cent of liberals agreed that "effects of global warming are already occurring," whereas only 30 per cent of conservatives concurred (Jones, 2010). Public opinion surveys of 1,024 Americans in 2012 revealed that 42 per cent contend that climate change claims are "generally exaggerated" and that political conservatives are more sceptical of climate change than liberals (Saad, 2012). Besides the US, big actors Germany and China have been pro-active during much of the climate negotiations (working, of course, towards different goals), formally acknowledged the need for fairness and laid out specific plans to fight climate change that included strategies to slash their GHG emissions (Khan, 2016: 15). But, countries are not doing enough and have shown little interest in operationalizing their commitment (Worland, 2018; Nilsson and Pitt, 1994). Together, the EU is another potentially important player. On 28 November 2018, the European Commission presented its strategic long-term vision for a prosperous, modern, competitive and climate-neutral economy by 2050 (McGrath, 2018). To date, however, a trio of central European countries has blocked the EU from inching closer to a net-zero carbon emissions target for 2050 (Rankin, 2019). Hence, regional cooperation under realism is impossible to achieve and sustain. Though realism, (Eckersley, 2010: 260) may have served it reasonably well in serving for relations between the superpowers mainly during the Cold War, it has struggle to make sense of complex international politics, including those associated with the climate change negotiations. While countries are repeatedly trying for regional cooperation, they find it difficult to get it through. As realism assumes that states are the primary actors in the anarchic realm of international politics, they are restricted to seeing threats only from the power maximization strategies of other states (Habib, 2011). Devoid of a central authority, states will fear that others may cheat on any climate change agreements. The reason being that very often they lack sufficient information to know that each states have common interests.

(ii) Liberalism

Liberalism differs from realism as it assumes a much more benign international environment and the general willingness of states to cooperate for mutual gain. It rejects realism's proposition on the centrality of states (Wallerstein, 1979: 1–37; Holsti, 1970; Jervis, 1976). Liberalism thinks universal peace is feasible, if only human beings could discover the reasoning capacities where they all share so as to devise effective mechanisms of international governance (Laferrière and Stoett, 1999: 108). Liberalism put forth that states having strong mutual self-interest can work together for achieving certain mutual benefits (Martin, 1995: 39–51; Haas, Keohane and Levy, 1994; O'Neill, 2009). According to liberalism therefore, the states view one another not as enemies but as partners which need to secure greater comfort and well-being for their own people (Grieco, 1988: 489). Liberalism inclines to focus on political economy or gains arising from mutual economic cooperation rather than conflict arising from the competitive acquisition of power (Ganesan, 2010: 5). In contrast to realism, liberalism has a tendency to see International Relations in optimistic terms. It believes humankind can transcend conflict through the pacifying influence of international institutions and the spread of liberal democratic political systems. Hence, it believes human beings are good and are capable of achieving more cooperative and less conflictive. This is possible (Rourke, 2005: 18) either through current government structures or through new models of governance to strive for understanding throughout their existence. For liberalism, peace is the normal state of affairs and it can be perpetual because the laws of nature dictate harmony and cooperation between people (Russett and O'Neal, 2001). When states can be jointly benefitted from such cooperation, it is expected that governments will attempt to construct such institutions as for them war is unnatural, irrational and undesirable and it is not a product of some peculiarity of human nature (Gardner, 1990: 23–39; Hoffmann, 1995: 159–177). Similarly, neo-liberalism argues that nations benefit from cooperation in an atmosphere of peace and harmony (Khan, 2016: 16).

While liberalism claims peace is the normal state of affairs, it is found that liberal states are also prone to make war, such as the US (Doyle, 1986: 1151–1169). It is seen the US led coalition forces launched incursions and bombardments quite often. Liberal governments who represent their countries in international environmental negotiations are not entirely free to formulate policy positions. It may be possible for government's representatives to ignore domestic constituents in the pre-negotiation phase, but national governments in democratic states ultimately rely on majorities in legislatures or in public referenda in order to ratify international agreements. Climate change has emerged into the mainstream of public consciousness in the Western Liberal democracies for more than two decades. Climate change has become a routine part of the media and street level discourse too. But when it comes to strategy to mitigate climate change both street-level 'common sense' and political 'wisdom' sit somewhere quite separate from the scientific consensus (Calder and McKinnon, 2012: 1). The ratification of global environmental agreements cannot guarantee that they will be implemented successfully while industries, courts, and interest groups often find sufficient leeway to delay and, potentially, circumvent the implementation of international obligations at the domestic level (Luterbacher and Sprinz, 2001: 67; Wolinsky, 1994). Government positions are likely to incline to the domestic pressure groups in anticipation of the challenges posed by ratification. The reason is liberal democracy is totally incompatible with attempts to dictate people's tastes and preferences (Wissenburg, 1998: 7). Nationally, domestic groups always pursue their interests by pressuring the government to adopt favourable policies and politicians seek power by constructing coalitions among those groups. Globally, national governments seek to maximize their own ability to satisfy domestic pressures, while minimizing the adverse consequences of foreign developments. According to Robert D. Putnam (1998: 427–460) neither of the two games can be ignored by central decision-makers, so long as their countries remain independent and sovereign. In contrast to earlier presentations of liberal institutionalism, it is also found that the neo-liberalism accepts the realism's arguments that states are the major actors in world affairs and are unitary-rational agents. It also claims to accept realism's emphasis on anarchy to explain state motives and actions (Neff, 1990). Thus, Robert Axelrod (1984:3–6) seeks to speak about this question: "under what conditions will cooperation emerge in a world of egoists without central authority"? Similarly, there is another argument which says "there is no common government to enforce rules, and by the standards of domestic society, international institutions are weak" (Grieco, 1988: 492). As such, many governments, including those who claim themselves to be democratic defy the decree given by global institutions. So, it is witnessed as member of the UNFCCC the Clinton Administration signed the Kyoto Protocol, but it never submitted it to the Senate for ratification because President Bill Clinton knew the Protocol would not be ratified (Kahn, 2003: 570). This is due to his knowledge the fate of the Kyoto Protocol rests in the hands of one hundred individuals in the U.S. Senate as their actions regarding its ratification would be crucial. According to the Kyoto Protocol, developing countries – including China, India, Brazil and South Africa– would face no restriction on their emissions but were encouraged to adopt policies to promote greener growth. Only the industrialized countries would be legally obliged to cut their GHG emissions i.e., 5 per cent on 1990 levels by 2008–2012. Further, to show the US's rejection on the Kyoto Protocol, on March 27, 2001, the Bush Administration officially rejected the Protocol when Christine Whitman says "we have no interest in implementing that treaty" (The New York Times, March 28, 2001; Weekly Compilation of Presidential Documents, 1997: 1629–1634) and Condoleezza Rice, the National Security Advisor went a step further when she says "Kyoto is dead" (Kluger, 2001). Further, on June 11, 2001 speech on global climate change, President George W. Bush asserted that the Kyoto Protocol is "fatally flawed in fundamental ways" because developing countries are not bound by it (President George W. Bush, 2001; Vespa, 2002). Speaking at the UN climate summit in New York (2014), President Barack Obama said, "We will do our part, and we will help developing nations do theirs. But we can only succeed in combating climate change if we are joined in this effort by every nation- developed and developing alike. Nobody gets a pass" (King, 2015). Although huge emitters of GHG such as the US, China and other powerful developing countries have realize to contribute- to lessen their emission, yet distinction between rich and poor countries remains. Presently, it appears that it is not easy to push nations to accept positions beyond their own proclaimed self-interest. The US withdrawal from the Paris Agreement is a clear example (Hai-Bin *et al.*, 2017:220–225). Similarly, many vulnerable developing countries cannot be viable partners without funding for adaptation from the rich nations. With this understanding, adaptation funding is viewed as inducing developing countries to go for mitigation (Buob and Stephan, 2013). In this way, national self-interest pressurize many nations towards free riding so that we are currently failing by a wide margin to do what may be required for long-run stability of GHG emissions (Luterbacher and Sprinz, 2001: 5).

Another assertion says liberal democracy as a political model may be incapable of responding to climate change as the short election cycles of liberal democratic political systems create an inherent in activity which prevents government from tackling long-term problems (Held and Hervey, 2011: 89–110; Shearman and Smith, 2007: 23). The policymakers who face periodic elections generally tend to underrate the future (Shearman and Smith, 2007: 23). The lack of a cooperative mechanism at the global level is significant example in this regard. The environmental adaptation which has been on the political agenda since the publication of the Brundtland Report or World Commission on Environment and Development (WCED, 1987) is still a convergence of opinions amongst different actors (Gupta, 2010b: 68). It is illogical to say the liberals are pacifists, for they do recognize self-interest will unleash passionate quests for power and wealth on the part of some states (Sand, 1990: 36). The presence of the so-called democratic/ liberal states in climate negotiating table means large potential range different positions and proposals to present, discuss and reconcile. Perhaps, it is theoretically feasible a huge number of liberal negotiating parties will share similar views, the fact that such negotiations taking place may denote disagreements when liberal negotiators have several opinions of their own (Depledge, 2006: 8).

(iii) Constructivism

Another theory for regional cooperation in International Relations is constructivism. This theory focuses on regional awareness and regional identity, on the shared sense of belonging to a particular regional community and on what is called "cognitive regionalism" (Hurrel, 1995: 84; Karns and Mingst, 2005: 50). For constructivism, the world can be changed by the change of ideas. Human ideas which include vision, determination and interest can play a vital role as designer and constructor of the structure of social norm and identity and vice versa. Regional organization is from this point of view a social crystallization to sustain the collective identity of a particular region which is via the realization of states and in turn such organization will shape the states' interests. It is due to this rationale identity is based on social norm and mutual interest of ideation and wellbeing. Therefore, an establishment of any regional organization means an identification of 'oneness' versus 'other-ness' or 'outsiders' in order to construct and instruct member states (Karns and Mingst, 2005; Palmujoki, 2001).

For constructivism, anarchy is not an unavoidable feature of international reality. Rather it gathers on the hypothesis where actors are shaped by the socio-cultural milieu in which they live (Wendt, 1998: 102–117). It has a tendency to believe that relations between states are always sincere and states actually endeavour to convey and appreciate each others' motives and intentions deliberately if not clearly (Wendt, 1992a: 183). Differences to constructivism are understood not as a collision between forces or entity (i.e., states conceived as units), rather as a disagreement or misunderstanding (Jackson and Sorensen, 2003: 253–258) or lack of communication between the conscious agents. For such a reason, regional cooperation can be a form of political community. Alexander Wendt's (1992b: 397) 'identity move' assumes that actors acquire identities, in order to participate in collective meaning. He says (Wendt, 1999: 224) identities are the property of international actors that generate motivational and behavioural dispositions, which means identities are rooted in an actor's self-understandings. This confirms that constructivism may be able to demonstrate the differences between states which have been resolved through normative discourse. However, very often disagreements about norms are determined by power and interests, not discourse. It is seen the US has differed with many other countries (Krasner, 2009: 11) on the appropriate response to the allegation of the disobedience of Saddam Hussein to the international community. It went to war against Iraq (Baylis and Smith, 2005: 178) citing the previous Security Council resolutions and relies on a coalition of the willing that was forged in the immediate aftermath. Thus, there is a possibility that even when mutual concern exists between neighbouring states, the powerful states will always violate the autonomy and the integrity of weak ones (Krasner, 1999). There is a pervasive element of deception in the relations between many states. Even Alexander Wendt (1992b: 396–399) has the opinion where the way international politics is conducted; it is made, not given, because identity and interests are constructed and supported by inter-subjective practice. In fact, everything which is inter-subjective is uncertain.

The saga "anarchy is what states make of it" is connected with a branch of the constructivism (Wendt, 1992b: 391–425). Constructivism also argues identities and interests in international politics are not stable- they have no pre-given nature. This is true for the identity of the sovereign nation-state as it is for the identity of international anarchy. The important thing is to look at how identities and interests are constructed- how they are made or produced in and through specific international interactions (Onuf, 1989; Wendt, 1994: 384–396). According to Cynthia Weber "we know from our own individual experiences that today we are not exactly who we were yesterday, and we are unlikely to be exactly the same tomorrow. Our identities- who we are- change, as do our interests- what is important to us. But the success of constructivism on climate change depends on an important move". The saga "anarchy is what states make of it" means that states decide what anarchy will be like- conflictual or cooperative. By making the state the key decision-maker about the "nature" of international anarchy, constructivism contradicts its own argument that identities and interests are always in flux. It allows the interests of states- conflictual or cooperative- change. But by making the character of international anarchy dependent on what states decide to make it, constructivism produces the identity of the state as decision-maker, and this identity cannot be changed. If the identity of the state as decision-maker were questioned, the constructivist myth "anarchy is what states make of it" would not function (Weber, 2001: 60; Wendt, 1992b: 391–425).

Neo-realism is also not prepared to accept the suggestion given by constructivism where states can easily become friends due to their social interaction. It believes such a goal may be desirable in principle, but not feasible in practice as the composition of the international system influences the states to behave as egoists. Anarchy (Mearsheimer, 1995: 82–93), offensive capabilities and uncertain intentions combine to leave states with little alternative but to compete aggressively against each other. It matters whether conflicting countries place themselves primarily in national or regional terms. The reason (Paterson, 2009: 164) is at the most basic level institutions construct not only how states interact but also what states are. In the field of climate change in particular, (Young, 1989: 349–375) states face considerable ambiguity about its outcomes. While concerns about climate change are mounting, differences between countries are also well-known. Seeking alternative to address this problem through constructivism may be a huge challenge and could be an impossible thing to achieve.

(iv) Functionalism

Functionalism is another theory which emerged during the inter-war period as a working philosophy which visualizes a gradual evolution of peaceful, unified and cooperative world. Rather than the self-interest (Rosamond, 2000) that realism sees as an appealing factor in International Relations, functionalism focuses only on common interests shared by member states. Functionalism (Griffiths and O'Callaghan, 2002: 116) is the idea in which international cooperation should begin by dealing with specific transnational problems. Functionalism hopes in the possibility of specifying technical and non-controversial aspects of governmental conduct and of weaving an ever-spreading web of international institutional relationships on the basis of meeting such needs (Haas, 1964: 6). Its argument says common interests across national boundaries appear as technical and social problems. In the case of climate change, this seems rather straightforward, as every nation presumably has an interest in averting the damaging effects of a destabilized climate. International cooperation is expected to begin as a solution to these transnational problems (Mitrany, 1943: 6; Long, 1963: 368) or the binding together of those interests which are common to the states. There are scholars (Haas, 1964: 6–7; Imber, 1984: 110) who have stated that, to functionalism, 'function means task'. It means an 'organizational task' (McLaren, 1985: 142). David Mitrany (1975: 27) the main advocate of functionalism has visualized the limitations of nation-state in the realm of mankind because government cannot protect them from such transnational problems. He emphasizes the scope of international organization to deal with common problems existing across the national borders (Groom and Taylor, 1975: 112–113). The functional logic which he defends most certainly play a key role in ushering in the new era of international organization, turning to technical experts and to codified law for a solution in containing problems.

Functionalism attributes power to the organization itself and bypassing the states (Wallerstein, 1974). But it does not aim to create a world federal structure; rather it stresses the sharing of sovereignty instead of its total surrender. It proposes a gradual approach towards regional unity which will aim to isolate, and in the end will obsolete the stubborn institutional structures of international organization (Tanter, 1969: 398–401). So, it approves the growth of non-political cooperative organizations involving economic and socio-cultural aspects (Tanter, 1969: 398–401). To functionalism these aspects are regarded as functional areas in which it is achievable and desirable by the states to cooperate. Such functional organizations working in these aspects are assumed to be less opposed by national governments when non-political issues are being addressed because it will be jointly advantageous for the states who involved. The basic calculation running behind such assumption is that it is easier to establish a functional organization than to develop pretentious political institutions which risk the national sovereignty of nation-states (Couloumbis and Wolfe, 1981: 305). These aspects are regarded necessary as well as desirable for states to cooperate. Such institutions as A. J. R. Groom (1991: 66) views will be global, subnational and transnational according to their needs for the development of a transnational community. Therefore, when an internationally cooperative venture works for mutual advantage (Couloumbis and Wolfe, 1981: 305–306) then it will inspire for the participating states to enter into cooperative ventures in some related functional areas, such as economic in addition to other aspects.

For this reason, David Mitrany (1975) enthusiastically hoped functionalism would yield a new sense of peace, not a peace that would keeps nations quietly apart but a peace that would bring them actively together, not the old static and strategic view but a social view of peace. States must put their faith (Mitrany, 1975: 92) not in a protected peace but in a working peace. Governments should relinquish the sole right to make legislation (national sovereignty) over a range of matters, (Mitrany, 1975: 92) in favour of joint decision making with other governments (pooled sovereignty). Henceforth, decisions that are previously taken by national government alone would be taken together with other governments. Initially, functionalism stressed for international integration or for collective governance and 'material inter-dependence' between states where it developed its own internal dynamic as states integrate in limited functional, technical, and/or economic areas (Mitrany, 1933: 101). From the discussion it is found that functionalism primarily focuses on international integration. This volume adopts function-alism as a suitable approach as it assumes international integration is the first step towards regional cooperation.

In terms of climate change, the demand for an international institution originates from the fact that this problem is a transnational dilemma and it poses exceptional challenge to the states and the world communities. A stable climate change can be regarded as transnational good. The demand for a cooperative institution originates from the fact that climate change is a transnational problem and it poses exceptional challenge to the states. Hence, functionalism is an appropriate framework since it argues that states agree to arrangements/agreements to solve common problems for common good as they cannot solve them unilaterally (Simmons, 1998). Since all nations in South Asia are facing the similar risks from climate change, there is a feel-ing amongst these nations to work together on this issue because it is likely they will benefit from a regional cooperation. It is within the framework of functionalism which seeks cooperative ventures based on mutual advantage and in which participating states have common problems to address as well as provides for re-evaluating their interests in the context of mutual benefit that this work titled: *Climate Change in South Asia: Politics, Policies and the SAARC*, has been undertaken. The next chapter will focus on climate change and its effects in the countries in South Asia- Afghanistan, Bangladesh, Bhu-tan, India, the Maldives, Nepal, Pakistan and Sri Lanka.

3 Climate change danger in South Asian countries

I Introduction

South Asia's geographical expanse includes a variety of mountains, plateaus, dry and marshy areas, river basins, as well as beaches. It extends from the Himalayan mountain ranges in the north to the Indian Ocean in the south. It has many rivers like the Indus, Ganges, Brahmaputra, Meghna, Godavari, Mahanadi, and Narmada (SAEO, 2009: 2) that support millions of people in the region. Its total land area comprises only about 3 per cent of the world's land mass (Obaidullah, 2010: 1–2). But it is inhabited by around 1.8 billion, which accounts for 24 per cent of the total world population (Sawe, 2018). The countries within it share geographical, socio-cultural, and civilizational pasts. A home to one of the oldest civilizations of the world, South Asia consists of eight sovereign states of different sizes and capabilities (Afghanistan, Bangladesh, Bhutan, India, the Maldives, Nepal, Pakistan, and Sri Lanka) who are negligible contributors to GHG emissions, yet climate change is manifesting itself in frequent disasters – both to gradual changes in temperature and sea level rise, to the increase in climate variability and extremes, more intense floods, droughts, erratic rainfall, and cyclones (Mani et al., 2018; World Bank, 2018).

The majority of the world's poor (56 per cent) live in this region and the bulk of its population are surviving with under $1.25 per day where agriculture continues to be an important sector for their livelihoods (UNDP, 2015; Chatterjee and Khadka, 2011: xiii). The World Bank (2015) recently reported that around 23 per cent (336 million) of the people who are consistently hungry live in South Asia. According to the Human Development Index 2010 and Global Hunger Index (GHI) 2014 (Islam et al., 2016: 209), the development indicators in South Asian countries are comparatively low. The United Nations Development Group (2014) estimates that more than 50 per cent of the world's undernourished children are in South Asia, predominantly in rural areas. South Asia's vulnerability to climate change's disasters is severe (World Bank, 2009b: 4), largely due to overpopulation and the extreme poverty of the people. Since the poorer households devote more of their money to buying food, they are the most sensitive to weather-related shocks that can make daily staples unaffordable.

Specifically, Afghanistan is a landlocked mountainous country having an arid and semi-arid continental climate with cold winters and hot summers (Savage *et al.*, 2009a). Despite having numerous rivers, large parts of this country are dry (ICARDA, 2002). Bangladesh is largely a delta plain of one of the largest river systems of the world (MoEF, 2005). Only a part of its south-eastern area is hilly. Floodplains occupy 80 per cent of the country (Rashid, 1991). Thus, it is exposed to a number of natural hazards like cyclones, floods, and riverbank erosion which have habitually displaced a large number of people (Karim, 2013: 39; Seraj, 2013). Known as the Land of the Thunder Dragon, the landlocked and isolated mountain kingdom of Bhutan is home to a variety of climates and ecosystems (Thinley, 2009: 1–17). It has a wide range of altitudinal zones and micro-climatic conditions which have created highly diverse ecosystems and a complex pattern of climatic conditions (RGB, 2006; Armington, 2002). India, the largest country in the region, comprises not only the bulk of the Indian subcontinent but also the islands of Andaman and Nicobar as well as the archipelago of Lakshadweep. Its climate is strongly influenced by the Himalayas and the Thar Desert, both of which drive the economically and culturally pivotal summer and winter monsoons (Chang, 1969: 373–396). Nepal is landlocked and located in the Himalayas. It is one of the ten poorest countries in the world (ADB, 2017; NPC, 2010; IFPRI, 2010). Its climate varies from the tropical to the arctic within the 200 km span from south to north. Much of Nepal (Dutt and Geib, 1998) falls within the monsoon region with regional climate variations.

Two-thirds of Sri Lanka is lowland. While heavy rains occur during the monsoon in the northeast, the region is otherwise hot and dry (Burt and Weerasinghe, 2014: 242–263; Abhayasinghe, 2007: 52–53; Basnayake, 2007: 54–55). This country faces grave risk from various climate change effects, such as sea level rise, floods and droughts, variability and unpredictability of rainfall patterns (SLMOE, 2010). Pakistan has a diverse array of landscapes with a great diversity in temperature and precipitation. Its territory encompasses portions of the Himalayas, Hindu Kush, and Karakoram mountain ranges, making it home to some of the world's highest mountains, including K2, the world's second-highest peak. Most of Pakistan has a dry climate and it receives less than 250 mm of rain per year, while the northern and southern areas have noticeable climatic differences (Chaudhry, 2017: 1). Numerous environmental problems threaten the economy and its population's health (Chaudhry, 2017: 1; Salma, Rehman and Shah, 2012: 37). The Maldives is an archipelago located 500 km (300 mi) southwest of the southern portion of India and consists of about 1,190 low-lying coral islands, of which only 200 are inhabited. These islands are scattered over 859,000 km^2 (MoEEW, 2012). Being low-lying small islands, sea level rise will seriously damage the Maldives economy or even wipe it out entirely (Shaiq, 2006; MoEEW, 2007, 2012; Kench *et al.*, 2006: 177–180; Majeed and Abdulla, 2004: 243–255).

As a region, South Asia is perceived as having plentiful water resources, including the magnificent Himalayan snows, a vast network of perennial rivers, high monsoon rainfall, and rich groundwater aquifers. The three largest river systems of the Indus, Ganges, and Brahmaputra are partially fed by snow and glaciers, and the southwest monsoon accounts for 70 per cent to 90 per cent (Mirza and Ahmad, 2005: 1–4) of the annual rainfall over most of the region. This makes half of the land area of the region arable (Kumara, Mittal and Hossain, 2008: 145–172) or suitable for crop production. The challenges of climate change (Stern, 2006; Byravan and Rajan, 2009: 134; Islam, Nazrul and Afroz, 2010) are however hugely demanding, interconnected, and increasingly viewed as the foremost problem of the 21st century. South Asia's geography coupled with its high levels of poverty and population density (Hirji, Nicol and Davis, 2017: 4; Sterrett, 2011) render the countries of the region fragile to the multitude of climate change–related hazards, such as cyclones, floods, landslides, drought, extreme temperature, erratic rainfall, heat waves, storm surges, and GLOFs among others. Floods affect India, Bangladesh, Nepal, Pakistan, and Sri Lanka. Drought affects Afghanistan, India, Pakistan, and parts of Nepal and Sri Lanka. Cyclones often hit Bangladesh, India, and Sri Lanka. Landslides occur in mountainous regions of India, Pakistan, Nepal, Bhutan, and Sri Lanka. The Maldives, Bangladesh, and Sri Lanka are subject to coastal erosion, sea water and salinity intrusion (Kafle, 2017: 2). The AR4 and theAR5 of the IPCC (Solomon *et al.*, 2007; Denman *et al.*, 2007; IPCC, 2007; Le Treut *et al.*, 2007; Philander, 2008; Edenhofer *et al.*, 2014) highlight that climate change would bring about the following challenges to South Asia:

- Melting of glaciers in the Himalayas would increase flooding and this in turn would alter long-term water resources and its availability in South Asia.
- Climate change would compound pressure on natural resources and the environment owing to rapid urbanization, industrialization, and economic development.
- Crop yields in South Asia would likely decrease up to 30 per cent by the middle of the 21st century.
- Periodic floods and droughts would impact on the health of the population.
- The rising sea level would exacerbate inundation, storm surges, soil erosion, and other coastal hazards.
- Human settlements and infrastructure, heat-related deaths, threaten economic growth and human security in complex ways.

> (Parry *et al.*, 2007; Solomon *et al.*, 2007; Denman *et al.*, 2007;
> Le Treut *et al.*, 2007; Philander, 2008; Edenhofer *et al.*, 2014;
> Jimenez Cisneros *et al.*, 2014)

In addition to natural causes these effects are assumed to have been compounded by deregulated human activities (Gore, 2007: 55; Philander, 2008:

210). The points previously noted state that South Asian countries are in tremendous danger from climate change–related effects. The effects of climate change on the region are due to several factors: the size of the population (over a billion people), the level of under-development (nearly 600 million people in South Asia still survive on less than US$1 a day), over-dependency on agriculture (60 per cent of employment is linked to agriculture), and the rapid rate of largely unplanned urbanization (in 2008, 464 million people lived in urban areas and nearly 40 per cent in slums) (Islam, Hove and Parry, 2011). Owing to ever-increasing development, countries in the region found it difficult to meet the targets of the MDGs in 2015. If the climate change–related hazards continue in the same manner, it will still be a serious challenge to the South Asian countries to achieve the SDGs – the targets developed by the UN General Assembly to succeed the MDGs which ended in 2015. The SDGs set the stage for developing countries to reduce extreme poverty and the problems that accompany it like hunger, poverty, good health and well-being, education, gender equality, water availability and sanitation, clean energy, reduction of inequalities, sustainable development etc. which are to be eliminated by 2030.

All the South Asian countries share more or less similar socio-economic characteristics and are facing comparatively more or less similar problems in the parameters of development. The South Asian region faces pockets of deep poverty, rising inequality, severe energy shortages, and high rates of malnutrition (Sarkar *et al.*, 2013: 2475–2476; IMF, 2011). Table 3.1 shows the broad human development indicators of the South Asian countries.

II Climate change–related disasters in the SAARC region

Statistics show (Edenhofer *et al.*, 2014; Jimenez Cisneros *et al.*, 2014) South Asia suffers a high number of climate change–related disasters. As per the International Emergency Disaster Data-Base (EM-DAT, 2011) global database of disasters, 2009 had witnessed 42 natural disasters in South Asia. A total of 3,379 persons have been killed (South Asia Disaster Report, 2009, 2010: 6–8; South Asia Disaster Report: 2007–2011, 2011). In 2010 alone, 3,863 died due to climate change–related events (South Asia Disaster Report: 2007–2011, 2011: 3). Quoting global statistics, while presenting a draft SAARC Agreement on Rapid Response to Natural Disasters to the Intergovernmental Meeting in Colombo (in May 2011), the former secretary general of SAARC pointed out that over the past 40 years, South Asia faced as many as 1,333 disasters that killed 980,000 people, affected 2.4 billion lives, and damaged assets worth US$105 billion (Memon, 2012: 8; Sharma, 2011). This loss is perhaps the highest among the recorded disasters in various geographical regions (Mall and Kumar, 2014: 25; South Asia Disaster Report: 2011, 2013). The 7.9 magnitude April 2015 quake in Nepal (The Times of India, June 2, 2015; UNDP, 2015) affected around 2.8 million Nepali. This earthquake destroyed the oldest neighbourhoods of Kathmandu, and was

Table 3.1 Human Development Index (HDI) Ranking of SAARC countries 2018

Sl. No.	Country	World Ranking	Human Development Index (HDI) (Value)	Life expectancy at birth (years) SDG3	Expected years of schooling (years) SDG4.3	Mean years of schooling (years) SDG4.6	Gross national Income (GNI) per capita (PPP $) SDG 8.5
1	Afghanistan	168	0.498	64.0	10.4	3.8	1,824
2	Bangladesh	136	0.608	72.8	11.4	5.8	3,677
3	Bhutan	134	0.612	70.6	12.3	3.1	8,065
4	India	130	0.640	68.8	12.3	6.4	6,353
5	Maldives	101	0.717	77.6	12.6	6.3	13,567
6	Nepal	149	0.574	70.6	12.2	4.9	2,471
7	Pakistan	150	0.562	66.6	8.6	5.2	5,311
8	Sri Lanka	76	0.770	75.5	13.9	10.9	11,326
World		–	0.728	72.2	12.7	8.4	15,295

Source: UNDP Report 2018.

strong enough to be felt across parts of India, Bangladesh, Pakistan, and the region of Tibet (The Times of India, April 26, 2015). Table 3.2, taken from the *Global Climate Risk Index 2019*, shows the countries of SAARC in the last two decades with their average weighted ranking (CRI score) and the specific results relating to the four indicators analyzed (Eckstein, Hutfils and Winges, 2018: 32–35).

In the year 2017 alone, Sri Lanka was the most affected of the SAARC countries (Eckstein, Hutfils and Winges, 2018: 5), and it ranks second globally. This was followed by Nepal (rank 4), Bangladesh (rank 9), and India (rank 14) (Eckstein, Hutfils and Winges, 2018: 5–7). The main reasons for this risk are the following: in May 2017 heavy landslides and floods occurred in Sri Lanka, which killed more than 200 people and about 600,000 were displaced (Prior, Athas and Mckirdy, 2017; Gettleman, 2017). In the same year, massive rainfall which flooded Nepal, Bangladesh, and India affected more than 40 million people: 1,200 people lost their lives and millions were displaced throughout the region. The floods were so strong that they caused landslides across the Himalayan foothills leaving tens of thousands of houses and vast areas of farmland and infrastructure destroyed (Epatko, 2017; Gettleman, 2017). Similarly, more than 300 people died and millions more were displaced after severe flooding hit parts of India, Nepal, Bangladesh, and Pakistan in 2019 and at least 700,000 were forced to flee their homes (Baynes, 2019). With the rapid growth of population, environmental problems are expected to become more complex and multifaceted (Bloom and Rosenberg, 2011). Speedy climate change (Dessler and Parson, 2006: 1–2) will represent an added threat to other environmental issues such as air and water quality, food shortages, floods and droughts, rise in temperature, endangered ecosystems and biodiversity, and threats to coastal zones and wetlands. The important issues considered for the study are the following: (i) rise in temperature, (ii) floods and droughts, and (iii) food shortages as they pose major problems confronting the region on a regular basis. When the future consequences are projected to be even more severe, it is imperative to study the significance of climate change in the region and investigate its drastic effects.

III Rise of temperature

South Asia is the only region in Asia to receive the status of "highly vulnerable" to the effects of climate change (Solomon *et al.*, 2007; USAID, 2010). The AR4 of the IPCC (Parry *et al.*, 2007) states that temperatures are increasing in every subregion of Asia, but in South Asia, the situation is even stronger. Rising temperatures and glacial melt (GIRA, 2009; GOI, 2009; BMEF, 2009) are noted as reasons for concern. There is escalating worry about the force of climate change on the high mountains because recent satellite observations have confirmed that glaciers of the HKH region are melting, presumably as a result of the steady rise of temperature (Berthier *et al*, 2007:

Table 3.2 Climate Risk Index (CRI) for 1998–2017 (SAARC countries)

CRI Rank	Country	CRI Score	Fatalities (annual average)		Fatalities per 100 000 Inhabitants (annual average)		Losses in US$ million (PPP) (annual average)		Losses per unit GDP (annual average)	
			Avg.	Rank	Avg.	Rank	Avg.	Rank	Avg.	Rank
26	Afghanistan	44.33	288.300	15	1.004	17	100.763	85	0.218	66
7	Bangladesh	26.67	635.500	9	0.433	41	2403.839	11	0.640	29
105	Bhutan	98.17	1.650	139	0.244	66	4.994	152	0.133	83
14	India	36.50	3660.600	2	0.316	48	12822.708	3	0.263	59
176	Maldives	168.83	0.000	173	0.000	173	0.549	174	0.013	160
11	Nepal	33.50	235.300	18	0.896	20	230.830	57	0.438	43
8	Pakistan	30.17	512.400	10	0.315	49	3826.028	7	0.567	33
31	Sri Lanka	48.33	60.750	39	0.305	53	491.048	35	0.294	55

(Avg. = average figure for the 20-year period, e.g. 37 people died in Albania due to extreme weather events between 1998 and 2017; hence the average death toll per year was 1.85.)

Source: Eckstein, Hutfils and Winges, *Global Climate Risk Index 2019*, 2018: 32–35.

327–333; Kaltenborn, Nellemann and Vistnes, 2010: 13; Bolch *et al.*, 2008a: 1329–1340; Bolch et al., 2008b: 592–600). It is assumed that there is an interaction of warming ocean waters and increasing air temperatures which is contributing to the thinning and breaking up of the ice sheets on the mountains. With its total area of 34,660 km² under glaciers, the Himalayas have the region's largest glacier cover (UNEP, 2009: 6). When the summer temperature rises, heat waves are projected to become more frequent and of longer duration, heating the Himalayan glaciers. This will ultimately result in the slow-moving assemblages of ice. Melting of ice will cause lakes to form at the bases of the glaciers (UNEP, 2009: 6; Shrestha, Mool and Bajracharya, 2007; Bajracharya and Shrestha, 2011), which eventually increases the potential of GLOFs. Many glacial lakes will then burst at the banks as the temperature continues to intensify and finally have devastating effects downstream. The large glaciers of Gangotri, Yamnotari, and Go-Mukh, which feed the Ganga and Yamuna rivers, have started to melt at an alarming pace. With the glaciers melting at an ever-increasing rate, the sea level will continue to rise. It is assumed that the rise in the sea level (Church and White, 2011: 585–602; Hunter, 2010: 331–350; Mengel *et al.*, 2016: 2597–2602) is due to thermal expansion and melting of the glaciers, ice caps, and polar ice sheets. There is an assessment which says the rate of global average sea level rise has increased from 1.8 mm/year to 3.1 mm/year from 1961 to 1993 (Dasgupta and Meisner, 2009; Meier *et al.*, 2007: 1064–1067). Different projections predict the sea level will rise between 0.18–0.59 m (Marquina, 2010: 198), or between 1.7 and 3.2 ft by 2100 (Muggah, 2019) or even up to 2 m (6.6 ft) at the end of the 21st century depending on anthropogenic activities and the impacts of warming air and the melting of ice bodies (Pfeffer, Harper and O'Neel, 2008; NOAA, 2012). Since most of the Himalayas' glacial cover is retreating (Kaltenborn, Nellemann and Vistnes, 2010: 13–14; Xu, 2009), it will immensely impinge on the people in the region in two fundamental ways.

First, coastal areas of South Asia (Whyte, 2008: 188–189; Roy, 2007; Harrabin, 2007) will be susceptible to sea level rise. Along the coastal region of Asia, including South Asia (Meehl and Stocker, 2007: 747–845), the rate of sea level rise is reported to be greater than the global average, rising at a rate of 1 to 3 mm per year. In South Asia, this process (Meehl and Stocker, 2007: 747–845; Islam, Hove and Parry, 2011: 9–11) is expected to lead to coastal flooding, intrusion of saltwater into freshwater resources. The disruption of fisheries (Mall *et al.*, 2006) is likely to affect large numbers of poor people. The Andaman and Nicobar and the archipelago of Lakshadweep Islands will not be spared. Bangladesh has (Ayeb-Karlsson, 2017; Banerjee, 2015; Harris, 2014) already faced salt-water intrusion and large fertile ponds have been submerged in the Bay of Bengal. Other than Bangladesh, coastal countries such as India, the Maldives, Pakistan, and Sri Lanka have each emphasized the vulnerability of their coastal zones to the effects of flooding, sea level rise, and coral bleaching (Banerjee and Juneja, 2015; Islam, Hove and Parry, 2011: 12). The Maldives will face the danger of total submergence (Zahir *et al.*, 2016: 4; MMHAHE, 2001), owing to rising sea levels as over 80 per cent of its land area is less than 1 m above mean sea level. Such a scenario will put at risk (Zahir *et al.*, 2016; MMHAHE, 2001; World Bank Group, 2009; Myers, 2002: 609–613; Pfeffer, Harper and O'Neel, 2008: 1340–1343; Anthoff, Nicholls and Tol, 2010: 321–335) the existence of this country. Millions of additional people living in the low-lying areas of Bangladesh, India, Pakistan, Sri Lanka, and the Maldives are expected to be flooded on an annual basis, leading potentially to large migrations.

Second, the frequent numbers of GLOFs (Harrabin, 2007) have caused an increase in the rise and volumetric flow rate of the most important rivers of Pakistan, India, Bhutan, Nepal, and Bangladesh, which in turn led to the impact on lives and infrastructure, as well as threatening the survival of humankind. The reason is that many of the rivers originating in the Himalayas receive (Kaltenborn, Nellemann and Vistnes, 2010) a substantial water contribution from the melting of glaciers. On the other hand (Ives, Shrestha and Mool, 2010: 1; Hasnain, 2002: 412–423; Hasnain, 2000: 8; Nandy, Dhyani and Samal, 2006: 108), glaciers in many parts of the HKH region are currently thinning and retreating, presumably as a result of the current climate change. It is expected (Raina, 2009: 17) that any fluctuation in the water contribution is also bound to have a major effect on the hydroelectric power potential and the irrigation potential of these rivers. Similarly, the rise in temperature (Hales, Edwards and Kovats, 2003) leads to the increase in heatstroke and respiratory diseases that will eventually result in high mortality rates, particularly the rural poor, rickshaws pullers, and outdoor labourers. Thus far, a number of heat wave–related deaths (Hindustan Times, May 31, 2015; Lal, 2003: 1–34; Glum, 2015; Mansoor, 2015; Pachauri and Meyer, 2014) have been reported in India and Pakistan. South Asian cities, especially in India, have witnessed unusually high temperatures, which caused a spike in heat-related deaths (Guleria and Gupta, 2018; Chandra, 2019). The phenomenon of this heat wave was not without prediction. In fact, as early as 2001 (Yassi *et al.*, 2001) it was predicted that heat waves would exacerbate the existing urban heat, which could result in susceptibility of some urban environments to heat-related mortality. By 2100, heat waves in South Asia could be too deadly to survive (Venkatesh, 2018a). The rise of temperature due to climate changes (WHO, 2008; *The Statesman*, 2008) has resulted in a change in the profile of various diseases leading to an increase of diseases in tropical countries. Table 3.3 shows the principal glacier-fed river systems of the Himalayas.

In collaboration with partners in different countries, the International Centre for Integrated Mountain Development (ICIMOD) has embarked on the preparation of an inventory of glaciers and glacial lakes, and identification of potential sites for glacial lake outburst floods (potentially dangerous glacial lakes), in the Hindu Kush-Himalayan region (excluding Afghanistan), using desk-based studies and systematic application of remote sensing. Table 3.4 summarizes the main findings of the studies in Bhutan, Ganges sub-basins, India, Nepal, and Pakistan (excluding Afghanistan).

Table 3.3 Principal glacier-fed river systems of the Himalayas

River	Mountain area (km²)	Glacier area (km²)
Indus	268 842	7890
Jhelum	33 670	170
Chenab	27 195	2 944
Ravi	8 092	206
Sutlej	47 915	1 295
Beas	12 504	638
Jamuna	11 655	125
Ganga	23 051	2 312
Ramganga	6 734	3
Kali	16 317	997
Karnali	53 354	1 543
Gandak	37 814	1 845
Kosi	61 901	1 281
Tista	12 432	495
Raikad	26 418	195
Manas	31 080	528
Subansiri	81 130	725
Brahmaputra	256 928	108
Dibang	12 950	90
Lohit	20 720	425

Source: UNEP (2009: 5).

IV Floods

Glacial melting in the HKH region (You *et al.*, 2017: 141–147; Singh *et al.*, 2011) may give rise to the increase of flash floods in the mountainous regions and other places located in the foothills. Flash floods can be described as rapid flooding of geomorphic low-lying areas: washes, rivers, dry lakes, and basins. It can be due to heavy rain associated with a severe thunderstorm, hurricane, tropical storm, or melt water from ice or snow flowing over ice sheets or snowfields. Flash floods may occur after the collapse of a natural ice or debris dam, or a human structure such as a man-made dam. This is because of the reason that the region's river systems are highly flood prone. In fact, South Asia's geography (Dewan, 2015: 36–42; Ghatak, Kamal and Mishra, 2012: 2; Rahman, 2011) makes it particularly susceptible to flooding. Floods are considered as catastrophes, because when floods occur, they cause wide-ranging damage. Both floods and sea level rise will have an adverse effect in the region. The World Bank (WB) has identified the Top High Impact Coastal Areas in South Asia which are open to floods. The top South Asian coastal areas vulnerable to climate change are Cox's Bazar, Khulna, Chittagong, Bakerganj, and Chandpur of Bangladesh; Jamnagar, Vadodara (Baroda), Thane, Bhavnagar, and Kolkata of India; Karachi of Pakistan; Moratuwa of Sri Lanka. These cities are identified as among the

Table 3.4 Summary of glaciers, glacial lakes, and lakes identified as potentially dangerous in selected parts of Bhutan, India, Nepal, and Pakistan

River basin	Glaciers			Glacial lakes		
	Number	Area (km²)	Ice Reserves (km³)	Number	Area (km²)	Potentially dangerous
Bhutan						
Amo Chu	0	0	0.00	71	1.83	0
Wang Chu	36	49	3.55	221	6.47	0
Puna Tsang Chu	272	503	43.27	980	35.08	13
Manas Chu	310	377	28.77	1383	55.51	11
NyereAma Chu	0	0	0.00	9	0.07	0
Northern basins	59	388	51.72	10	7.81	0
Total	677	1317	127.31	2674	106.77	24
India: Himachal Pradesh						
Beas	358	758	76.40	59	236.20	5
Ravi	198	235	16.88	17	9.16	1
Chenab	681	1705	187.66	33	3.22	5
Satluj	945	1218	94.45	40	136.46	136.46
Sub-basins	372	245	11.96	11.96	0.18	2
Total	2554	4161	387.35	156	385.22	16
India: Uttaranchal (Uttarakhand)						
Yamuna	124	173	17.88	20	0.17	0
Bhagirathi	393	1034	143.41	32	0.44	0
Alaknanda	540	1675	191.36	54	1.37	0
Kali	382	1178	122.78	21	0.51	0
Total	1439	4060	475.43	127	2.49	0
Tista river basin (Total)	285	577	64.78	266	20.20	14
Nepal						
Koshi River	779	1410	152.06	1062	25.09	16
Gandaki River	1025	2030	191.39	338	12.50	4
Karnali River	1361	1740	127.81	907	37.67	0
Mahakali River	87	143	10.06	16	0.38	0
Total	3252	5324	481.32	2323	75.64	20
Pakistan (Indus river basin)						
Swat	233	224	12.22	255	15.86	2
Chitral	542	1904	258.82	187	9.36	1
Gilgit	585	968	83.34	614	39.17	8
Hunza	1050	4677	808.79	110	3.21	1
Shigar	194	2240	581.27	54	1.09	0
Shyok	372	3548	891.80	66	2.68	6
Indus	1098	688	46.38	574	26.06	15
Shingo	172	37	1.01	238	11.59	5
Astor	588	607	47.93	126	5.52	9
Jhelum	384	384	6.94	196	11.78	5
Total	5218	15041	2738.50	2420	126.32	52
Grand Total	13425	30480	NA	7966	716,64	225

Source: Ives *et al.* (2010: 7).

top 25 cities in the world with the most severe human effects from storm surges (Dasgupta *et al.*, 2009: 22–24). Such a scenario (Dasgupta *et al.*, 2007: 44; Najam, 2003: 59–73) will pose unprecedented environmental problems to the national integrity, threatening the survival of millions of the poor.

V Droughts

Afghanistan, India, Pakistan, and Sri Lanka have reported droughts at least once in three years in the past five decades, while Bangladesh and Nepal also suffer from drought frequently. From the mid-1990s (FAO, 2002), prolonged and widespread droughts have occurred in consecutive years in Afghanistan, India, and Pakistan, while the frequency of droughts has also increased in Sri Lanka, Nepal, and Bangladesh. The growing urbanization, changing consumption patterns, and inefficient water use (Jaitly, 2009: 1) will add to the decline in physical water availability in the near future. When the Ganga, Indus, and Brahmaputra rivers become seasonal, which is likely to happen, (Cruz *et al.*, 2007: 469–506) Afghanistan, Bangladesh, India, Nepal, and Pakistan will face water shortages or drought. As the region's population continues to grow, the demand for water will continue to increase, surpassing the available resources (IPCC, 2013; Anglia Ruskin University, 2014). In other words, the water sector will face multiple challenges such as from population increase, urbanization, shifting agriculture and livelihoods, privatization of water rights, over-extraction and resource degradation (Eldho, Sreeja and Madhusoodhanan, 2016: 724–741). The AR5 of the IPCC also confirms (Edenhofer *et al.*, 2014: 476) the future drought situation in the region. The other immediate impact of drought is land degradation. This reality may lead to a decline in food availability. It is further predicted (Lal *et al.*, 2011) that constraints on water resources may lead to other problems like health, human settlements, power generation supply, and other related sectors.

VI Food shortages

A large number of the population in the region survives on under US$1.25 per day (Sterrett, 2011: 4; Canuto, 2013; Ghosh, 2014). This number is highest in countries like Afghanistan (36 per cent), Bangladesh (43 per cent), India (33 per cent), Nepal (25 per cent), and Pakistan (21 per cent) (6th SAES Theme Paper, 2013; Proceedings of the 7th SAES, 2014). Agriculture is a key source (Chatterjee and Khadka, 2011: xiii) of employment for the majority of the population. It contributes 22 per cent of the region's Gross Domestic Product (GDP) (World Bank, 2013). Approximately 85 per cent of Afghans and Nepalese depend on rain-fed agriculture for their livelihoods while in Bhutan and India the figures are 80 per cent and 64 per cent respectively (USAID, 2010; NMOE, 2010; BMEF, 2005; RGB, 2008). But agricultural production in South Asia is also prone to high risks from large variations in weather. Climate change–related effects such as temperature rise, erratic rainfall, the regularity and severity of extreme droughts have largely affected crop yields and threatened food security (Aryal *et al.*, 2019: 1–31). The definition of food security used in this volume is the one adopted at the 1996 World Food Summit held in Rome. Food security exists when all people, at all times, have physical and economic access to sufficient safe and nutritious food that meets their dietary needs and food preferences for an active and healthy life. The decline of crop yields has threatened food security for the overwhelming majority of the population of South Asia (Pachauri and Meyer, 2014: 48; IFPRI, 2010; NMoE, 2010; BMEF, 2005; ESCAP, 2012: 61–62), which is estimated to be the home of about 40 per cent of the world's hungry population. In reality, the status of food security in the region has deteriorated due to declining agricultural growth (1993–2006), both in terms of its contribution to GDP and its share of the labour force (Mittal and Sethi, 2009; Chatterjee and Khadka, 2011; Ludi, 2009). The future projections of climate change indicate that South Asian agriculture is likely to be affected by warming during this century. It is predicted that a rise in temperature may reduce yields of rice, wheat, other cereals, and certain cash crops significantly (Chatterjee and Khadka, 2011). A further change in socio-economic conditions, such as population expansion, growing urbanization, changing patterns in consumption will affect the agricultural sector, the food supply and demand system in the near future (World Bank, 2009b: 76; Iqbal and Amjad, 2012: 376). The International Food Policy Research Institute (Nelson *et al.*, 2009) records for South Asia say the yield in terms of rice (23 per cent), wheat (57 per cent), and maize (36 per cent) are projected to decline. At the regional level, the 2018 Global Hunger Index (GHI) scores for South Asia stand at 30.5, that is dramatically higher than the other regions of the world. This score indicates a serious level of hunger, standing in stark contrast to those of East and Southeast Asia, the Near East and North Africa, Latin America and the Caribbean, Eastern Europe and the Commonwealth of Independent States, which range from 7.3 to 13.2 and indicate low or moderate hunger levels. Further, South Asia has the highest child-stunting and child-wasting rates of any region, followed by Africa south of the Sahara. In terms of undernourishment and child mortality, it is next only to Africa south of the Sahara which has the highest rates (UNICEF, WHO and World Bank, 2018; von Grebmer *et al.*, 2018: 12). Understanding the effects of climate change on agriculture is a subject of huge importance as it links directly to food shortages and the lives of millions in the region. Table 3.5 shows the 2018 GHI scores (2000–2018) for South Asian countries.

In a region where famines like floods and droughts occur frequently, it is natural to assume that these crises will likely worsen poverty while pushing food prices and other resource costs higher. In the context of South Asia, therefore, climate change is not only the biggest environmental threat but

Table 3.5 2018 Global Hunger Index scores (SAARC countries)

Country	2000 '98–'02	2005 '03–'07	2010 '08–'12	2018 '13–'17
Afghanistan	52.3	43.2	35.0	34.3
Bangladesh	36.0	30.8	30.3	26.1
Bhutan	–	–	–	–
India	38.8	38.8	32.2	31.1
Maldives	–	–	–	–
Nepal	36.8	31.4	24.5	21.2
Pakistan	38.3	37.0	36.0	32.6
Sri Lanka	22.3	21.2	17.9	17.9

Note: Data are not available or not presented. Some countries did not exist in their present borders in the given year or reference period.

Source: von Grebmer *et al.* (2018: 50).

Table 3.6 Potential effects of climate change in SAARC countries

Country	Potential Effects
Afghanistan	Earthquake, drought, floods, landslides, extreme winter conditions, avalanches, sand and dust-storms, agriculture paste.
Bangladesh	Cyclone, floods, saline water intrusion, drought, epidemic.
Bhutan	Glacial lakes outburst floods, flash floods, landslides, droughts, forest fire, epidemic.
India	Floods, droughts, earthquake, cyclone, tsunami, sea water intrusion, epidemic, landslides and forest fire.
Maldives	Tsunami, sea water intrusion, floods, cyclone, earthquake.
Nepal	Floods, landslides, earthquake, epidemic, glacial lakes outburst floods, avalanche, fire.
Nepal	Earthquake, floods, landslides, sand-storm, drought, avalanche.
Pakistan	Tsunami, floods, landslide, drought, cyclone.

Source: Kafle (2017: 2).

also the likely cause of extraordinary social, economic, and political problems in the near future. It is projected (Ratna, 2015) to trigger conflicts, violence, and migration due to water scarcity, food insecurity, lack of shelter, and livelihood constraints. Table 3.6 shows the potential effects of climate change to SAARC's individual countries.

After a quick look at Table 3.6, it can be stated that the soaring rates of population growth, high population density, and continuing high rates of poverty will make South Asia one of the most vulnerable regions to the effects of climate change. The succeeding paragraphs will discuss the effects which climate change brings to the individual countries of South Asia.

VII Climate change hazards to individual countries of South Asia

Afghanistan

Afghanistan is frequently ranked among the countries most vulnerable to climate change (Kreft *et al.*, 2015, 2016). The country is regularly hit by extreme weather or climatic events, causing substantial economic damage and loss of lives showing that even today Afghanistan is not sufficiently adapted to the current climate (Kreft *et al.*, 2015, 2016; Aich *et al.*, 2017). Despite this very alarming situation, climate data and information is scanty in Afghanistan (World Bank, 2014). Large parts of the historical datasets have been lost during the political turmoil in the country. Nevertheless, despite the absence of good long-term climatic records and available data, those who have modeled climate change projections for Afghanistan (McSweeney, New and Lizcano, 2008) note that since the 1960s, the mean annual temperature has increased by 0.6°C and by 0.13°C on average per decade. During the same period, the frequency of hot days and hot nights has increased in every season. According to Savage *et al.* (2009a), the mean annual temperature in Afghanistan is projected to increase by 1.4 to 4.0°C by the 2060s, and 2.0 to 6.2°C by the 2090s. However, the INDC report mentions an expected warming of 1.5°C until 2050 and of approximately 2.5°C until 2100 under an "optimistic scenario" (GIRA, 2015). According to the Government of Afghanistan, the "pessimistic scenario" projects a 3°C increase until 2050, with further warming up to 7°C by 2100 (GIRA, 2015). It is worth mentioning that the Ministry of Transportation installed the first meteorological weather stations in Afghanistan in 1953 (GIRA, 2009: 65) only in selected locations. The Meteorological Department of the Ministry of Transportation has collected and monitored hydro-meteorological data (GIRA, 2009: 29), as did a number of other ministries and projects, including the Ministry of Agriculture, Irrigation and Livestock (MAIL). As of September 2007, AgroMet has installed (GIRA, 2009: 29) many weather stations and is managing 89 sites, all of which put on record the rainfall and snowfall across Afghanistan. Climate change is expected to intensify regional contrasts in the precipitation pattern: dry areas are expected to become drier, and wet areas become wetter. In contrast to temperature projections, the uncertainty of model projections for precipitation is higher, and regional and seasonal differences are more distinct. The amount of rainfall over the country has decreased by 0.5 mm per month or 2 percent per decade in the past 50 years (GIRA, 2015).

Due to its mountainous topography, the Hindu Kush region is particularly prone to natural disasters such as landslides, avalanches, floods, flashfloods, and substantial erosion of fertile soil, rendering the poor susceptible to climate change (Savage *et al.*, 2009a; UNEP, 2004). In addition, the overflow of rivers due to ill-timed and heavy rainfall may lead to the collapse of irrigation canals, destruction of agricultural lands, loss of crops and live-stock, collapse of dwellings, spread of epidemic diseases, and destruction of infrastructure such as roads and bridges and damage to the national economy (Aich *et al.*, 2017: 8; GIRA, 2009: 70). Afghanistan has been hit with the worst flooding in recent years. In June 2014 flash floods killed around 80 people in northern Afghanistan, washed away hundreds of homes, and forced thousands to flee (The Guardian, June 7, 2014; Mirajnews, June 8, 2014). In 2018, flash floods killed at least 34 people in several Afghan provinces and caused serious damage to property and livestock (Sediqi, 2018). In 2019, thousands of homes were swept away as rains hit the country after a devastating drought (Janjua and McVeigh, 2019). Other climate change–related effects such as rising temperatures, droughts, and floods present the greatest hazards to ecosystem services and means of livelihood. Means of livelihood like irrigated agriculture, livestock herders, and dry land farmers are also considered the most susceptible to climate change–related effects (NEPA, 2012; GIRA, 2009).

The Hindu Kush region which is approximately 7.5 per cent of the land is covered with permanent snow and ice and receives the highest amount of precipitation. Therefore, it is a major water source, feeding main rivers like the Amu Darya (FAO, 2016a; Aich *et al.*, 2017: 5). From 1960, Afghanistan has experienced immense drought in 1963–1964, 1966–1967, 1970–1972, and 1998–2006 (GIRA, 2009: 66). Severe drought conditions between 1998 and 2001 were the worst in the last five decades (Savage *et al.*, 2009a). The report says that even the Amu Darya's watercourse that runs along Afghanistan's border with Tajikistan, Uzbekistan, and Turkmenistan has been flowing less. In the late 1980s this river, which provides water to Afghanistan and its entire region, failed to reach the Aral Sea (Glantz, 2002). There are studies (Hagg *et al.*, 2013: 62–73; Jaramillo and Destouni, 2015: 1248–1251; Jarsjö *et al.*, 2012: 1335–1347) which assert that climate change is now influencing the water resources in the Amu Darya basin, including the Aral Sea as part of this basin. Another study (Blood, 2001) says the Afghans are facing climatic hazards such as drought due to increasing temperatures and the 120-day winds. Prolonged droughts due to rising temperatures (Field *et al.*, 2012; Aich *et al*, 2017; GIRA, 2009) present the greatest hazards to ecosystem services and livelihood security in Afghanistan. Such worsening climatic conditions will bring hardship to the inhabitants of the desert and the steppe lands and affect Afghan's socio-economic growth.

Despite drought in the last few years, the use of water keeps increasing but without any proper water reservoir and dams to store water necessary for irrigation purposes. The growing effects of more frequent and intense droughts on reservoirs and groundwater threaten the water supply of the entire population in its most arid regions. Accessibility to clean water is a serious challenge (PAN, September 17, 2011, cited in Ahmadzai, 2013: 196) for the people in all parts of Afghanistan, leading to a range of humanitarian crises which is likely to be felt in hydro-electricity production.

Around 80 per cent of the population make a living from agriculture, though only 3 per cent of its area is covered with forests (PAN, September 17, 2011, cited in Ahmadzai, 2013: 10). The arid countryside, which contributes to the low level of agricultural production, is a grave problem for Afghanistan (GIRA, 2009; Aich et al., 2017). Much of the country is at very high altitude with average summer temperatures not exceeding 15°C, and winter temperatures below zero in the highest regions (Mackenzie, 1993; Azimi and McCauley, 2003). On the contrary, the lowland plains in its southern part experience extreme seasonal variations in temperature, with average summer temperatures exceeding 33°C and winter temperatures of around 10°C (Peikar, 2011: 1). With the rise of temperature, Afghanistan will become drier, resulting in an increase in levels of the incidence of diseases and a decline in farming (Philander, 2008: 11; Peikar, 2011). It is reported that there was a 15 per cent decrease in wheat production i.e. 3.3 million tonnes in 2010–2011 (PAN, September 17, 2011, cited in Ahmadzai, 2013: 196).

Water is the lifeline of the people of Afghanistan, not just for living but also for the economy, which has been dominated by agriculture (FAO, 2013: 87; MoIWRE, 2004: 2; Ghiasy, Zhou and Hallgren, 2015). Agricultural development in this country is largely dependent on weather conditions in any given year (Ghiasy, Zhou and Hallgren, 2015; ICARDA, 2002: 9). About 82 per cent of water for agriculture is derived from surface water sources, which rise or fall depending on rainfall (FAO, 2013: 95). But now the availability of, and access to, water by farmers is severely affected by changes in mean annual and seasonal temperatures and precipitation over the last five decades (APPRO, 2014: 12). The amount of rainfall has also decreased by 2 per cent per decade in the past 50 years (McSweeney, New and Lizcano, 2008). Thus, the impact of increased rainfall-related drought risk on food security and livelihoods has been most severe in the north and central parts of Afghanistan over the last 30 years. These are places where farmers are highly dependent on rainfall. The productive rain-fed mixed farming areas of Kunduz and Baghlan provinces in the north, as well as parts of Paktia, Logar, and Nangahar in the east, have witnessed the decrease in agricultural productivity because of declining spring rainfall (Zaher, 2016: 33).

The decline of rainfall especially in these zones has significant impacts on food security and livelihoods well beyond these immediate areas. While agricultural products form a core component of Afghanistan's national economy, this country needs over 6 million tonnes of wheat every year (World Bank, October 22, 2012). Studies in the last few years in Afghanistan have confirmed that the rise in the cost of food has led to a switch in consumption from nutrient-rich foods, such as vegetables, meats and other proteins to nutrient-poor staples, such as rice and wheat (World Bank, October 22, 2012). Thus, Toby Lanzer, the UN Humanitarian Coordinator for Afghanistan, has also raised concerns over a possible drought in Afghanistan in 2019, pledging the UN's all-out support to Afghanistan in case there is a wheat shortage in the country (Akbari, 2018). It can be stated that the rising temperature and decreasing rainfall affects food availability in Afghanistan. This problem covers both the access to and the availability of food. Presently, however, food availability is relatively good because imported foods are available. But high prices coupled with low employment may decrease the purchasing power of the people. The Human Rights Report of the United Nations in Afghanistan says "poverty kills more Afghan than those who die in a direct result of the arm conflict" (UN News, 30 March 2010, cited in Ahmadzai, 2013: 196). The last three decades of conflicts and wars, droughts coupled with damaged irrigation systems, have also badly affected the productivity of the farmer.

Bangladesh

Bangladesh remains one of the world's poorest and most densely populated countries (UNICEF, 2017; NIPORT *et al.*, 2016; Census, 2011; Shiekh, 2013: 8; BMEF, 2009). According to reports (World Development Indicators, 2015; BMEF, 2009; Khatun, 2002: 2), the highest poverty rate recorded in South Asia was that of Bangladesh. Its population is expected to reach 170 million with the density set to increase to 1,200 people per square kilometre by the year 2020 (Khatun, 2002). Since 2009 Bangladesh's economy has grown by 188% and its per capita income has surpassed $ 1,909 (Hasina, 2019) and poverty has been cut by more than half. Though it is now well positioned to achieve most of its MDGs, it remains a low-income country with substantial poverty, inequality, and deprivation. At least 45 million of its people (Paralkar, 2017) or almost one-third of the population live below the poverty line, and a significant proportion of them live in extreme poverty. Bangladesh is susceptible to floods, cyclones, tornadoes, and tidal bores (Hossain, 2013: 127; BMEF, 2005; Agrawala *et al.*, 2003a: 14) on account of its somewhat unique location, topography, and the low capacity of its society to cope with such extreme events.

Almost every year (Younus and Harvey, 2013: 1–32; CIF, 2010) Bangladesh experiences climate change–related disasters. Most of its landmass lies less than 10 m above sea level with considerable areas at sea level, and this scenario leads to frequent and prolonged flooding during the monsoons (CIF, 2010). Extreme precipitation in the monsoons together with the physical (Mirza *et al.*, 2001: 37–48; Kripalani *et al.*, 2007: 133–159) settings of the river basins have caused severe floods on the flat deltaic terrain. Similarly, the low deltaic terrain combined (Thompson and Sultana, 1996: 1) with extreme rainfall in the nearby hills and high flow from the large catchment means that over 20 per cent of Bangladesh is inundated in a normal flood year. Other studies say (Denissen, 2012; Agrawala *et al.*, 2003a; Cash *et al.*, 2013) flood can engulf between 30 per cent and 70 per cent of the country in each year, rendering 70 per cent of Bangladeshis at risk. Recent studies (Ashfaq *et al.*, 2009; Lenton *et al.*, 2008: 1786–1793) have noted that the deteriorating monsoon leads to more and more flooding over Bangladesh, which is a grave concern. Extreme flooding means coastal zones (Dasgupta *et al.*, 2010: 27) will be subjected to inundation risk. The most frightening situation is that Bangladesh's scientists (Bansal and Datta, 2012: 226) have estimated that up to 20 per cent of the country's land may be lost to flooding by 2030. If this is so, it is expected that the poor people who lack adequate means to take protective measures and who also have very little capacity to cope with it would be the most vulnerable. They will lose their property and their income too will be affected (UN, 2009; Brouwer *et al.*, 2007: 313–326; Parvin and Rajib, 2013: 165–184). As such, it will increase poverty and affect the country's development (Paul and Routray, 2010: 489–508; Parvin and Rajib, 2013: 165–184; Parvin *et al.*, 2016: 1–13). So, persistent floods are dangerous hazards to the people of Bangladesh, particularly the poor and those living in its coastal areas.

On the other hand, Bangladesh is a country (Ramamasy and Baas, 2007: 13) which is affected by country-wide droughts every five years. Every year, it experiences a dry period for 6 months, i.e., from November to April when rainfall is normally low (Adhikary *et al.*, 2013: 2918). Its northwestern region (Shahid and Behrawan, 2008: 391–413) is a severely drought-prone region due to the high variability in rainfall. When all the river systems are parched during the dry season, the people are compelled to entirely depend on groundwater even for irrigation (Adhikary, 2013: 2917–2918). There were approximately 20 drought conditions in Bangladesh in the last five decades (Ramamasy, 2007). By 2050, it is expected that around 8 million people will be affected by drought annually (MoEF, 2005, cited in Islam, Hove and Parry, 2011: 51). In comparison with floods, drought develop slowly over time and its impacts can be underestimated. However, it can have drastic and long-term effects on vegetation, animals, and the people as well.

Bangladesh's economy is primarily agricultural where the majority of its population depends on agriculture for their livelihoods (FAO, 2016a; von Grebmer *et al.*, 2018: 35). When the agricultural land is dominated by flood plains, flooding will be of catastrophic proportion. The overall result is that it has too much water in summer (Pathmarajah, 2012: 107). Floods cause food shortages too. For example, if floods occur in Khulna district, Bangladesh will suffer from food shortages, as this area is one of the main vegetable-growing regions of the country (USAID, 2017). From 1949–1991, Bangladesh experienced more than 20 episodes of drought. Some of the worst droughts occurred in 1995, 1994, 1992, 1989, 1982, 1981, 1979, 1978, and 1973 (Rakib *et al.*, 2015: 2). Besides droughts, unpredictable rainfall (Karim *et al.*, 1990) and the lack of soil moisture during the dry season is already a problem to Bangladesh's agricultural sector. Droughts have affected water supplies and plant growth leading to the loss of production, food shortages, and starvation for the 35 per cent of its people who are suffering from malnourishment (Magnani *et al.*, 2015; Rahman *et al.*, 2008). In 2012, Bangladesh achieved self-sufficiency in rice as it could produce enough rice to meet the domestic consumption (FAO, 2016a), yet poor access to food was an ongoing problem with 15.2 per cent of the population undernourished, with insufficient access to calories (Compact2025, 2016). In 2018, Bangladesh's hunger and undernutrition situation remained worrying; its GHI score is only 26.1, which is considered seriously alarming (von Grebmer *et al.*, 2018: 32). Despite significant economic progress and poverty reduction, about 35 per cent of Bangladesh's population remains food insecure, with around 10 per cent of ever-married women reported as moderately or severely food insecure (NIPORT *et al.*, 2013). The rates of malnutrition in this country remain the highest in the world, with an estimated 6 million children chronically undernourished (Richards, 2015). Hence, the malnutrition burden is significant to Bangladesh. The loss of arable land, rising sea levels, frequent flooding, and extreme weather patterns, due in part to climate change, compound the threats to food security (Magnani *et al.*, 2015). Perhaps, Bangladesh's food security will get worse in the near future because climate change is likely to have more drastic effects.

Due to rising temperatures Bangladesh is likely to experience coastal flooding and sea level rise when the Himalayan glaciers melt (MoEF, 2009). The rise of sea level will cause more coastal flooding as storm surges increase (Yu and Wang, 2009: 5). When the rising sea levels physically reduce the amount of land in Bangladesh, the large deltaic region will be at risk of submergence (CIF, 2010). Along the Bay of Bengal, the coast of Bangladesh will have about 11 per cent of its total land area submerged and 75 per cent of the Sundarbans mangrove forests. As a result, the effects on the poor who rely on the mangrove forests for fish, fuel wood, timber, and other raw materials will be severe. Many shrimp fisheries are likely to disappear by the emergence of new estuarine fisheries (Asaduzzaman, 1994). About 35 million Bangladeshis live in coastal areas and as many as 27 million could be at risk of sea level rise by 2050 (Antos, 2017: 3). It is also predicted that due to sea level rise, around 13 per cent to 15 per cent of Bangladesh's population will be displaced in the next 60 years. When the total area of the country is reduced due to submergence, people will no longer gain a secure livelihood in their home regions (Karim and Mimura, 2008: 490–500). So, they will be forced to move away. This will generate millions of climate migrants or environmental refugees (Myers, 2002: 609–613; Myers, 1995; Imam, 2008). Such a displacement of several million poor people will be dramatic, particularly from the country's coastal zone. A number of environmental refugees will be forced to flee to other places within the country or to neighbouring countries illegally or legally, either temporarily or permanently. A point to be noted here is that the movement of the Bangladeshi from coastal regions into neighbouring upland areas like Assam, Tripura of India and the Chittagong Hill Tracts in Bangladesh is a classic case of conflict-generating migration (Smit and Vevekananda, 2007; Weiner, 1978; Phadnis, 1989; Hazarika, 1993; Islam, 1991: 8–9). Perhaps millions more will be displaced by climate change–related phenomenon (Bose, 2013: 61–82). Such sudden movements of people will cause a drastic change in the ethnic balance of the neighbouring areas (Macdonnell, 1995: 107). There is the possibility that such people who migrate across the borders will worsen both the intra-regional and inter-regional tensions (Mintzer and Leonard, 1994: 11). The situation may in turn lead to growing insecurity and conflict in the region (Ansorg and Donnelly, 2008: 3; Hossain, 2013: 126). The issue of climate migrants from Bangladesh will soon haunt the tribal-dominated state of India called Meghalaya in the next few years because it shares 443 km of international border with Bangladesh. This is a neighbourhood of India where cross-border movement is now taking place because the border is open and porous.

Bhutan

Being a mountainous country, Bhutan has numerous snow-clad mountains and glaciers, which supply water to the rivers and streams (Dorji, 2013: 183). Mountainous regions on earth are dominated by the presence of glaciers. But the increase of temperature which is likely to rise by 1°C from 2010 to 2039, and by 2°C from 2040 to 2069 can give rise to GLOFs because the permanent snowline in Bhutan has moved significantly (BMCI, 2016: 32–33). These GLOFs will be devastating, as the glacial lakes will impose potential risks from sudden outbursts and consequent floods (UNDP, 2012; Mool, Joshi and Bajracharya, 2001). Its significant effects will include perturbation in the quantity of river water used for hydropower generation; destruction of settlements, infrastructure, and agricultural lands; loss of biodiversity and even human lives downstream (Shrestha, Gautam and Bawa, 2012; RGB, 2009a: 24; Alam and Regmi, 2004: 38; Regmi *et al.*, 2008). The GLOFs are not a new phenomenon in Bhutan. The frequency has risen in the past three decades, which resulted in disasters (Tirwa, 2008). Significant GLOFs have occurred in 1957, 1960, 1968, and 1994, devastating lives and property downstream (Rinchen, 2008; Armington, 2002). The October 1994 GLOF in Bhutan underscores the nature of the risk. It occurred at 90 km upstream from Punakha Dzong, the outburst flood from Lugge Tsho led to massive flooding on the Pho Chhu River, damaging the Dzongchu and causing casualties (BMCI, 2016: 34). It is predicted that by the year 2100, the temperature will increase by at least 2°C and glaciers will retreat by 49 cm (Vince, 2019; World Bank, 2012c; Tesfaye *et al.*, 2017: 959–970). The scenario is poised to make many lakes burst their banks and send millions of gallons of floodwater downstream (DoGM, 2009). Hence, it will be destructive for Bhutan considering that its entire northern highlands are either covered with glaciers or snow (Namgyel, 2003; ADB, 2003). However, the Bhutan Department of Energy says the majority of rivers in Bhutan are more susceptible to fluctuation with changing rainfall patterns than to flooding which is directly attributable to glacier or snowmelt (Pelden, 2010). The heavy monsoon floods in 2000 affected the economic growth of Bhutan by more than 2 per cent (RGB, 2002; Tshering, 2007: 75). In May

2009, cyclone Aila, which originated in the Bay of Bengal, resulted in incessant rainfall causing one of the worst disasters in Bhutan. According to the Thimphu Meteorology Department record-breaking rainfall, measuring up to 76 mm over a 24-hour period was recorded as one of the highest in the last five years (BMCI, 2016: 35). The future increase in frequency of intense monsoon rains will cause flash floods and landslides. It may also increase liability to health hazards (Mozaharul and Tshering, 2004: 15). Whatever may be the causes – GLOFs or erratic rainfall – the dire consequences will be severe in Bhutan. Even so, the populace of Bhutan lacks awareness of climate change's effects, especially the floods from glaciers (Chhetri, 2010).

Bhutan may never record water shortages, yet it is very likely that climate change may render it highly exposed to water scarcity (RGB, 2006a). The basis for saying this is that its territory is very sensitive to the changes in temperature and precipitation (Anglo *et al.*, 1996). Due to the rising temperatures glaciers and snow cover will decline in the course of the 21st century. During warm and dry periods, this will reduce water supply to the rivers, which depend on glaciers. Such an episode will not only reduce the potential of catchments to retain water, but also cause water quality to deteriorate (Shrestha, Gautam and Bawa, 2012; Bates *et al.*, 2008; RGB, 2009a: 23–24). The winters of 2005 and 2006 were unusually dry winters with no rain and snow (BMCI, 2016: 37). In 2012, Pemagatshel district was facing drought where vegetation was visibly dry, stream and river flows declined, water levels in lakes and reservoirs fell (Bhutan Observer, June 16, 2012). In this way, erratic rainfall has been a problem in recent years in Bhutan (Gelmo, 2015).

Bhutan is one of the ten global biodiversity "hotspots" and to date its rich biodiversity resources still make a large contribution to the economy (BMCI, 2016: 24; Gleick, 1989: 333–339). With the rise in temperature there will be enormous effects on the forest resources and biodiversity of Bhutan says the Report of the ICIMOD (ICIMOD, 2016: 58–64). While warming may have positive effects on the growth of some trees, it may also reduce tree survival by benefiting insects or pests. Warmer winters will imply reduced snow cover and less carryover of water to the growing season, which can lead to drought-induced forest decline (ICIMOD, 2016: 37, 60; RGB, 2011). As a result, some species will die out or become extinct as soon as temperatures exceed their tolerance limits. Their extinction may directly be due to the changing physiological responses of species, or indirectly by changing the relationships between species (ICIMOD, 2016: 64). Alas, human activities are also threatening its existing ecosystems (ICIMOD, 2016: 17). The combination of climate change with the pressures of deforestation, land use changes, habitat degradation and fragmentation presents a significant threat to biodiversity (RGB, 2009a: 23).

Approximately 80 per cent of its population depends on farming for their livelihoods (RGB, 2006). The majority of agricultural production in Bhutan is smallholder subsistence farming (MoAF, 2016) and about 69 per cent of the Bhutanese population is employed in the agricultural sector (FAO, 2015).

Given the mountainous and extreme biophysical conditions, less than 3 per cent of the country is used for agriculture (Meenawat and Sovacool, 2011: 515–533). This will severely constrain agricultural production and also exposes the country to the risk of food insecurity (Mozaharul and Tshering, 2004: 21). It is worth noting here that Bhutan's upland crop production is practicing close to the margins of viable production which can be highly sensitive to the variations of climate and flash floods (ICIMOD, 2016: 59). The Third Assessment Report of the IPCC in 2001 (McCarthy *et al.*, 2001) states that climate change variation in water quantity and quality will affect food availability. Incessant rainfall can lead to swelling of rivers and streams causing dangerous floods. This can be seen in 2010 when landslides and flash floods damaged more than 2,000 acres of agricultural land, affecting some 4,165 households over 20 *Dzongkhags*, and damaged farm roads and irrigation channels affecting 529 households (BMCI, 2016: 36). Therefore, in 2015, Chimi Rinzin, the chief agriculture officer with Bhutan's Ministry of Agriculture, said the October rains and hailstorms affected farmers from eight districts and hundreds of acres of paddy crops in Bhutan. He also says "If we look back Bhutanese farmers never experienced such rainfall especially during harvest time and also no drought cases were reported during summer" (Gelmo, 2015). On seeing this situation, he attributes these impacts to climate change (Gelmo, 2015). In this way, the rural populations who depend on the agricultural sector become susceptible to food security. Their livelihoods will be at grave risk. An increase in rainfall will enhance soil erosion and may affect vegetation (BMCI, 2016: 37; RGB, 2002).

From the discussion it shows that climate change will have its own consequences in Bhutan. It will put the Bhutanese in jeopardy and impair the vision of Jigme Thinley who traveled across the world promoting the happiness measure (the concept of happiness has often been explained by its four pillars: good governance, sustainable socio-economic development, cultural preservation, and environmental conservation), which made him a popular figure among Western academics. Therefore, Prime Minister Tshering Tobgay says, "rather than talking about happiness we want to work on reducing the obstacles to happiness" (Harris, 2013); perhaps this will include mitigation of climate change effects (Harris, 2013).

India

Kerala floods reveal the horror of climate change's latest act in India (Krishnakumar, 2018; Faizi, 2018). Much of India's state that is Kerala was submerged due to flooding with large areas cut off completely from the rest of the world, without power, drinking water, food, or communication links, by the time the rain's fury began to subside after the second week of August 2018. The Government of Kerala described the horror as "the most intense floods to hit the region in the last hundred years" as it ravaged 774 villages, directly affecting over 5.4 million people; at least a million people had to be evacuated; there were 483 deaths and US$3 billion in damages (Krishnakumar, 2018; Sangomla, 2018; Faizi, 2018; Venkatesh, 2018b). In June 2005 the state of Gujarat was affected by floods where about 900 villages in the three districts were marooned by floodwaters (Dasgupta, 2005; Bunsha, 2005). On July 26, 2005, Mumbai was hit by 94 cm of rain – the heaviest rainfall since India began keeping weather records in 1846. Heavy rains flooded Mumbai like never before and the city came to a standstill due to flooding. Around 150,000 people were stranded on the roads, or lost their homes while many died (Sturcke, 2005; Katakam, Bavadam and Bunsha, 2005). Both events have caused mayhem in India.

India is frequently facing the brunt of climate change. It is expected that this country will also see severe stress on water resources and foodgrain production in the future, increasing the risk of armed conflict. According to German Watch Report (2017), India is ranked as the fourth most vulnerable country to climate change. The probable future effects of climate change, identified by the Government of India in its Initial National Communication to the UNFCCC, will be on many fronts. These are:

i) Decreased snow covers, affecting snow-fed and glacial systems such as the Ganges and Brahmaputra; 70 per cent of the summer flow of the Ganges comes from snowmelt;
ii) Erratic monsoons with serious effects on rain-fed agriculture, peninsular rivers, water and power supply;
iii) Decline in wheat production by 4–5 million tonnes with as little as a 1°C rise in temperature;
iv) Rising sea levels causing displacement along one of the most densely populated coastlines in the world and threatening freshwater sources and mangrove ecosystems;
v) Increased frequency and intensity of floods; increased vulnerability of people in coastal, arid and semi-arid zones of the country.

(GOI, 2008: 14–18)

Parts of India, particularly the east and west coasts, are expected to fall under "high water stress" in the near future as a result of current withdrawals and availability, coupled with future climate impacts (Antos, 2017: 3; Eldho, Sreeja and Madhusoodhanan, 2016: 724–741). When agriculture is the mainstay of India's economy, floods and droughts may have immense potential to harm it in a variety of ways. India ranks first among countries that rely on rain-fed agriculture, in both size (86 million hectares) and value of production (Sharma *et al.*, 2010: 23–30). Rain-fed agriculture alone accounts for roughly 44 per cent of total food grain production in India (Sharma *et al.*, 2010: 23–30). While large parts of its arable land are rain-fed, the productivity of its agriculture depends on monsoon and rainfall patterns (Skymet Weather Team, 2017; MEoF, 2004; Dasgupta, 2008). So, a volatile monsoon season could be damaging to India's agriculture and food supply (Skymet Weather Team, 2017). Information is offered that semi-arid regions of western India are expected to receive higher-than-normal rainfall as temperatures soar, whereas the central parts will experience a decrease between 10–20 per cent in winter rainfall by the 2050s (Lal *et al.*, 2001: 1205). The Economic Survey 2017–2018 reveals that on an average annual rainfall in India has declined by about 86 mm in the last three decades (Rattani, 2018: 7). Such a shift in rainfall patterns will create a severe threat to its agriculture, the economy as well as food scarcity. The National Communications Report of India to the UNFCCC also says that climate change is likely to affect all its natural ecosystems as well as socio-economic systems (MEoF, 2004).

Future climate change projections indicate that in the coming years the effect of temperature rise will be significant on Indian agriculture (CDKN, 2014; Kumar and Parikh, 2001a, 2001b: 147–154; Kumar, 2008). If global temperatures rise by 2°C, India's monsoon season would be deemed highly unpredictable (Pachauri and Meyer, 2014). The UN Report in *The Times of India* (Mohan, 2014) says that the effects of climate change will be felt severely in the Indo-Gangetic Plains in the near future, affecting the poor people. It further states that the areas, which are facing frequent floods currently, may face drought-like situations in the distant future. Further, the IPCC Report says by the 2090s, climate change may double the frequency of extreme drought, causing problems to India's agriculture (Parry *et al.*, 2007). The result is projected to increase the risk of crop failure and lower crop production (CDKN, 2014). In addition, B. Venkateshwarlu, former director at the International Central Research Institute for Dryland Agriculture (CRIDA), Hyderabad, says climate change affects all three aspects of food security: availability, access, and absorption (Goswami, 2017). According to him, climate change has about 4–9 per cent impact on India's agriculture each year, which results in a loss of about 1.5 per cent in the GDP annually (Goswami, 2017). Important crops like rice and wheat are likely to see about a 6–10 per cent decrease in yields by 2030 (Goswami, 2017; The World Bank, 2014). Though the rising frequencies of floods, storms, and cyclones is likely to increase agricultural production variability (Swaminathan, 2008), unusual weather patterns such droughts, rising temperatures, heat waves, and river recession are among the environmental scenarios which have already harmed large number of Indians (Parik, 2009). Rising temperatures will increase fertilizer requirements for agricultural production (Shivay and Rahal, 2008: 19). The Indian Agriculture Research Institute has estimated that with every 1°C rise in temperature, India might lose 4 to 5 million tonnes in wheat production (Sharma, 2008).

The rise in temperature will see glaciers melting in the Himalayas (Bolch *et al.*, 2012: 310–314; Ladha *et al.*, 2000). With the rise in temperature the residents living in the Ganges Delta are particularly at risk from accelerated global sea level rise and will face flood risks (Ericson et al., 2005:

63–82; Rahmstorf *et al.*, 2007: 709). The reason is that the water bodies will rise alongside the Indian coast (Cruz *et al.*, 2007: 469–506; Rahmstorf *et al.*, 2007; Nicholls *et al.*, 2007: 315–356). With the rise of ocean water, India may lose a large percentage of its total land area, including lots of its famous beaches and tourist infrastructure. Coastal flooding will not only kill people and cause destruction; it will also affect tourism areas of India. Goa and Kerala will be the worst-hit areas (Mohan, 2014). Mumbai's northern suburbs like Versova beach, India's coastal cities like Kolkata, Surat, and Chennai, and other populated areas will be defenceless to loss of land and increased flooding (Mohan, 2019; Castle, 2002; EPW, 2009). The Andaman and Nicobar Islands and the coral atolls of the Lakshadweep archipelago will become vulnerable areas (Bruce, Lee and Haites, 1996: 211; Nicholls *et al.*, 2007). As a result, the supplies of freshwater may become scarcer because of inundation. Numerous species living along the coastline will be endangered. Further, it has been estimated that even a 100 cm sea level rise would lead to coastal welfare loss of US$1,259 million (Roy, Ghosh and Baruah, 2006). In addition, there are possibilities that the problems of India's coastal areas may induce people to migrate from low-lying and risky areas elsewhere (McGranahan, Balk and Anderson, 2007: 17–37; Chella, 2008). This may put a greater pressure to human crisis especially when there is lack of fresh water and food. On seeing the scenario, India should be concerned about the crisis as the phenomenon will have substantial adverse effects on its coastal regions and the outcome will be intolerable.

Records show that 2017 was the third consecutive year in which global temperatures rose by a degree above late nineteenth-century levels. The year 2017 was also the warmest non–El Niño year ever marked by extreme weather events around the globe (NASA, 2018; Rattani, 2018: 7). In India, the temperature is expected to go up by the end of the century (Kumar *et al.*, 2006: 334–345). When this is the consequence, India will considerably suffer the brunt of heat waves as the majority of its population is poorly equipped to cope effectively with the adversities (Das, 2015; The Telegraph, June 3, 2015). It will increase mortality and become a cause of public health concern. As a result, the threats of climate change are alarming for India as it is already threatening lives, food, and health across the country.

The Maldives

As the flattest country on earth, the Maldives comprises only small islands, surrounded by the ocean (MoEEW, 2012). Due to its small size and proneness to natural hazards, the Maldives are subject to external shocks that can leave any island nation including the Maldives defenceless to climate change (Stojanov *et al.*, 2017a; Houghton *et al.*, 2001: 881). If this is the case, the small coral islands that make up the Maldives are likely to suffer the most from its adverse effects. In reality, the Maldives is among the countries most vulnerable to existing natural hazards and future climate change impacts (Sovacool, 2012: 295–300; Julca and Paddison, 2010: 717–728; Ghina, 2003: 139–165). A few of its islands could even become uninhabitable (Hunter, 2005: 2; Eklund, 2009). This makes the Maldives a typical case that requires the help and attention of the international community. "This is what will happen to Maldives if climate change is not checked" says Mohamad Nasheed, the former president of Maldives, while holding his first Cabinet meeting underwater with all his ministers in scuba gear (Dyer, 2015).

The Maldives has no safe sanitation. The majority of its islands do not have a functioning water supply and distribution network that can ensure sufficient safe freshwater during dry periods, except for in Malé, Vilingili, and Hulhumalé, which are home to over a third of the total population (Lubna, 2012). Only 11 per cent of the Maldives' inhabited islands had potable groundwater before the 2004 *tsunami* in the Indian Ocean (OCHA, 2007; UNEP, 2005). Water scarcity can pose serious constraints to the people (Pandey, 2004). The inundation of seawater during the event contaminated these groundwater supplies as well as the soils. Sea level rise is also likely to put more stress on the Maldives' scarce freshwater resources. Though the Maldives provides drinking water to about 87 per cent of its population by collecting rainwater, groundwater is still required for non-drinking purposes and for drinking water during the dry season months. But groundwater aquifers on the islands are shallow, and high extraction levels have made them vulnerable to inundation by saltwater (Stojanov *et al.*, 2017a: 2; UNEP, 2005). In December 2014, the Maldives ran out of water. When taps ran dry the thirsty residents of Malé were receiving bottled and desalinated water provided by India, China, Bangladesh, and Sri Lanka via public taps and mobile vehicles (Aljazeera, 2014; EPoA, 2014). Whilst almost everyone even in Malé depends for their water supply on shallow wells, there is growing concern that climate change is likely to break down human health causing adverse kinds of diseases such as malaria, dengue, food- and waterborne diseases, and other climate-sensitive diseases (Stojanov *et al.*, 2017a: 2; Kruijk, 2010: 15).

With limited natural, human, and economic resources, the country's populations are dependent on marine resources to meet their basic needs. Though its population is small, many of the inhabitable islands have high population densities. Its population is distributed among various islands, but more than a quarter of them live in an area less than 2 km² (Shaljan, 2004: 1835–1840). Internal migration to urban areas such as Malé has been creating severe imbalances in the distribution of population. At present, nearly one-third of the population live in Malé (UNDESA, 2015). The overcrowding in Malé has caused lots of pressing environmental problem (Stojanov *et al.*, 2017a: 2; Shaljan, 2004: 1835–1840).

Rice, tubers, and millet, being the essential part of island diets, may be available. But the reason for concern is that the soil in this country is very new and it is chemically alkaline as a result of the excess of calcium from basic coral rock and sand. The soil is also very porous, so it needs to be irrigated many times in a day (Suja, 2017). These soil characteristics limit agricultural production and food security (Stojanov *et al.*, 2017a: 2). Being a tourism-based country, the Maldives' potential for agriculture is limited by the scarcity of farmable land. Only about 23 per cent of soil is used for agriculture (CIA, 2015), which is too little to produce enough food to the Maldivians. Thus, the country is heavily reliant on imports, with 90 per cent of all food coming from outside (FAO, 2011: 2). Hence, by no means will everyone be able to afford food in the near future. Outside of Malé fishing is the main source of food security and livelihood for the vast majority of the people (FAO, 2011: 1). But fishery is expected to suffer from climate change effects. Many of its islands are already exposed to acute food shortages especially when fishing seasons proved short or storms devastate subsistence crops (Kruijk, 2010: 15–16).

Tourism accounts for 34 per cent of the GDP of the Maldives, which is more than any other economic sector (Shareef *et al.*, 2015). Rapid expansion of coastal and tourism activities like scuba diving, sport fishing, and snorkeling may threaten coral reefs and other marine resource (Kundur, 2012: 4; Scheyvens, 2011: 148–164). Sea level rise and related beach and coastal erosion, coral bleaching, and salinity intrusion have already been reported to directly and indirectly affect tourist resorts in the country (MoEE, 2015). Therefore, climate change is central to all discussions about food security in this country because it is adversely affecting crops and fish stocks and reducing land area as the sea level rises (FAO, 2011: 1).

The islands of the Maldives are reef-based. The coral reefs serve as natural breakwaters. But its dependence on these coral reefs is very high as they are mined for building materials (Schultz, 2017; Bruce, Lee and Haites, 1996: 213). With their damage comes the bigger danger of losing the natural protection of the islands from the ocean waves and currents. This makes the country more open for beach erosion and more susceptible to inundation by uncontrolled waves reaching the shore due to its relatively large coastline (MEEW, 2007; Ghina, 2003). Gradual sea level rise has aggravated the existing problem of beach erosion. In the recent past, 62 per cent of all inhabited islands and 45 per cent of tourist resorts have reported severe beach erosion (Shaig, 2006). Additionally, given mid-level scenarios for global warming emissions (the scenarios referred to here are the middle-emissions pathways known as A1B and IS92a from the IPCC), the Maldives is projected to experience sea level rise on the order of half a meter (1.5 ft) and to lose some 77 per cent of its land area by around the year 2100 (Tol, 2007: 741–753; Woodworth, 2005: 1–18). If sea level were instead to rise by 1 m (3 ft), the Maldives could be almost completely inundated by about 2085 (Anthoff, Nicholls and Tol, 2010: 321–335). This situation will force people to flee to take refuge elsewhere (GACGC, 2006; MHAHE, 2001). Perhaps the worst affected country in terms of environmental degradation in the 2004 *tsunami* was the Maldives. While referring to the effects of 2004 *tsunami*, Tom Bergmann-Harris who is the former head of the UN Children's Fund in the

Maldives, put forth this statement, "some of the places I have visited look like they have been hit by a nuclear bomb" (UN, 2005). Some scholars have however suggested that the increased risk of flooding during the 21st century for the Maldives is overstated (Morner, Tooleyb and Possnertc, 2004: 177–182). Others say that a rise in sea level of approximately 50 cm during this century remains the most reliable scenario, which is indeed an authentic concern for the Maldivians (Woodworth, 2005). With about three-quarters of its land area, which lie less than a meter above the sea level (Woodworth, 2005), the slightest rise in seawater will prove threatening to the very existence of the livelihood of the Maldivians (Karthikeyan, 2010: 343–350). This is indeed frightening information. But migration to foreign countries is not an option for the Maldivians as they are aware that they have to take into account all the aspects; they may lose their identity, culture, history and face many problems in different countries. One environmental activist from the Maldives says "people have to do strange things when they are in danger. But, are we willing actually to lose our nation, our culture, tradition and history? If we immigrate due to sea level rise, we lose our nation, our history. . . . But, at one stage in future, we may have to leave" (Stojanov *et al.*, 2017b: 26).

The Maldivian economy is heavily dependent on tourism and fishing industries. Due to sea level rise, beautiful and precious coral, the beaches, which attract lots of tourists, will face devastation. Much of the islands' sand at the beaches and shoreline are washed off regularly (MHAHE, 2001b: 9). Thus, tourism will be disrupted due to the loss of beaches, coastal inundation and degradation of coastal ecosystems, saline intrusion, damage to critical infrastructures, and the bleaching of coral reefs (MoEC, 2004). The tourism industry may also suffer from climate change mitigation measures, such as levies on aviation emissions, which will increase the cost of air travel (Hunter, 2005: 23). The revenue it collects from tourism will be reduced. The statement by H. E. Maumoon Abdul Gayoomin, former president of the Maldives (December 4, 1997), said:

> The Maldives is one of the small states. We are not in a position to change the course of events in the world. But what you do or do not do here will greatly influence the fate of my people. It can also change the course of world history.

And, the address by His Excellency Mr. Maumoon Abdul Gayoom (October 19, 1987) said:

> As for my own country, the Maldives, a mean sea level rise of 2 metres would suffice to virtually submerge the entire country of 1,190 small islands, most of which barely rise over 2 metres above mean sea level. That would be the death of a nation. With a mere 1 metre rise also, a storm surge would be catastrophic and possibly fatal to the nation.

Another statement by Mr. Abdullahi Majeed (December 11, 2003) said:

> The Maldives is an archipelago consisting of tiny islands scattered in a vast expanse of the Indian Ocean. Over 80 percent of the land area have less than one meter above mean sea level, climate change and its associated sea level rise would undoubtedly be a catastrophe and threaten the livelihood of the islanders in the Maldives alike many thousands of others in low-lying island states. Sixteen years ago in April 1987, Maldives experienced unusual high waves causing extensive damage to the islands. Two thirds of the whole Maldives, including the capital island, Malé, was inundated for two days causing extensive damage to the infrastructure. Malé International Airport, the only gateway to the Maldives, was closed for two days, causing delays in receiving the relief assistance from the international community, cancellation of tourist arrivals and lot more
>
> (Statement by Mr. Abdullahi Majeed (Maldives)
> 11 December 2003, Milan, Italy.

The entire discussions about the Maldives validate that climate change's calamity will be catastrophic as it has the potential to wipe out the entire country from the face of the earth in the near future. Therefore, huge challenges arising from climate change are waiting.

Nepal

Nepal lies in the HKH expanse. Due to the limited number of scientific studies conducted in this region, Nepal is described as a climate change white spot (ICIMOD, 2012: 5). This country is one of the 11 countries globally that is most at risk of climate change–induced poverty. The reason is its economy is highly exposed to climate change–related hazards, the direct economic costs of which have been estimated in the range of 1.5 per cent to 2 per cent of the country's current GDP, which is equivalent to almost US$270 million to US$360 million per year, in 2013 prices (IDSN, 2014; Shepherd *et al.*, 2013).

According to climate models, temperature of Nepal is rising at the annual rate of 0.04–0.06°C per year, which is more than global rate (Shivakoti *et al.*, 2015; Dhakal, Silwal and Khanal, 2010). This country is also likely to experience an increase in temperature up to 1.4°C by 2030 and 2.8°C by 2060 (NAPA, 2010). This increase in temperature could be highly significant for a society largely reliant on agriculture and mountain ecosystems (NAPA, 2010; Bartlett *et al.*, 2010: 2). In other words, Nepal will be exposed to the danger of climate change in many ways (Rai and Gurung, 2005: 316–320). A significant threat in the Himalayas that is directly correlated to rising temperatures is GLOFs. The GLOFs which amass water into the glacial lakes subsequently burst and rush downstream, and the immediate impacts will be of havoc (ICIMOD, 2011: 9–13). Such havoc will inflict turmoil at the downstream communities and damage valuable infrastructure such as hydropower facilities and roads (Shrestha, Mool and Bajracharya, 2007; ICIMOD, 2011: 14). The speedy melting of glaciers during the last half-century has created many new glacier lakes and expansion of existing ones (Mool, Joshi and Bajracharya, 2001). At present, there are few glaciers that have shown signs of retreat. In addition, more than 1,466 glacial lakes, with a total geographic area of 64.78 km², have been identified in the Koshi, Gandaki, Karnali, and Mahakali basins of Nepal (Mool, Bajracharya and Shrestha, 2005: 80–82; Gaan, Acharya and Mohapatra, 2013: 85). Floods, both riverine and those linked to GLOFs, are frequent in many parts of Nepal. When these glaciers melt, there is a possibility that they will have serious effects on Nepal's pristine Himalayan mountain range and the surrounding communities (Butler, 2009: 8–31). The over-dependence of the people on natural resources, its weak governance capacity, and its poor economy, combined with its location in the heart of the HKH region, leave Nepal vulnerable to climate change impacts.

The entire HKH expanse of Nepal is prone to severe floods. (Jianchu *et al.*, 2006: 1; Shrestha and Bajracharya, 2013). In addition, floods in the Tarai region are growing in severity and regularity (Rabbani, Shafeeqa and Sharma, 2016: 162). During the monsoon the flow of rainwater to the rivers is unpredictable. These rivers may wreak havoc in the valleys when they overflow from the mountain tops. Being a mountainous country, Nepal's risk of landslides due to degradation of forests and faster surface runoff will increase (NCVST, 2009). In fact, landslides have been rising in Nepal (Chaudhary, Jimee and Basyal, 2015). From 1971 to 2013 (Chaudhary, Jimee and Basyal, 2015), there were 3,220 landslide events, which resulted in 4,691 human deaths, 18,902 homes destroyed and nearly 34,126 damaged, and almost 22,576 hectares of arable land lost. In 2018, Nepal experienced flash floods and landslides across the southern border, amounting to US$600 million in damages (Eckstein, Hutfils and Winges, 2018: 7). In the high Himalayan Mountains of Nepal, floods may not cause much damage to human settlements as the upper mountainous areas are sparsely populated. In the southern plains region such as Terai, floods occur regularly and cause considerable damage because it is a densely populated area. It also affects human settlement (Hare *et al.*, 2011: 1–13; WWF, 2005). So, the scale of the effect of any flood in Nepal depends on natural conditions and the density of population.

More than 6,000 rivers and rivulets in Terai with a total of 45,000 km in length support irrigated agriculture and other livelihoods (Dixit, 2010). Rains are increasingly unpredictable in Nepal (Vidal, 2006). On the one hand, Nepal faces extreme flooding caused by heavy rains (IDSN, 2014). On the other hand, due to unpredictable rainfall, studies foresee that freshwater flow in its rivers will diminish, lowering the levels of reservoirs (groundwater and river flow) after 2035 (Dutt and Geib, 1998; Springate-Baginski and Blaikie, 2007). There has been an increase in severity and frequency of droughts in Nepal. The winter droughts of 2008–2009 greatly affected Nepal's agriculture as agricultural productivity in this country is dependent on weather patterns (Rabbani, Shafeeqa and Sharma, 2016: 162; WFP, 2009; Gurung and Rai, 2005). The farmers will be unable to properly feed themselves because this variability of rainfall suggests that agriculture in Nepal will face immense challenges (Oxfam, 2009: 1–26; MOE, 2010; Shrestha *et al.*, 1999: 2775–2787). Thus, food shortages will become the end result, mostly to the hill and mountain populations which is already a chronic problem (Gill, 2003a; Sharma *et al.*, 2009: 15–17). Reports have confirmed that changing weather patterns have dramatically affected crop production in Nepal (Oxfam, 2009: 1–26; Pettengell, 2010). Its 40 districts, mostly in the western part, are facing food deficits (Dixit, 2010; MOE, 2010). Worse is that about 6.9 million Nepalese are living with chronic food insecurity (Clewett, 2015). The consequences of drought will not only be felt in agriculture but more on a daily basis too as water sources dry up.

In addition, the Nepalese are dependent on tourism for their livelihood (Ghimire, 2009). But a change in climatic conditions may disturb Nepal's tourism industry. Unfavourable climate change phenomena can have an impact on trekking and mountaineering tourism in Nepal. For instance, in October 2014, more than 32 people were killed by a sudden snowstorm in Annapurna Conservation Area of western Nepal. Hundreds of trekkers were trapped at more than 5,000 m altitude from sea level in the Thorong La Pass area (Anup, 2017: 35–38; Burke and Walker, 2014). In April 2015, the earthquake caused snow avalanches in the Mt. Everest region at about 7,000 m altitude from mean sea level, killing 57 foreign tourists and more than 1,000 were reported missing (Gayle, 2015; Beaumont, 2015). At the same time in 2015, a devastating earthquake caused avalanches in the Langtang region. This event buried 116 houses and killed ten army men (Anup, 2017: 35–38; Callaghan and Thapa, 2015).

Natural disasters, such as floods, landslides, and GLOFs, affect migration trends in Nepal. Migration of people from the Hill region to the Tarai region, due to water shortages, further confirms climate change forced people to migrate (Dash, 2016: 7). Nowadays, Nepalese are moving to urban areas and many of them cross the border to India on a seasonal basis to earn money when workloads in the farms/villages are low (Gill, 2003b; Sharma and Thapa, 2013). The most recent 2015 Gorkha Earthquake is estimated to have damages and losses over NPR 7.5 billion in monetary terms with more than 500,000 houses fully destroyed and around 250,000 partially damaged (NPC, 2015a). A total of 649,815 families were displaced because of this earthquake and the aftershocks (MoHA and DPNet-Nepal, 2015:18; NPC, 2015b). Rural to urban migration and displacement further escalated during the aftermath of the earthquake and because of the high risk posed by landslides and floods (MoHA and DPNet-Nepal, 2015). Another study says (Ginnetti, 2015) the risk of climate change–induced displacement is quite high in Nepal with approximately 4,152 Nepalese per million at risk of being displaced each year. Therefore, migration may not be a new phenomenon to the Nepalese but climate change–related disasters in various parts of the country will certainly force individuals and families to move out of their homes and shift elsewhere and to foreign countries. That is why Nepal is on the front line of climate change.

Pakistan

With a population of 191.71 million (as of 2015), Pakistan is now the sixth most populous country in the world (Pakistan Bureau of Statistics, 2015: 3; MoF, 2014–15: 199). The data reveals that the population is estimated to reach up to 265.6 million at mid-2030 and 344 million at mid-2050 (UN Population Division, 2012; Warraich, 2017). At its current rate of population growth, Pakistan will be the fifth most populous country in 2050 (GOP, 2013). This country was ranked 21st by the Global Climate Risk Index (GCRI) in terms of exposure to extreme weather conditions for the period from 1993 to 2012 (Kreft and Eckstein, 2014). It is also listed as the twelfth most highly exposed country to climate change by the WB (Nomman and Schmitz, 2011: 1–12). Recently, the German Watch Report (2017) GCRI ranked Pakistan number seven in the Long Term Climate Risk Index (1996–2015). The total losses assessed in the index for the 15-year period for Pakistan stand at US$ million 3823.17 Purchasing Power Parity (PPP). The average costs for annual adaptation and mitigation to climate change for Pakistan were estimated to range annually from US$14 billion to US$32 billion leading to 2050 (GOP, 2017: 13–19).

Large numbers of its population live in low coastal areas or river deltas. These are fertile areas which will also be heavily devastated by sea level rise and flooding due to heavy rainfall and climate shifts like the rising temperature (Farooqi, Khan and Mir, 2005: 12). Based on the Pakistan Meteorological Department (PMD) station data from 1951 to 2000, a rising tendency in the annual mean surface temperature was observed throughout the country (Farooqi, Khan and Mir, 2005: 11–21). From 1960 to 2007, it was found that the annual temperature in Pakistan has increased by 0.87°C (maximum) and 0.48°C (minimum) (Chaudhary *et al.*, 2009: 1–43). Its glacial-covered regions too have experience the heat (Chaudhary *et al.*, 2009: 1–43; Zahid and Rasul, 2010:85–96). As a result, it is leading to glacial retreat (Barnett, Adam and Lettenmaier, 2005: 303–309). Specifically, glaciers in Pakistan are retreating, and their retreat speed is faster than the rest of the world (You et al., 2017: 141–147). The retreat of glaciers has resulted in the formation of new lakes and extension in volume and size of existing lakes (Rasul *et al.*, 2011: 1–5). The hasty release of water from these glaciers has caused an increase in the frequency of GLOFs (Ashraf, Naz and Roohi, 2012:113–132). These GLOFs are of catastrophic proportion for the mountain population due to flash floods. The communities downstream too are not spared from the dangers as huge loads of debris and mud flowing downstream sweep the infrastructure and agricultural lands, causing tremendous hazards mainly to its coastal populations along the coasts (Rasul *et al.*, 2011: 1).

Pakistan has a 1,046 km-long coastline along the border of the Arabian Sea in the south provinces of Sindh and Balochistan (Chaudhry, 2017: 34). The coastal areas of Sindh are considered more vulnerable to sea level rise than those of the Balochistan coastal areas due to its tidal flat topography and higher population concentration with marked industrial activities along coastal areas, such as Karachi. In this province, a 2 meter rise in sea level is expected to submerge 7,500 km^2 in the Indus Delta (Chaudhry, 2017: 34). Fearing this unprecedented rise of sea level the Senate's Standing Committee on Science and Technology of Pakistan issued a letter to Prime Minister Nawaz Sharif in 2015, expressing fears about the seriousness of sea intrusion along the coastal areas of Sindh and Balochistan which can result in the sinking of Badin and Thatta in a period of 30 years, followed by Karachi (Butt, 2015). The other low-lying coastal areas of Balochistan such as Pasni may also be affected by sea level rise as the mean sea level in the coastal town of Pasni is about 1.4 m (Khan and Rabbani, 2000). In Pakistan, under a high emissions scenario, and without large investments in adaptation (WHO, 2016: 3), an annual average of 1,207,700 people are expected to be affected by flooding due to sea level rise between 2070 and 2100. The rise in sea level is also expected to threaten the streams in the delta regions such as Hajamaro, Ghoro, Kaanhir, and Kahhar as the erosion rate in these places is ranging from 31 m/year to 176 m/year. The south side of the mouth of Ghoro Creek shows the highest erosion frequency of 176 m/year with a retreat rate of 425 m from 2006 to 2009 (WWF Pakistan, 2012). Furthermore, due to lack of sedimentation, the delta region is both shrinking and sinking (Syvtiksi *et al.*, 2009: 681–686). It is expected, therefore, the sea level may have severe impacts on the coastal areas of Pakistan as is already seen from the inundation of low-lying areas, degradation of mangrove forests, declining drinking water quality, and decrease in fish and shrimp productivity (GFDRR, 2011).

A significant increase in the number of heatwaves has been observed from the 1980s till the first decade of the 21st century in Pakistan (Zahid and Rasul, 2012: 883–896). But, from 2015 to 2018 the increase in frequency of heatwaves has been steep. Among all the climate-related disasters that are confronting cities, heatwaves are the deadliest. For instance, a week-long scorching temperature in June 2015 took the death toll to more than a thousand in Karachi and hundreds more in Sindh Province (Mansoor, 2015; Imtiaz and Rehman, 2015). In May 2018, Pakistan weather left people collapsed in the streets and begging for water as a stifling 45°C heatwave killed at least 65 people (The Sun, May 22, 2018; Sayeed, 2018; Safi, 2018) in the southern city of Karachi.

While addressing at a ceremony to mark World Environment Day, the former Federal Minister for Environment of Pakistan, Hameed Ullah Jan

Afridi, said that the harmful effects of climate change are showing in the form of extreme weather disasters like storms, floods, and droughts (Khalid, 2009; WBGU, 2008: 143–146; GOP, 2012). It is true that water remains an issue of high priority for Pakistan as it frequently experiences floods and droughts (Eckstein, Hutfils and Winges, 2018: 4). Its water availability declined from 5,300 cm per capita in 1950 to 1,105 cm per capita in 2005. There is also expectation that climate change will affect water availability in Pakistan, a country whose 34 per cent of its energy supply is based on hydropower (MOE, 2003).

In this country, about 59 per cent of the annual rainfall is due to monsoon rains, but precipitation is scarce in most parts of Pakistan (Farooqi, Khan and Mir, 2005: 12–13). Due to the increasing evapotranspiration (a combined process of water evaporation from the earth's surface and transpiration from vegetation), climate change is likely to increase salinization of shallow groundwater both in semi-arid and arid areas (Ahmad, Bari and Muhammad, 2003; Ahmad *et al.*, 2004; Bates *et al.*, 2008: 43). Excessive groundwater harvesting in Baluchistan will exhaust the aquifers (Saddiqui, 2010: 74). As stream flow is anticipated to decrease in many semi-arid areas, the risk of getting more desertification will increase (Oldeman, Hakkeling and Sombroek, 1991).

On the contrary, millions who live in low-lying lands are suffering from flooding (Parry *et al.*, 2007). This is caused by the Indus River Basin (Ali, 2013). Sometimes it is leading to the loss of land, displaced coastal villages, and population displacement. The year 2010 started with a drought in January and February, followed by a heat wave in March and tropical cyclone in May. June records another heat wave with temperatures which rise to 53.7°C, the highest in the last 50 years. On July 24, heavy rainfall got going and continued till July 29, which resulted in devastating floods (Habib and Nawaz, 2010: 129). In addition, from 1992–2010 more than five megafloods occurred in Pakistan whereas frequent droughts also occurred from 1999 to 2002 (Dawn, July 22–23, 2010; The Times of India, August 1, 2010). These situations have revealed the vulnerabilities of the Indus River Basin irrigation system as it has been going on through unprecedented variations (Ahmad, Bari and Muhammad, 2003).

Agriculture is a key economic sector that contributes 21 per cent to the country's GDP, employs 45 per cent of the total workforce, and contributes about 60 per cent to exports (MoPDR, 2015b). Climate change is also likely to have a huge blow on Pakistan's agriculture. The possible effects will include liability to heat stress, water availability, and changes in productivity (MOE, 2003). The former environment minister Mukhdoom Syed Faisal Hayat says agricultural production in Pakistan gets affected by the changes in land and water regimes (Bansal and Datta, 2012: 228). For a case in point, about 5 million acres of land were affected by the 2010 flood alone (Preliminary Assessment, 2010). The direct cost in terms of damages to infrastructures, livelihood, and agriculture has been estimated to be about

3 trillion rupees (Habib and Nawaz, 2010: 131). While its population is depending heavily on agriculture, its densely populated deltas are threatened by potential risk of soil degradation (Farooqi, Khan and Mir, 2005:12–14). Extensive logging in northern areas has resulted in slope instability with the potential for soil erosion and water pollution. Though trees are essential resources for rural communities, only 5 per cent of Pakistan's land has forest cover (FAO, 2005). Over-grazing results in desertification and soil salinity (Saleh, 2007; Zia *et al.*, 2004).

Again, Pakistan's agriculture production system is mostly based on irrigated flows of water it derives from glacial melt-off of Himalayan glaciers together with monsoons during the summer period. Despite having the world's largest glaciers, however, Pakistan is now among the world's 36 most water-stressed countries (Reig, Maddocks and Gassert, 2013; Ghani and Muhammad, 2017: 1; Kundi, 2017; Khan 2009: 5). According to the Water Risk Security Index 2010 given by Maplecroft (WSRI, 2010) for global risk analysis, Pakistan is ranked at seventh position in the list of countries having an extreme risk of water shortage (Asif, 2013: 8). According to National Drinking Water Policy (2009), 35 per cent of the Pakistani population is deficient in getting access to safe drinking water. Khan (2009: 82) argued that this country went from being relatively water-abundant in 1981 to water-stressed by about 2000 and will be expected to be water-scarce by 2035. Water shortage is now a problem as the demand for freshwater resources is extremely important due to the agrarian nature of its economy. Even the Indus Valley, which is considered as the cradle of Pakistan's agriculture, is presently threatened by water shortage as agricultural irrigation in these areas is often not practised in a sustainable fashion (MEA, 2005). For the past several years, drought and water shortage have reached a critical stage in the food-growing areas, especially in Sindh and Punjab provinces which threatened food production (Khan, 2015). Water crises endangered both the winter crops that are about to be harvested, and the summer crops which would be sown immediately afterwards (Asif, 2013: 9). That is why while Pakistan was a net exporter of wheat and other cereal grains in the 1980s and 1990s, it suddenly became a great net importer of food products during the 2001–2003 periods (FAOSTAT, 2006). Pakistan has been reported to have one of the highest levels of prevalence of child malnutrition compared to other developing counties (Di Cesare *et al.*, 2015: 229–239). According to the National Nutrition Survey (2011), 33 per cent of all children were underweight. With its soaring population, the demand for food is also expected to increase further (Ringler and Anwar, 2013). With the rise of temperature (+0.5°C–2°C), agricultural productivity will decrease by around 8–10 per cent by 2040 (Dehlavi *et al.*, 2015). As a result, there is warning that if the situation worsens, people may raid storage facilities for food (Saleem, 2008). Unless climate change trends are reversed and water availability returns to normal, Pakistan will be in great danger (Rana, 2013). Hence, the Pakistan Council of Research in Water Resources

(PCRWR) delivered a grave warning, saying that if the government does not take action, the country will run out of water by 2025 (Kundi, 2017). From the discussion Pakistan is getting plagued by multiple ecological crises.

Sri Lanka

Sri Lanka, a tropical country, falls under various climate change–related threats, such as sea level rise and severe floods, droughts, and erratic rainfall patterns (McCarthy *et al.*, 2001; SLMOE, 2010; Mendelsohn, Munasinghe and Niggol, 2005: 581–596). Other effects include (i) a faster increase in mean monthly and annual temperatures compared to the average global rate of warming; (ii) winter temperatures increasing more than summer temperatures; and (iii) more frequent extreme weather events such as heat waves (SLMOE, 2010). There is also a study which has attempted to project the future effects of climate change on Sri Lanka's agriculture, water resources, sea level, the economy, and health (Eriyagama *et al.*, 2010: 51).

Some data suggest that atmospheric temperature is steadily rising almost everywhere in Sri Lanka (Chandrapala, 2007a: 56–57; Eriyagama *et al.*, 2010; Nissanka *et al.*, 2011; Basnayake, 2007). It has been reported that mean daytime maximum and mean night-time minimum air temperatures have also increased (Basnayake, 2007; Zubair *et al.*, 2005). Owing to rising temperatures large areas of Sri Lanka will suffer from a periodic deficit in water supply and increased water demand for domestic and industrial needs. This condition will worsen water scarcity in the country and may even lead to periodic drought (SLMFE, 2000: 64). Other studies (MoERE, 2011; Eriyagama *et al.*, 2010) suggest that climate change could alter natural systems connected to water cycles, eco systems, and biodiversity of the country.

Temperature rise depletes soil moisture and premature aridness of crops, increases pests and diseases. The reason is that if the temperature increases by another 1–2°C during the flowering stage of a rice crop, the production will decrease (Weerakoon and De Costa, 2009: 6–7; Human Development Report, 2006; Punyawardena, 2007). It was in 2009 where Sri Lanka's agricultural sector was affected by drought, and variable rainfall affected the production of paddy and tea (Central Bank of Sri Lanka, 2010; Eriyagama *et al.*, 2010: 50–55). Subsequently in 2013, the northeast monsoon that is a source of irrigation for the main rain-fed agriculture (*Maha*) season across the paddy-producing areas in Sri Lanka was delayed (OCHA, 2014). By April 2014, the Department of Agriculture stated the decreasing rainfall had damaged 83,746 hectares of paddy-planted area, resulting in an estimated production loss of 280,000 Mt of rice (15 per cent of forecasted production) (OCHA, 2014). The discharges of water from major rivers are decreasing, resulting in a setback for irrigation and crop production (Weerakoon and De Costa, 2009). Besides agriculture, scarcity of water will affect the health of the population leading to various types of diseases (UNDP, 2006).

For Sri Lanka, monsoon is the main source of water which is now unpredictable (Jaitly, 2009: 20). There are studies which suggest that average rainfall is showing a decreasing trend in Sri Lanka (Basnayake, 2007; Chandrapala, 2007b: 58–59; Jayatillake *et al.*, 2005: 62–87). But other studies observed that heavy rainfall events have become more frequent especially in Yala season (Punyawardena and Premalal, 2013: 1–12; Chandrapala, 2007b; Eriyagama *et al.*, 2010; Punyawardena, Dissanaike and Mallawatantri, 2013). Higher rainfall intensities have however resulted in damages to the country's infrastructure (Eckstein, Hutfils and Winges, 2018: 6). It generates higher surface runoff too. In due course, it will possibly lead to flooding. Floods triggered by heavy rainfall cause heavy landslides, mudslides, and soil erosion, which finally leads to road and railway track devastation (Baba, 2010: 4–16). As such, this situation affects the people. In December 2014, Sri Lanka was affected by heavy flooding where thousands of people were displaced and a million were affected (Khaleej Times, December 31, 2014; The Economic Times, December 21, 2014). In May 2017, heavy landslides and floods occurred in Sri Lanka after strong monsoon rains in the southwestern regions of the country. The monsoons killed at least 150 people, displaced more than 600,000 of them from their homes, and 12 districts were affected. The inland southwest district of Ratnapura was most affected where over 20,000 people faced flash floods (Leach, 2017; Eckstein, Hutfils and Winges, 2018: 6–7).

The intensity and the frequency of the extreme events such as droughts have increased during recent times (MoMDE, 2015: 18; Imbulana, Wijesekara and Neupane, 2006; Ratnayake and Herath, 2005: 192–205). Droughts can disrupt public life and damage property and crop harvests (MoE/ADB, 2010). Insufficient rain or dry spells affected paddy cultivation during the 2016–2017 *maha* season (September–March), which provides two-thirds of the annual national domestic supply of rice, Sri Lanka's main staple food (FAO, 2017a: 32). According to a World Food Programme (WFP) document food shortages brought on by extreme weather events have resulted in almost a quarter of Sri Lanka's 21 million people becoming malnourished (Parera, 2016).

With the rise in sea level which is the outcome of rising temperature, Sri Lanka will be affected in many ways (Pachauri and Reisinger, 2007). The 2004 tsunami has shown that low-lying plains in the coastal zone are vulnerable to any future rise in sea level. With the rise of sea level cyclones are expected to increase their strength (Parry *et al.*, 2007). Inundation caused by sea level rise will result in coastal erosion, loss or damage to the country's infrastructures such as boat landing sites, fisher folk settlements and shrimp fishing under coastal aquaculture. It will also affect tourism, freshwater availability, and the livelihoods of people in low-lying coastal areas (MoERE, 2011; Senaratne, Perera and Wickramasinghe, 2009; Baba, 2010: 4–16; UNDP, 2006). The intrusion of saltwater on low-lying agricultural land may result in huge loss through degradation of arable land. Such a case will likely expose people to grave environmental risk and they may face food scarcity (Rabbani, Rahman and Islam, 2010: 25). Besides, a substantial share of foreign income is earned through export crops which are highly sensitive to fluctuations of weather (MoMDE, 2015: 11).

Besides, floods, droughts, sea level rise, and food shortages, climate change–related effects can have a toll on the health of the people (MoE, 2010). For instance, sea level rise may affect people's health due to seawater incursion and storm damage, fisheries, and the effects of saltwater intrusion on agricultural irrigation and drinking water. Temperature rise will cause heat-related illnesses. Drought, floods, and increased rainfall will have implications for housing damage and a number of water-borne diseases including diarrhoea, typhoid, dysentery, and dengue, and viral hepatitis (MoE, 2010; WHO, 2015: 24). Hence, the bigger question of national importance is what Sri Lanka's climate will look like in 50 or 100 years and what preparation the country is ready to make.

From the entire discussion it has been found that all the countries of South Asia have experienced some of the deadliest furies of climate change. The rise of temperature will melt huge glaciers or result in the loss of land-based ice. As a result, it will result in the volumetric flow of GLOFs. Its coastal areas will be susceptible to sea level rise. Perpetual inundation will affect coastal countries such as Bangladesh and the Maldives (Tallmeister, 2013). Other coastal areas of South Asia will be no exception from this threat. The other worst effect will be the emergence of environmental refugees (Brammer, 2009: 87). People who flee to neighbouring countries may alter the socio-economic, cultural, and demographical balance in the region or maybe implicated in social conflict and instability (Nordas and Gleditsch, 2007: 627). This will create a non-military threat, as climate refugees will have the potential to create law-and-order problems in the adjoining countries and imposed economic and social burdens (Hussain, September 5, 2009; The Times of India, August 24, 2014).

With the melting of the Himalayan glaciers, which act as natural water reservoirs, there will be extreme consequences for water supply in the perennial rivers, which sustained water availability (Lovett, December 21, 2011). Erratic rainfall causes drought. When water becomes a scarce commodity, people will have to relocate to a place where they can grow food find accessible water to drink. This could be the reason the former president of India, A. P. J. Abdul Kalam, once cautioned that there is a possibility that the neighbouring countries of India could "create problem" for India to meet the demand for water for their increasing population because their demand for water is also increasing (The Economic Times, July 17, 2010).

The Indo-Gangetic Plains could become significantly heat-stressed by the 2050s, potentially causing losses of 50 per cent of its wheat-growing area (Porter *et al.*, 2014: 485–533). When food production degrades, the availability of food will decline. This situation will be further intensified by the escalating rise of food prices due to scarcity of food. When income levels are low for the poor and food expenditures are high, they will be in a tricky position to afford food. The rich will also face the problem to access food due to the decline in supply (Vidal, 2013). A decline of food supply could escalate food insecurity to hundreds of millions of people who are malnourished. Such problems could be the basis for the mass of hungry and impoverished people to become desperate due to their hunger. More than that climate change may hamper the achievement of many of the SDGs, including those on poverty eradication, child mortality, malaria, and other diseases, and environmental sustainability. Considering the escalating frequency, intensity, and rising danger from climate change events as shown by the numerous studies discussed previously, there is a need to enhance regional cooperation in South Asia. This is not the only region which has experienced the extremes of climate change. Other countries of the world have also experienced similar problems which will be emphasized in the subsequent sections.

VIII Climate change in China and the United States

Besides South Asia, the impacts of climate change have been felt in China, which is the most populous country and second-largest economy in the world (Zhang *et al.*, 2010: 183–189; Ding, Qian and Yan, 2010: 1452–1462). Over the past several decades, China has experienced some devastating climate change extremes (Zhang *et al.*, 2010: 183–189; Ding, Qian and Yan, 2010: 1452–1462; Wang, Ren and Zhang, 2014: 3054–3065; Wei and Chen, 2009: 153–158) like heavy heatwaves, floods, droughts, and rainfall.

Extremely hot summers in 1988, 1990, 1994, 1998, 1999, 2002–2008, and 2013 together have resulted in thousands of human casualties (Kurukulasuriya *et al.*, 2016; Grieser, 2013; Ma *et al.*, 2015: 103–109). From July to August 2013, China faced long-lasting high temperatures, in a zone extending from north China southward to the lower reaches of the Yangtze River and to the south of the lower reaches of the same river (Ding and Ke, 2015: 651–665). On July 24, 2015, the temperature reached 50.3°C (122.5°F) near Ayding Lake in Xinjiang Province, which was the highest temperature ever recorded in China (Sandalow, 2018: 25). On July 20, 2017, Shanghai had its hottest day ever, with the temperature reaching 40.9°C (105°F) (Phys. org, July 21, 2017). Altogether, heat extremes during 2013–2017 have had consequences in China. Around 1,610 million people and 18,720 km² of agricultural lands in Hunan Province were affected (Duan, Wang and Feng, 2013: 633–640). Over the same period, scanty precipitation, high solar radiation, and high evaporation in the south of the Yangtze River have had a devastating impact on many aspects of social life (PRC, 2007; Sun *et al.*, 2014: 1082–1085).

China is sometimes severely afflicted by floods particularly when its major rivers are flooded with rainwater. Of the two major rivers – the Yellow River and the Yangtze River – the former is accountable from 70 per cent to 75 per cent of all China's floods (Butler, 2007). This river is known to regularly overflow its banks and flood huge plains of land along the river basin during the monsoon season (PRC, 2007; Ren, 2007). For several years flooding in the middle and lower reaches of this river and southeastern China has intensified (PRC, 2007). In July 2007, the worst rainstorms in 115 years hit Chongqing, causing dozens of deaths and extensive property damage. In July 2012, the heaviest rainfall in 60 years hit Beijing, leaving 37 people dead (Nan, 2012). From June 30 to July 6, 2016, heavy rainfall caused a series of extensive floods, claiming 132 lives across 13 provinces (Lyu et al., 2018: 2). Reports predict rainfall is expected to increase from 2 to 3 per cent by 2020 and 5 to 7 per cent by 2050 across China (Ren, 2007; Zhang and Wang, 2007). This dramatic rainfall pattern will have serious impacts on the socio-economic development of China and on the welfare of its population (Ren, 2007; Zhang and Wang, 2007).

Over the last two decades, drought anomalies have become more severe and frequent, affecting water supply mostly in the north and the northwest part of China (Lyu et al., 2018: 4; Wang et al., 2009: 141–158). Millions of Chinese have been affected by droughts. This can be seen from the statistics of the Ministry of Civil Affairs published in May 2011, which showed around 35 million people were affected by the drought, 4.24 million suffered from acute water shortages, and 370,000 hectares of harvestable crops destroyed (BBC China, 2011a). To worsen the matter, average precipitation at the middle and lower streams of the Yangtze River was reduced to only 194 mm since spring 2011, which meant 51 per cent of normal precipitation, a record low in 60 years (BBC China, 2011a; Jelenkovic, 2012: 49). As a result, around 3.3 million people and 22,000 hectares of land in Jiangxi were affected by drought, whereas 6.1 million people in neighbouring Hunan Province faced acute water shortages (BBC China, 2011b). In Hunan and Jiangxi provinces, where villagers are allowed to own small farmlands as well as fishponds, drought deprived people of their livelihood as their ponds and lands were drying up (BBC China, 2011c). So many villagers fled to the cities in search of job and livelihood opportunities (BBC China, 2011c).

The extreme heat too has had its own consequences for water shortage. Extreme heat from 2013 to 2017 has (Duan, Wang and Feng, 2013: 633–640) affected the supply of drinking water to 353.5 million people and 1.417 million livestock. Water shortages may not end here. The scenario may intensify as several studies have predicted the probability that the arid area in western China will become larger and the risk of desertification will increase (Ren, 2007; Zhang and Wang, 2007).

Collectively, heatwaves, rainfall, floods, and droughts raise serious concern over the long-term development of regional agriculture, the economy, and environmental health of the Chinese people. First, China holds up to 22 per cent of the world's population but only 7 per cent of the world's arable land (Zhang et al., 2010: 183–189). The share of irrigated land varies significantly across regions due to diverse environmental conditions, ranging from more than 70 per cent in the east to only about 20 per cent in the northeast (Wang, Huang and Rozelle, 2010: 5). Therefore, it is doubtful whether China can feed its entire population adequately by the year 2030, given that food supply is a problem, particularly to those living in poor and remote areas of the country (Huang and Rozelle, 2009; Tian *et al.*, 2016; Zhang *et al.*, 2010: 183–189; Ye, 2013). Thus, climate change is a source of concern for agriculture and human life in China especially in its north-western parts (Zhai and Zou, 2005: 16–18; Zhai, 2010: 649–662). Another study predicts that full-blown heatwaves are likely to increase the frequency of isolated hot days, further impairing health and productivity for millions of working people (Zhao *et al.*, 2016: 4640–4645). Akin to South Asia, climate change may hinder China in meeting SDG 1 and SDG 2 which stress on tackling poverty and hunger respectively. So, it is vital for China to implement the Agenda 2030, especially when it comes to SDG 13.

The US is already experiencing the effects of climate change (Karl, Melillo and Peterson, 2009). In the near future heat waves, wildfires, extreme weather events, and rising sea levels may cost the country hundreds of billions of dollars due to the reduction of crop yields, health problems, and crumbling infrastructure (Gramling and Hamers, 2018). The Fourth National Climate Assessment of the US that was released November 23, 2018, predicts the US economy will shrink by as much as 10 per cent by the end of the century if rising temperature continues apace (USGCRP, 2018). As of 2013, coastal shoreline counties of the US are home to 133.2 million people, and are economic engines that support jobs in defence, fishing, transportation, tourism industries and contribute substantially to the GDP of the US and also serve as hubs of commerce, with seaports connecting the country with global trading partners (Kildow *et al.*, 2016: 31; Talley and Ng, 2017: 175–179). With the rapid ice loss from Greenland and Antarctica revealing higher scenarios, an extreme scenario of global sea level rising upwards of 6–8 ft by 2100 is a possibility (Pfeffer, Harper and O'Neel, 2008; NOAA, 2012; Fleming *et al.*, 2018: 322–352; Sweet *et al.*, 2017: 75; DeConto and Pollard, 2016: 591–597; Wong, Bakker and Keller, 2017: 347–364). Under this rise, flooding from rising sea levels and storm surges is likely to destroy, or make unsuitable for use, billions of dollars of property by the middle of this century, with the Atlantic and Gulf coasts facing greater-than-average risk compared to other regions of the country (Fleming *et al.*, 2018: 322–352; Neumann *et al.*, 2015: 337–349).

Future climate change impacts on hydrology, floods, and drought for the US have been discussed in the Third National Climate Assessment (Georgakakos *et al.*, 2014: 69–112) and in the USGCRP's Climate Science Special Report (Easterling *et al.*, 2017: 207–230; Wehner *et al.*, 2017: 231–256). Increasing temperature has resulted in a shift in the timing of snowmelt runoff to earlier in the year (Lall *et al.*, 2018: 152). Shifts in the hydrological regime owing to glacier melting will alter stream water volume, water temperature, and runoff timing (Lall *et al.*, 2018: 152). As temperatures continue to rise (McDonald and Girvetz, 2013; Lall *et al.*, 2018: 152), the stress on water supplies will adversely impact water supply in the US. In some parts of the US, annual rainfall is expected to decrease (Stanton and Ackerman, 2007). Decreasing rainfall and hotter and drier conditions during the growing season will threaten the agricultural sector in Great Plains states such as the Dakotas (Karetnikov *et al.*, 2008; Wolfe *et al.*, 2007: 555–575; Gowda *et al.*, 2018: 391–437). Water shortages are projected to constrain electricity production in Arizona, Utah, Texas, Louisiana, Georgia, Alabama, Florida, California, Oregon, and Washington state by 2025 (Bull *et al.*, 2007). Long-lasting droughts and warm spells may also compromise earth dams and levees as a result of the ground cracking due to drying, a reduction of soil strength, erosion, and subsidence (Vahedifard, AghaKouchak and Robinson, 2015: 799–799; Vahedifard, Robinson and AghaKouchak, 2016). On the other hand, extreme precipitation events are projected to increase in a warming climate and may lead to more severe floods and greater risk of infrastructure failure in some states (Ragno *et al.*, 2018: 1751–1764).

Under a high-emissions scenario, many cities in the US can expect 60 or more days above 90°F by 2100 and 14–28 days above 100°F, with some of the hottest temperatures expected in large cities like Philadelphia and New York (Frumhoff *et al.*, 2007: 92–96; Twilley *et al.*, 2001; Stanton and Ackerman, 2007; Medina-Ramon and Schwartz, 2007: 827–833). High temperatures in the summer are conclusively linked to an increased risk of a range of illnesses and death, particularly among older people, pregnant women, children, and the poor (Sarofim *et al.*, 2016: 43–68). In the absence of adaptation, exposure to the mosquito *Aedes aegypti*, which can transmit dengue, *Zika*, *chi-kungunya*, and yellow fever viruses, is projected to increase by the end of the century due to climatic, demographic, and socioeconomic changes, with some of the largest increases projected to occur in North America (Ebi *et al.*, 2018: 572–603; Butterworth, Morin and Comrie, 2017: 579–585).

From the discussion it is clear the impacts of climate change will be felt across the globe. It is greatly affecting lives and communities, disrupting national economies and countries today and even more tomorrow. The developing countries of South Asia are more likely to experience its negative effects although their GHG's contribution is marginal. In the next chapter, the study will focus on the various declarations, policy initiatives, and action plans that SAARC has undertaken so far in order to tackle the brunt of climate change within the region.

4 SAARC policy initiatives for cooperation in climate change

I Introduction

Generally, nation-states have largely ignored cooperation for climate change. This is due to the fact that the controversy about climate change became a subject of political debates only in recent times. Precisely, climate change first surfaces as a political issue in the 1980s, and soon after it attracted extensive international attention (Bodansky, 2001: 46). Reasons for the emergence of climate change, how much it has occurred in modern times, what will its effects be, whether any action should be taken to curb it, and if so what that action should be have been discussed (Oreskes, 2004: 1686; DiMento and Doughman, 2014). In 1988, partly under US leadership, the UN established the IPCC to conduct scientific research on the phenomenon of climate change. To support the UNFCCC, the IPCC produced the reports which are the main international treaty on climate change. The eventual objective of the UNFCCC as stated in Article 2 is to "stabilize greenhouse gas concentrations in the atmosphere at a level that would prevent dangerous anthropogenic interference with the climate system" (Parry *et al.*, 2007: 766). The IPCC report provides "the scientific, technical and socio-economic information relevant to understanding the scientific basis of risk of human-induced climate change, its potential impacts and options for adaptation and mitigation" (Principles Governing IPCC Work, Art. 2). Thus, it can be stated that climate change as an issue started to come to the political scene during the second half of the 20th century.

It has been discussed in the previous chapter that the SAARC region has frequently experienced various climate change–related disasters such as floods, droughts, rise of temperature, food shortages, etc. When climate change inflicts these catastrophes on the people in the region on a regular basis, climate change therefore presents the opportunity for the members of SAARC to formulate new interstate policies and initiatives (Underdal, 2001: 1–47). These policies and initiatives are crucial as they build the structure and enforce uniform standards throughout a regional cooperation group such as SAARC. So, this chapter focuses on the declarations, policies, initiatives, action plans, and arrangements that exist within the SAARC agenda since its inception; the vision of this organization was to work together towards finding solutions in certain agreed areas of cooperation.

II The SAARC Summits

The SAARC organization which presents the Summits at the apex is supported by the Council of Ministers, acting as the decision-making wing, and the Standing Committee that serves as the executive wing. A network of sector-based Technical Committees provides the base of the organizational pyramid (Khatun and Hossain, 2012). It is this systematized set-up which drives SAARC as an organization. The willingness of SAARC members to mitigate the effects of climate change began due to the concerns over the region's heavy dependence on environmental resources and its susceptibility to the natural disasters. These factors have motivated the members of SAARC to commission a study on the subject at its Third Summit in 1987 (Sharma, 2011a). Back then, SAARC decided, inter-alia, to commission a study that was titled "Protection and Preservation of Environment and the Causes and Consequences of Natural Disasters" in a well-planned comprehensive framework (Kathmandu Declaration, 1987). It suggested an appropriate institutional mechanism for coordinating and monitoring the implementation of its recommendations in the form of a SAARC Committee on Environment (Kathmandu Declaration, 1987).

The Fourth SAARC Summit in December 1988 coincides with the SAARC Heads of State's decision to undertake a joint study titled "Greenhouse Effect and Its Impact on the Region". This study was initiated to provide a basis for an action plan for meaningful cooperation among its member states (Islamabad Declaration, 1988). It recommended regional measures in sharing experiences, scientific capabilities and information on climate change, sea level rise, and technology transfer (Shaw, 2010). At the Fourth Summit in Islamabad, its members expressed their deep sense of sorrow and profound sympathy at the loss of valuable lives and extensive damage to property they suffered due to climate change–related disasters. In this connection, they recalled their earlier decision at Kathmandu in November 1987 to strengthen regional cooperation with a view to strengthening their disaster management capabilities and took note of the recommendations of the meeting of the SAARC Group of Experts on the Regional Study on the Causes and Consequences of Natural Disasters and the Protection and Preservation of Environment that met in Kathmandu in July 1988 (Islamabad Declaration, 1988; SAARC, 1992a, 1992b). All the members voiced their conviction that the identification of measures and programmes as envisioned by the Group of Experts would supplement national, bilateral, regional efforts to deal with their increasing problems. The member countries also urged the regional study be completed within the shortest period of time to provide a basis for them to draw up an action plan for a meaningful cooperation amongst the member states (Declarations of SAARC Summits, 1985–1998, 2001: 28).

The Fifth Summit (Malé, November 21–23, 1990) directed that the study "Greenhouse Effect and its Impact on the Region" (SAARC, 1992b) be completed for consideration at the Sixth Summit. In fact, by then its members knew that the methodology for undertaking the Regional Study on the Greenhouse Effect and its Impact on the Region (SAARC, 1992b) was likely to be finalized soon. At this summit, SAARC's Heads of State were alarmed at the unprecedented climatic changes predicted by the Assessment Report of the IPCC (Pielke, 2005: 952–954; Goodwin, 2009; Oppenheimer *et al.*, 2007: 505–506; The Fifth Summit Malé Declaration, 1990). Knowing their countries' poverty therefore, the SAARC members urged the international community to mobilize additional finances for them and to give them appropriate technologies so as to enable them to face the new challenges arising from climate change–related problems such as sea level rise etc. It was during this Summit that the SAARC leaders decided to observe 1992 as the "SAARC Year of Environment" (The Fifth Summit Malé Declaration, 1990; Pattanaik, Bisht and Bommakanti, 2015: 2).

It was at the Sixth Summit in Colombo on December 21, 1991 (The Colombo Declaration, 1991; Pattanaik, Bisht and Bommakanti, 2015: 2) that the members of SAARC expressed their satisfaction at the completion of the Regional Study on the Causes and Consequences of Natural Disasters and the Protection and Preservation of Environment (SAARC, 1992a, 1992b) and they also endorsed the decision of the Council of Ministers to establish a Committee on Environment or Technical Committee on Environment to:

- examine the recommendations of the Regional Study
- identify measures for immediate action
- decide on modalities for their implementation.

Accordingly, a Committee on Environment was set up in 1992 (SAARC, 2015: 1).

This Regional Study on the Causes and Consequences of Natural Disasters and the Protection and Preservation of Environment is a synthesis of the studies on the environment undertaken by member states (Statements and Declarations of SAARC Summits, 2010: 110; SAARC, 1992a). The member states have understood that the protection of the environment is a common imperative for all countries. But they also have their opinion that it is the responsibility of the economically advanced countries to address the problems of global climate change for the reason that most of the pollutants emitted to the atmosphere originated from such rich countries. As developing countries they upheld their own development goals (Rajamani, 2012: 605–623; Tolba, 1988: 151). On the other hand, attempts have been made by SAARC countries to render their services towards a better environment through their regional negotiations. The Regional Study on the Causes and Consequences of Natural Disasters and the Protection and Preservation of

Environment (SAARC, 1992a) embodies a strong indication of the aware-ness of the environmental problems on the part of SAARC members. It was also at the Sixth Summit (Colombo, December 21, 1991) that the SAARC leaders expressed their satisfaction about the decision of the Standing Com-mittee concerning the time frame to finalize the Regional Study on the Greenhouse Effect and its Impact on the Region (The Colombo Declaration, 1991; SAARC, 1992b).

During the Seventh Summit (April 10–11, 1993, Dhaka) the SAARC lead-ers recognized that the completion of the Regional Study on the Greenhouse Effect and its Impact on the Region (SAARC, 1992b; The Declaration of the Seventh SAARC Summit, 1993) was a significant step forward in promot-ing regional cooperation in this vital area. The Regional Study has compo-nents for regional measures in sharing experiences, scientific capabilities, and information on climate change; and global collaboration in monitoring climatology, sea level rise, natural disasters, technology transfer, finance etc. (SAARC, 1992b; The Declaration of the Seventh SAARC Summit, 1993). The Seventh Summit also welcomed the outcome of the United Nations Confer-ence on Environment and Development (UNCED) held in Rio de Janeiro in 1992 and underscored the imperative need to ensure the flow of resources to successfully implement the wide range of suggested initiatives and actions contained in Agenda 21 (The Declaration of the Seventh SAARC, 1993: 5; Sitarz, 1993; text of Agenda 21). Agenda 21 is a comprehensive plan of action to be taken globally, nationally, and locally by organizations of the UN and governments in every area where humans impinge on the environ-ment. The SAARC's Heads of State have said that all international actions in the area of the environment should be based on common but differen-tial responsibilities, collective endeavours, and a balanced perspective. The principle of common but differentiated responsibilities is one of the basic principles in international environmental law; the core idea of this principle is that the developed countries and the developing countries should bear dif-ferent environmental protection responsibilities in all kinds of international environmental protection issues. In other words, the richer countries must shoulder the lion's share of responsibility for adaptation on the basis that they are historically responsible for the current levels of greenhouse gases in the atmosphere.

At the Eight Summit (The Eight Declaration, 1995) in New Delhi, the SAARC members realized the slow progress of its Regional Studies (SAARC, 1992a, 1992b). During this Summit (The Eight Declaration, 1995) the Heads of State or Government of SAARC stressed the importance of effective and speedy implementation of the recommendations of the Regional Study on the Causes and Consequences of Natural Disasters and the Protection and Preservation of Environment, and the Regional Study on Greenhouse Effect and its Impact on the Region (SAARC, 1992a, 1992b). They directed the Technical Committee on Environment to monitor the progress made in the implementation of the recommendations of the two Regional Studies.

At this Summit (The Eight Declaration, 1995) all the countries acknowledged the importance of international cooperation for building up national capabilities like the transfer of appropriate technology and promotion of multilateral projects and research efforts in natural disaster reduction. The transfer of technology pertains to taking fully into account that economic and social development and poverty eradication are the first and overriding priorities of the developing country. This shows that SAARC was aware of the fact that actions in the area of environment protection should be based on partnership and collective endeavours. Further, the members of SAARC recalled the decisions expressed at the Seventh Summit (The Declaration of the Seventh SAARC Summit, 1993) on the outcome of the UNCED of June 1992 and reiterated the urgent need to ensure the flow of new and additional resources that are adequate and predictable to successfully implement the programmes of Agenda 21 (The Eight Declaration, 1995).

The Ninth SAARC Summit (1997) in Malé welcomed the offer of the Maldives to host a meeting of the SAARC Environment Ministers to focus more directly on the Environment concerns of the region, including the formulation of the Plan of Action for immediate implementation of recommendations contained in the two Regional Studies on the Greenhouse Effect and its Impact on the Region, and the Regional Study on the Causes and Consequences of Natural Disasters and the Protection and Preservation of Environment (SAARC, 1992a; SAARC, 1992b). This Summit directed that the meeting may also consider the feasibility of drawing up a Regional Treaty on Environment. A decision was also taken that the meeting of SAARC Environment Ministers should be institutionalized, henceforth as an annual event. Consequently, the Third Meeting of the SAARC Environment Ministers (Malé, October 15–16, 1997) adopted the SAARC Environment Action Plan which is highlighted in **Annex I**, in a Brief on Regional Cooperation under SAARC (SAARC, 2015: 2). The Action Plan identified some of the key concerns of member states and set out the parameters and modalities for regional cooperation (SAARC, 2015: 11–12; SAARC Environment Ministers Conference, October 15–16, 1997; The Ninth SAARC Summit, 1997). More than that there was an expression of concern at the slow progress in the implementation of Agenda 21 at this Summit. Hence, its leaders called for the urgent implementation of the commitments, recommendations, and agreements reached at the UNCED held in Rio de Janeiro in June 1992 (The Ninth SAARC Summit, 1997).

The Declaration of the Tenth Summit (July 29–21, 1998) in Colombo expressed satisfaction at the positive outcome of the Third Meeting of the SAARC Environment Ministers, and called for the effective and early implementation of the SAARC Environment Action Plan (1997). The Fourth Meeting of the SAARC Environment Ministers held in Colombo, October 30–November 1, 1998, adopted the Colombo Declaration on a Common Environment Programme which is highlighted in **Annex II** (SAARC, 2015: 2; SAARC Environment Ministers Conference, October 15–16, 1997). It was at the Tenth Summit (The Declaration of the Tenth SAARC Summit, 1998) which was held in Colombo that the SAARC Member States welcomed the offer of the Maldives to prepare a feasible study on the establishment of a Coastal Zone Management Centre. This centre aimed at promoting regional cooperation in the management of the coastal zones, including research, training, and promotion of awareness in the region (SAARC Coastal Zone Management Centre, 2016). In addition, the governments of all SAARC members voiced their dedication to prepare their own National Environment Action Plans and State of the Environment Reports before the end of 1998. They further welcomed the adoption of the Kyoto Protocol to the UNFCCC (1997) and underscored the importance of the Protocol for the protection of the climate system (The Declaration of the Tenth SAARC Summit, 1998).

Further, the members of SAARC then expressed their warm welcome of the adoption of the Kyoto Protocol to the UNFCCC in December 1997 (Kyoto Protocol, 1997: 162; The Declaration of the Tenth SAARC Summit, 1998). This Protocol aimed at the stabilization of GHG concentrations in the atmosphere. Simultaneously, SAARC countries stressed the importance of the Kyoto Protocol for the protection of the climate system and advised the industrialized countries to ratify it and to undertake urgent and effective steps to implement the commitments to reduce their emissions of GHG. It is worthwhile to note that the Kyoto Protocol didn't become international law until more than halfway through the years from 1990–2012. By that point, global emissions had risen substantially (The Guardian, March 11, 2011).

The Eleventh SAARC Summit (2002) reiterated the need for the early and effective implementation of the SAARC Environment Action Plan (1997). The Summit also felt a strong need to devise a mechanism for cooperation in the field of early warning as well as preparedness and management of natural disasters, along with programmes to promote conservation of land and water resources.

The Twelfth SAARC Summit (2004) held in Islamabad underlined the importance of the early and effective implementation of the SAARC Environment Action Plan (1997) and also stressed on the early submission of national state of the environment reports to expedite the preparation of SAARC State of Environment Report and the commissioning of the work on drafting a Regional Environment Treaty.

The Declaration of the Thirteenth Summit (2005) in Dhaka decided to consider the modalities for having a Regional Environment Treaty in furthering environmental cooperation among the member states. The Summit decided to proclaim the Year 2007 as the "Year of Green South Asia" devoted to a region-wide afforestation campaigns (Mall, Attri and Kumar, 2011: 30; The Declaration of the Thirteenth SAARC Summit, 2005). The Summit also called for elaboration of a Comprehensive Framework on Early Warning and Disaster Management and underlined the necessity to put in place a permanent regional response mechanism dedicated to disaster preparedness, emergency relief, and rehabilitation to ensure immediate response (The Declaration of the Thirteenth SAARC Summit, 2005).

At the Thirteenth SAARC Summit the SAARC leaders expressed satisfaction at the progress of the implementation of the SAARC Environment Action Plan. Its leaders also hailed the decision of the Council of Ministers to establish a SAARC Forestry Centre in Bhutan (The Declaration of the Thirteenth SAARC Summit, 2005). The Forestry Centre intends to serve as a centre of excellence for forestry within the region and operate as a nodal point for research, information, and policy development (SAARC Forestry Centre Opened in Bhutan, 2008). The Summit endorsed the recommendation for elaboration of regional programmes and projects for early warning, preparedness for and management of *tsunami* and other likely calamities. In addition, the member states of SAARC called for a detail Comprehensive Framework on Early Warning and Disaster Management towards establishing and strengthening the regional disaster management system to reduce any possible risks (Dhaka Declaration, Thirteenth SAARC Summit, 2005; SAARC Disaster Management Framework, 2007; Jeganaathan, 2011). This Summit also sanctioned the decision of the Special Session of the SAARC Environment Ministers to move ahead with the capability of the existing SAARC Institutions. These include the SAARC Meteorological Research Centre in Bangladesh that constituted a framework of SAARC Institutions which addressed diverse aspects of environment, climate change, and natural disasters (Jeganaathan, 2011; The Declaration of the Thirteenth SAARC Summit, 2005) and the SAARC Coastal Zone Management Centre to carry out their mandated tasks. Through the SAARC Comprehensive Framework on Disaster Management the SAARC Heads of State emphasized the necessity of putting in place a permanent regional response mechanism which is devoted to disaster preparedness, emergency relief, and rehabilitation to ensure immediate response (The Declaration of the Thirteenth SAARC Summit, 2005). This is in view of the extensive loss of life and colossal damage to property as a result of the earthquake and the 2004 Indian Ocean *tsunami* (Kenneally, 2004; Lambourne, 2005) and other climate change disasters in South Asia. Such calamities have impressed upon them the need for a joint action in the area of environment, including water conservation, to encourage sustainable development (The Declaration of the Thirteenth SAARC Summit, 2005).

While noting the progress in the execution of the SAARC Environment Action Plan (1997), which was adopted by the Third Meeting of the SAARC Environment Ministers, Malé, October 15–16, 1997 (SAARC, 1999: 101–103), the Fourteenth Summit (2007) held in New Delhi also called for its timely implementation. In addition, this Summit expressed "deep concern" over global climate change and called for pursuing a climate resilient development in South Asia. During the Fourteenth Summit, the SAARC members voiced their contentment at the declaration of 2007 as the "Year of Green South Asia" (The Fourteenth Summit, 2007). Accordingly, a Ministerial

Meeting on Climate Change was held in Dhaka on July 3, 2008, which adopted the Dhaka Declaration and SAARC Action Plan on Climate Change. The Dhaka Declaration and SAARC Action Plan on Climate Change are highlighted in **Annex III** (SAARC, 2015: 3; The Fourteenth Summit, 2007).

The Twenty-ninth session of the SAARC Council of Ministers held in New Delhi on December 7–8, 2007, among others, adopted a SAARC Declaration on Climate Change which was presented by the President of Republic of Maldives at the High Level Session of the UN Conference on Climate Change held at Bali in December 2007 on behalf of SAARC. The Declaration is highlighted at **Annex IV** (SAARC, 2015: 3).

The Fifteenth Summit (2008) held in Colombo reiterated the need to intensify cooperation within an expanded regional environmental protection framework, to deal in particular with climate change issues. Expressing concern at the human loss suffered through natural disasters in the region, the Summit also stressed the need for timely provision of relief in humanitarian emergencies and directed that a Natural Disaster Rapid Response Mechanism be created to adopt a coordinated and planned approach to meet such emergencies under the aegis of the SAARC Disaster Management Centre.

The most significant step taken by SAARC was at the Sixteenth Summit (Thimphu, April 28–29, 2010). The Sixteenth SAARC Summit adopted climate change as its theme and its members reaffirmed their commitment to address this challenge. In this context, they adopted the Thimphu Statement on Climate Change (2010) and directed that the recommendations contained therein be implemented in earnest. The Thimphu Statement on Climate Change is highlighted in **Annex V** (SAARC, 2015: 3). The Thimphu Statement on Climate Change (2010) outlines a number of initiatives to be implemented at the national and regional level to address the adverse effects of climate change. The Inter-Governmental Expert Group on Climate Change (IGEG.CC) which was established by the Thimphu Statement on Climate Change is required to monitor and review the implementation of the Thimphu Statement on Climate Change (2010) and make recommendations to facilitate its implementation. The First Meeting of the IGEG.CC was held in Colombo on June 29–30, 2011, and the report of the meeting was approved by the Ninth Meeting (2011) of SAARC Environment Ministers held on September 29, 2011, at Thimphu (SAARC, 2012: 134–136). A matrix outlining the status of implementation of the Thimphu Statement (2010) as reviewed by the Second Meeting of the IGEG.CC (SAARC Secretariat, April 16–17, 2012) is highlighted in **Annex VI** (SAARC, 2015: 3).

A SAARC Convention on Cooperation on Environment (2010) was also signed during the Sixteenth Summit (Mall, Attri and Kumar, 2011: 31). This Convention provides for cooperation in the field of environment and sustainable development through exchange of best practices and knowledge, capacity building, and transfer of eco-friendly technology in a number of important areas. The Convention is highlighted in **Annex VII** (SAARC, 2015: 3) and it will enter into force upon ratification by all member states and issuance of the notification of its entry into force by the Secretariat. Fortunately, it was ratified by all member states and entered into force with effect from October 23, 2013. The Seventeenth SAARC Summit was held on November 10–11, 2011, at Addu City, Maldives (The Daily Star, November 12, 2011). Besides the signing of SAARC Agreements on Rapid Response to Natural Disasters, SAARC Seed Bank, this Summit adopted a 20-point Addu Declaration to forge effective cooperation among the member states in a host of areas including climate change, food security, and the need to ensure timely implementation of the Thimphu Statement on Climate Change (Mall, Attri and Kumar, 2011: 32; Khanom, 2012: 4). The Eighteenth Summit (Kathmandu, November 26–27, 2014), harped on the crises of climate change including taking into account the existential threats posed by it to its members (SAARC, 2014; The Economic Times, November 26, 2014). The Eighteenth Summit also harped on the need to mitigate climate change. During the Summit the members underscored the importance of promoting sustainable agriculture (Srinivasan, 2014). The Nineteenth Summit was scheduled to be held in Islamabad from November 9–10, 2016. But this Summit could not be held when India, Afghanistan, Bhutan, Bangladesh, Sri Lanka, and the Maldives declined to attend (Waheed, 2016; Yousaf, 2016) due to the clashes in the Indo-Pakistan border when militants attacked India's army camp in Uri in Indian-held Kashmir (IHK).

III Meetings of the SAARC Environment Ministers

Given the high priority attached towards enhancing regional cooperation in the areas of environment, climate change, and natural disasters, the SAARC Environment Ministers have been meeting regularly since 1992. The SAARC Ministerial Meetings (SAARC, 2015) provide a platform for taking stock of progress and providing direction and guidance towards further enhancing regional cooperation in the areas of environment, climate change, and natural disasters. To date, there have been nine meetings of the SAARC Environment Ministers (SAARC, 2015: 4). In addition, a Special Session of the Environment Ministers in the aftermath of the Indian Ocean *Tsunami* was held in Malé in July 2005 (The Malé Declaration, 2005; SAARC, 2012: 121–124) and a SAARC Ministerial Meeting on Climate Change was held in Dhaka on July 3, 2008 (SAARC, 2012, 2015: 4).

In pursuant of the declaration by the Heads of State during the Sixth Summit (Colombo, December 21, 1991), the First Meeting of the SAARC Committee on Environment took place in April 1992 in New Delhi (SAARC, 1999: 87). This meeting was convened to deliberate upon the issues related to the forthcoming UNCED (June 1992 in Rio de Janeiro). A brief essay on the First Meeting of a SAARC document states:

> Detailed statements were made by the Ministers of Bangladesh, Bhutan, Maldives, Nepal, Pakistan and Sri Lanka. The Ministers highlighted the

need for a coordinate approach by SAARC and welcomed this initiative to harmonise views on the approach to UNCED. The SAARC countries share most of the problems and concerns related to environment and development. The Ministers expressed their views on critical issues such as Agenda 21, financial resources, technology transfer and institutional structure. The text of the Ministers' statements are highlighted in **Annex-V to X**.

(SAARC, 1999: 87)

The Second Meeting of the Environment Ministers of SAARC countries, which was held in New Delhi from April 2–3, 1997, prepared a common SAARC position prior to the UN General Assembly (UNGA) Special Session on the Implementation of Agenda 21 scheduled to be held in June 1997, New York. Detailed statements were made by Bangladesh, Bhutan, India, the Maldives, Nepal, Pakistan, Sri Lanka which highlighted the comprehensive actions initiated by the SAARC countries to promote broad-based economic growth, accelerated eradication of poverty, and the problems of environmental degradation. They stressed that they had an abiding interest to achieve significant progress on all these issues and are henceforth committed to continue sustained efforts at the national and regional levels for the purpose. The Ministers and Leaders of delegations expressed their deep concern at the slow fulfilment of the commitments undertaken by the developed countries on important issues such as increases, and adequate and predictable flow, of financial resources to developing countries, transfer of environmentally sound technologies as well as assistance to be extended to developing countries to enhance their capacity to address major environmental issues (SAARC, 1999: 93–100, 2015: 4).

The next meeting of the SAARC Environment Ministers was held in New Delhi from April 2–3, 1997. It is to be noted that,

> The New Delhi meeting of the SAARC Environment Ministers culminated in the formulation and adoption of 'The 1997 New Delhi Declaration of Environment Ministers of SAARC Countries on a common SAARC position before the UNGA Special Session on the Implementation of Agenda 21' (SAARC, 1999: 94; New Delhi Declaration of Environment Ministers on a common SAARC position before the UNGA Special Session on the Implementation of Agenda 21). The copy of the declaration is highlighted in **annex VII**.
>
> (SAARC, 1999: 96–100)

On the slow progress in the implementation of Agenda 21 and the failure of the international community to bind all polluters into an effective global common framework, the Ninth Summit at Malé (1997) called for the urgent implementation of the commitments and agreements that was reached at the UNCED, in Rio de Janeiro (The Rio Declaration, 1992; Report of The United Nations, 1992). During this Summit, SAARC emphasized the importance of projecting its coordinated and collective position at the then forthcoming Special Session of the UNGA to Review and Appraise the Implementation of Agenda 21 (the Ninth Summit at Malé, 1997; The Rio Declaration, 1992; Report of The United Nations, 1992). Agenda 21 proposed an array of actions which were intended to be implemented by every person on earth especially the specific changes in the economic activities of all people (Agenda 21, 1992). In addition, Agenda 21 detailed a programme of policy goals and initiatives that emphasized environmentally sustainable paths to long-term economic growth and prosperity (Hillstrom, 2010: 665). In this regard, SAARC fully endorsed the concern of the New Delhi Declaration issued at the end of the SAARC Ministerial Conference on Environment held on April 2–3, 1997, as amended at the Eighteenth Session of the Council of Ministers (The Ninth SAARC Summit, 1997).

It was stated earlier that the Third Meeting of the SAARC Environment Ministers Conference (Malé, October 15–16, 1997) adopted the SAARC Environment Action Plan (1997) which is highlighted at **Annex I** (SAARC, 2015: 2, 1999: 101–104). The "SAARC Environment Action Plan" identified some of the key concerns of member states and set out the parameters and modalities for regional cooperation (SAARC, 1999: 104; SAARC Environment Action Plan, 1997). During the meeting the SAARC Environment Ministers called for its effective and early implementation (SAARC Environment Ministers' Conference, 1997). All the SAARC Member States participated at the Conference. At this Conference, the Ministers/Heads of the Delegation in their Countries' Statements highlighted their efforts made by their respective countries towards the protection and preservation of the environment and reiterated their commitment to further strengthening the efforts of regional cooperation in this important area (SAARC Environment Ministers' Conference, 1997; SAARC, 1999: 102).

In consideration of the implementation of the SAARC Study on Greenhouse Effects and its Impact on the Region (SAARC, 1992b), the Ministers while reviewing the recommendations of the SAARC Study on the Greenhouse Effect and its Impact in the Region:

> Decided to identify some priority areas for immediate action and provide the framework for enhanced partnership and cooperation. The measures adopted for their implementation are reflected in the Malé Declaration of SAARC Environment Ministers.
>
> (SAARC, 1999: 102)

While reviewing the recommendations of the SAARC Study on the Causes and Consequences of Natural Disasters and Protection and Preservation of Environment (SAARC, 1992a), the SAARC Environment Ministers:

> They were of the view that their follow-up had been very slow in spite of some efforts to implement them at the national levels. They therefore,

decided to identify some priority areas for immediate action and provide the framework for enhanced partnership and cooperation in some of the vital areas. The measures adopted for their implementation are reflected in the Malé Declaration of SAARC Environment Ministers.

(SAARC, 1999: 102)

Taking into account the recommendations made by the meetings of the Environment Ministers in New Delhi in April 1997 and the directives given by the Heads of State during the Ninth SAARC Summit in Malé 1997, the Third Meeting adopted the "Malé Declaration of SAARC Environment Ministers". Moreover, the Report along with the 1997 Malé Declaration of SAARC Environment Ministers on a SAARC Environment Action Plan and a Common SAARC Position on Climate Change to be presented at the Third Conference of Parties (COP 3) to the UNFCCC which would be held at Kyoto was also adopted unanimously (SAARC, 1999: 103–108).

The Fourth Meeting of the SAARC Environment Ministers took place in Sri Lanka (Colombo, October 30–November 1, 1998). It adopted the Colombo Declaration on a Common Environment Programme which it is highlighted in **Annex II**. The meeting decided to take further measures to expedite the implementation of the SAARC Environment Action Plan (SAARC Environment Action Plan, 1997) which was adopted in Malé 1997. In addition to some other measures, the SAARC Environment Ministers agreed to present a common position at the then forthcoming COP meeting of climate change in Buenos Aires, Argentina, November 1998 (Colombo Declaration on a Common Environment Programme, 1998; SAARC, 1999: 109–114).

The Fifth Meeting of the Ministers of Environment took place at Thimphu from August 10–11, 2002. During the meeting (The Fifth Meeting of Ministers of Environment, 2002: SAARC, 2012: 111–116) its members endorsed the Report of the Second Meeting of the Technical Committee on Environment, Meteorology and Forestry which was held in Thimphu on August 9, 2002. They also decided to have a collective SAARC position for presentation before the World Summit on Sustainable Development to be held in Johannesburg August 27–September 4, 2002. In addition to other decisions, these Ministers also endorsed and adopted the Thimphu Resolution to be presented at the World Summit on Sustainable Development to be held in Johannesburg on August 27–September 4, 2002 (SAARC, 2012: 111–116).

Prior to the Sixth Meeting, it was at the Twelfth SAARC Summit (Islamabad, January 4–6, 2004), where its member countries agreed to undertake and strengthen regional cooperation in the conservation of water resources, environment, prevention and control of pollution as well as preparedness to deal with any natural calamities arising out of climate change (The Twelfth Summit Meeting, 2004). It was at this summit that they welcomed the early establishment of the Coastal Zone Management Centre in the Maldives (The Twelfth Summit Meeting, 2004; SAARC, 2015: 2). The Centre engages in a continuous process of administering and of studying the nature of problems such as tidal surges, cyclones, and the greenhouse effect (SAARC Workshop, 2008; Mishra, 2013: 17; Thapa, 2013: 4). Together, these countries stressed the importance of the early and effective implementation of the SAARC Environment Plan of Action (1997). They also stressed the early submission of national state of the environment reports to expedite the preparation of SAARC State of Environment Reports and the commissioning of the work on drafting a Regional Environment Treaty (The Declaration of the Twelfth Summit, 2004).

Soon after the Twelfth Summit, the Sixth Meeting of Ministers of Environment took place at Thimphu (June 12–13, 2004). At this meeting,

> The ministers reviewed the recommendations made by the First Meeting of the Technical Committee on Environment and Forest held in Thimphu on 11–12 2004 and appreciated its work. They underscored the need for effective implementation of the decision made at the previous Ministerial and other meetings. They approved the recommendations of the First Meeting of the Technical Committee on Environment and Forestry on the decision adopted by the Ministerial Meetings on Environment.
>
> (SAARC, 2012: 118)

During the same meeting the ministers appreciated the presentation of the Thimphu Resolution adopted at the Fifth Meeting (Thimphu, August 10–11, 2002) which contained the collective SAARC position before the World Summit on Sustainable Development (Johannesburg, August 27–September 4, 2002) by Nepal's representative as the then Chairperson of SAARC (SAARC, 2015: 63). Its members also noted with appreciation the implementation of various activities under the SAARC Environment Action Plan (1997) and endorsed the recommendations of the Technical Committee thereon (SAARC, 2012: 111–118).

More than that, on June 25, 2005, Ministers of the Environment of the member countries of SAARC gathered in Malé for the Special Session of SAARC Environment Ministers in the Aftermath of the Indian Ocean *Tsunami* Disaster of December 26, 2004. The Special Session (Special Meeting of Ministers of Environment, 2005; SAARC, 2012: 121–122) was held due to the unprecedented loss of life and property in South Asia caused by the *tsunami*. After the assessment of the state of the environment of the SAARC countries affected by the Indian Ocean *Tsunami*, the ministers were eager to develop a mechanism to cooperate in *tsunami* disaster recovery (SAARC, 2012: 121–122). The outcome of the Special Meeting was the Malé Declaration on Environment in the Aftermath of the Indian Ocean *Tsunami* Disaster (SAARC, 2012: 123–124). The adoption of the Malé Declaration on Environment in the Aftermath of the Indian Ocean *Tsunami* Disaster was welcomed by the Thirteenth SAARC Summit (2005).

The Seventh Meeting of Ministers of Environment was held on May 24, 2006, in Dhaka (Seventh Meeting of Ministers of Environment, 2006;

SAARC, 2012: 125–127). This meeting noted that the detailed discussions on the observance of 2007 as Green South Asia Year is a step forward towards the implementation of the decision of the Thirteenth SAARC Summit (SAARC, 2012: 125–127). Further,

> The Meeting considered the Report of the Expert Group Meeting on Formulation of a Comprehensive Framework on Disaster Management held in Dhaka on 7–9 February 2006, including the Draft Framework on Disaster Management in South Asia: A Framework for Action 2006– 2015, as contained in Document No. **SAARC/ENV/MM.7/SOM/3**.
>
> (SAARC, 2012: 126)

During the Twenty-ninth session of the SAARC Council of Ministers (New Delhi, December 7–8, 2007), the Council felt that given the vulnerabilities, inadequate means, and limited capacities of the countries, there was a need to ensure rapid social and economic development to make SAARC climate change resilient (Area of Cooperation-Environment, 2012). Pursuant to this decision,

> The Twenty-ninth session of the SAARC Council of Ministers (New Delhi, 7–8 December 2007), among others, adopted a SAARC Declaration on Climate Change which was presented by the President of the Republic of Maldives at the High Level Session of the UN Conference on Climate Change held at Bali in December 2007 on behalf of SAARC. The Declaration is at **Annex IV**.
>
> (SAARC, 2015: 3; Quarterly Newsletter, 2008)

Another Ministerial Meeting on Climate Change was held in Dhaka on July 3, 2008. This Ministerial Meeting was connected to the Fourteenth SAARC Summit (2007) where the members of SAARC expressed their "deep concern" over global climate change and called for pursuing a climate-resilient development in South Asia. Hence the Ministerial Meeting that was held in Dhaka on July 3, 2008,

> Adopted the Dhaka Declaration on Climate Change and SAARC Action Plan on Climate Change. The Dhaka Declaration and SAARC Action Plan on Climate Change are at **Annex III**.
>
> (SAARC, 2015: 3)

In his inaugural speech on intensifying regional cooperation by means of climate change adaptation, H.E. Sheel Kant Sharma who was then SAARC Secretary General, placed emphasis on intensifying the regional cooperation to mitigate the negatives of climate change. He also highlighted the emphasis of SAARC to move from a declaratory to implementation phase, wherein the ten SAARC Regional Centres established to institutionalize regional cooperation could play enabling roles (SDMC, 2008: 2–3). He urged the SAARC Meteorological Research Centre, the SAARC Coastal Zone Management Centre, SAARC Disaster Management Centre (SDMC), and SAARC Forestry Centre to contribute synergistically with their respective mandates in enhancing the SAARC climate change resilience by pursuing the SAARC Action Plan on Climate Change (SDMC, 2008: 2–3). It is to be noted that the SDMC stands with the mandate to serve the member countries by providing policy advice and facilitating capacity to disasters like climate change within the region. It placed a very high priority on implementing the SAARC Action Plan on Climate Change (SDMC, 2008: 2–3). In fact, in its declarations, SDMC highlighted climate change adaptation as one of its priority areas for action (SAARC Workshop, 2008). The SAARC Workshops on Science & Technology Applications in Disaster Risk Reduction (January 2008) in New Delhi and Coastal and Marine Risks (May 2008) in Goa emphasized an exchange of information and research on the linkages between climate change adaptation and disaster risk reduction in the region. As such it opened up a range of possibilities for integration of climate change adaptation in disaster risk reduction strategies in the region (SAARC Workshop, 2008).

The Eight Meeting of Ministers of Environment (October 20–21, 2009; SAARC, 2012: 128–133) agreed to publish a compendium of SAARC National Plans of Action on Climate Change before the then forthcoming COP 15 in Copenhagen in December 2009. The meeting requested the member states to submit copies of their respective National Action Plans to the SAARC Secretariat (SAARC, 2012: 129–130). The meeting noted that cooperation in the number of areas pursuant to the directives issued by the Fourteenth SAARC Summit (New Delhi, April 3–4, 2007) have been identified in the draft "SAARC Treaty on Cooperation in the Field of Environment". It is worth noting that this meeting also considered and approved the Report of the Expert Group Meeting on SAARC Natural Disaster Rapid Response Mechanism (New Delhi February 5–6, 2009) and endorsed the recommendation of the Senior Officials that the Draft SAARC Agreement on Natural Disaster Rapid Response Mechanism (NDRRM) be finalized, so that it could be finally signed during the Sixteenth SAARC Summit (SAARC, 2012: 130). All these efforts reflected the determination of SAARC countries to minimize the adverse furies of climate change in the region.

Pursuant to the decision taken by the Eighth Meeting of the SAARC Environment Ministers (New Delhi, October 20–21, 2009; SAARC, 2012: 128–133), a common SAARC position on climate change was presented by Sri Lanka, as the Chair of SAARC, at the COP 15 in Copenhagen in December 2009 (Daily News, October 24, 2009). A Joint Statement on Climate Change was also issued by the Permanent Representatives of Member States of SAARC based in New York (The Daily Star, April 29, 2010). As decided by the Sixteenth SAARC Summit, a common SAARC position on climate change for COP 16 (Cancun, December, 2010) was finalized by the Inter-governmental Expert Group Meeting (Thimphu, August, 2010). The common position was presented by Bhutan, as then current Chair of SAARC, and it is highlighted in **Annex VIII** (COP 16/CMP 6, 2010; Bhutan Observer, December 17, 2010; SAARC, 2015: 4). Further, the Eighth Meeting of the SAARC Environment Ministers adopted the draft of the SAARC Ministerial Statement on Cooperation on Environment (Delhi Statement on Cooperation in Environment, 2009; SAARC Ministerial Statement on Cooperation on Environment, 2009). The statement identified many critical areas that needed to be addressed and reaffirmed the commitment of member states towards enhancing regional cooperation in the area of environment and climate change. The Delhi Statement is highlighted in **Annex IX** (SAARC, 2015: 4).

The Ninth Meeting of the SAARC Environment Ministers (2011) took place in Thimphu. The Meeting (2011) reviewed the implementation of directives/decisions of the Sixteenth SAARC Summit (Thimphu, April 28–29, 2010) and the Thirty-third Session of the Council of Ministers (Thimphu, February 8–9, 2011). The Ninth Meeting of Ministers of Environment (2011) endorsed the recommendation of the Fourth Meeting of the Technical Committee on Environment and Forestry (SAARC, 2012: 134–136). It also underlined the need to ensure timely follow-up and implementation of various decisions/directives (SAARC, 2012: 134–136). At the meeting, the environment ministers renewed their commitment towards a common regional stand on environmental issues (Business Bhutan, October 1, 2011). A range of views were brought out in the statements of the environment ministers which included the need to take a common stand at the COP 17 (Durban, South Africa, November 2011) and at the RIO+20 (Rio de Janeiro, June 2012). The purpose of their common stand at the World Summits would be to stress the need to demand accessing the global fund for climate change (Business Bhutan, October 1, 2011; SAARC, 2012: 135–136). Due to time constraints, the common statement, which was meant to be presented at Durban, could not be finalized.

Further, the Forty-ninth Meeting of the Programming Committee which comprises joint secretaries of the SAARC countries decided to merge the SAARC Forestry Centre in Bhutan, Disaster Management Centre in New Delhi, Coastal Zone Management Centre in the Maldives, and Meteorological Research Centre in Dhaka into a single body. Moreover, it also decided to set up a SAARC Environment and Disaster Management Centre. The implementation of the closure and merger of the centres was to be completed by December 2016 (The Indian Express, November 22, 2014).

The IGEG.CC established by the Thimphu Statement on Climate Change required to monitor and review the implementation of the Thimphu Statement on Climate Change (2010) and made recommendations to facilitate its implementation. The IGEG.CC is as well expected to meet every year and give its reports to the SAARC Environment Ministers. The First Meeting of the IGEG.CC was held in Colombo on June 29–30, 2011. The report of the meeting was approved by the Ninth Meeting of SAARC Environment Ministers (Thimphu, September 29, 2011). A matrix outlined the status of implementation of the Thimphu Statement as reviewed by the Second Meeting of the IGEG.CC (SAARC Secretariat, April 16–17, 2012) highlighted in **Annex VI** (SAARC, 2015: 3). This meeting reached a broad consensus on an agreement that aimed to put in place an effective mechanism for rapid response to natural disasters (Radhakrishnan, 2011).

IV Technical Committee on Environment and Forestry

Since the establishment of the Committee on Environment in 1992 till the present Technical Committee on Environment and Forestry, the mandate of the committee has undergone a number of changes. The Committee on Environment was set up in 1992 by the Council of Ministers to examine the recommendations on the Regional Study on the Causes and Consequences of Natural Disasters and the Protection and Preservation of Environment (SAARC, 1992a); to identify measures for immediate action; and decide modalities for the implementation. The Seventeenth Session of the Standing Committee (Dhaka, December 7–9, 1992) determined to choose the Committee on Environment as the Technical Committee on Environment and include within its preview the regional study "Greenhouse Effect and its Impact on the Region" (SAARC, 1992b). The Second Special Session of the Standing Committee (New Delhi, August 25–26, 1995) recommended the merger of the Technical Committee on Meteorology with the Technical Committee on Environment. The Twenty-first Session of the Council of Ministers (Sri Lanka, March 18–19, 1999) approved the recommendation of the Standing Committee on a new arrangement for the Integrated Programme of Action under which the Technical Committee on Environment and Meteorology was given the additional mandate of Forestry (SAARC, 2015: 5). Technical committees comprising representatives of member states are responsible for the implementation, coordination, and monitoring of the programmes related to environment.

Under the restructured Regional Integrated Programme of Action (RIPA) approved by the Twenty-ninth Session of the Standing Committee (Islamabad, December 31, 2003–January 1, 2004), the subject of Meteorology merged with the Technical Committee on Science and Technology (Areas of Cooperation, 2012). Since 2004, the Technical Committee on Environment and Forestry has met four times in June 2004, May 2006, January 2009, and September 2011. The sectoral mandate of the Technical Committee comprises environment, forestry, and natural disasters. In addition to the Terms of Reference (TOR) outlined under Article VI of the SAARC Charter (1985: 4), the Technical Committee is required to followup on the implementation of directives issued by SAARC Summits and Meetings of the SAARC Environment Ministers. The TOR is highlighted in **Annex X**. The coordination and monitoring of the implementation of the SAARC Environment Action

Plan and the SAARC Action Plan on Climate Change are also entrusted to the Technical Committee (SAARC, 2015: 5–6).

The Fourth Meeting of the Technical Committee on Environment and Forestry that took place in Thimphu, Bhutan, on May 17–18, 2011 (Arruda, 2012; The Kathmandu Post, April 27, 2010) made Apa Sherpa, an Everest Summiteer, the SAARC Goodwill Ambassador for Climate Change for two years (May 2010–May 2012). He was assigned to advocate and to raise the awareness about the negative effects of climate change and to garner support from various stakeholders so that they work together to address the threats that climate change posed to the people; to propagate the work being done by SAARC in the area of climate change; and to stress the need to ensure timely implementation of the Thimphu Statement on Climate Change (2010) as well as the Dhaka Declaration on Climate Change (2008) and SAARC Action Plan on Climate Change (2008). But due to his pre-occupations and busy schedule, the Secretariat has not been able to engage him in his capacity as SAARC Goodwill Ambassador for Climate Change (SAARC, 2015: 5).

V SAARC Environment Action Plan

The SAARC Environment Action Plan (1997) was adopted by the Third Meeting of the SAARC Environment Ministers in Malé (SAARC, 1999: 102). The SAARC Environment Action Plan (1997) is based on the recommendations of the two regional studies. The two regional studies are the Regional Study on the Causes and Consequences of Natural Disasters and the Protection and Preservation of Environment (SAARC, 1992a) and the Regional Study on Greenhouse Effect and its Impact on the Region (SAARC, 1992b). It seeks to assess the status of SAARC cooperation in the field of environment, identified some of the key concerns of member states at the regional and global levels, and set out the parameters and modalities for regional cooperation (Mishra, 2013: 19; Arif and Karim, 2015). Henceforth, regional cooperation in the field of environment including in disaster preparedness and mitigation has been largely focused towards the implementation of the various components of the SAARC Environment Action Plan (1997). However, the key responsibility for the implementation of the SAARC Environment Action Plan (1997) rests with member states.

A SAARC Coastal Zone Management Center (SCZMC) was established in 2004. The SCZMC engages in a continuous process of administering and of studying the nature of problems such as tidal surges, cyclones, and the greenhouse effect. Since its establishment, therefore, it has been serving as the focal institution of SAARC to promote cooperation in planning, management, and sustainable development of coastal zones, including research, training, and awareness in the region (SAARC Workshop, 2008). Likewise, a SAARC Forestry Center (SFC) was established in Thimphu in 2007. It began its operations from 2008 for the protection, conservation, and prudent use of forest resources by adopting sustainable forest management practices through research, education, and coordination among the member states. The SFC (2008) also intends to serve as a centre of excellence for forestry within the region and operate as a nodal point for research, information, and policy development (SFC). A South Asia Environment Outlook (SAEO) 2009 was also finalized by member states in collaboration with the United Nations Environment Programme (UNEP, 2009). The SAEO 2009 was launched during the Eighth Meeting of the SAARC Environment Ministers (New Delhi, October 20–21, 2009). A SAARC Convention on Cooperation on Environment (2010) as stipulated under Item 17 (Legal Framework) of the Action Plan was signed during the Sixteenth SAARC Summit (2010).

VI Dhaka Declaration on Climate Change and SAARC Action Plan on Climate Change

It is worth mentioning that during the Twenty-ninth session of the SAARC Council of Ministers (New Delhi, December 7–8, 2007), the Council felt that due to the vulnerabilities of the region and their inadequate means as well as limited capacities, there was a need to ensure rapid social and economic development to make SAARC climate change resilient. To pursue this decision, a Ministerial Meeting on Climate Change was held in Dhaka on July 3, 2008, and preceded by an Expert Group Meeting on Climate Change on July 1–2, 2008. On realizing the urgent need to make South Asia a climate change resilient region, the Ministerial Meeting (Dhaka, July 3, 2008) adopted the Dhaka Declaration on Climate Change (2008) as well as the SAARC Action Plan on Climate Change (2008).

The Dhaka Declaration on Climate Change (2008) requires member states to carry out activities to encourage advocacy programs and mass awareness on the crises created by climate change; cooperation in capacity building including the development of Clean Development Mechanism (CDM) projects and Designated National Authority (DNA) and on incentives for removal of GHG by sinks; and exchange of information of best practices, sharing of the results of research and development for mitigating the effects of climate change and undertaking adaptation measures; and for enhancing south-south cooperation on technology development and transfer, as per established SAARC norms; and to initiate and implement programmes and measures as per SAARC practice for adaptation for dealing with the onslaught of climate change to protect the lives and livelihood of the people. It also calls upon the Annex I countries to fulfil their commitments as per UNFCCC for providing additional resources (SAARC, 2015: 6). There are 43 Annex I countries which include countries of the EU. These countries are classified as industrialized countries and economies in transition.

The SAARC Action Plan on Climate Change (2008) identifies seven thematic areas of cooperation related to adaptation; mitigation; technology transfer; finance and investment; education and awareness; management of impacts and risks; and capacity building for international negotiations (The Ninth Meeting of the SAARC Environment Ministers, 2011; SAARC, 2012: 134–138; Mall, Attri and Kumar, 2011: 31; Climate Change Adaptation and Disaster. ., 2010: 13). The adoption of the SAARC Action Plan on Climate Change showed that the SAARC members are eager to contribute and to restore harmony with nature in order to make South Asia a sustainable region. The SAARC Action Plan on Climate Change also lists the areas of capacity building for CDM projects; exchange of information on disaster preparedness and extreme events; exchange of meteorological data; capacity building and exchange of information on climate change impacts (e.g. sea level rise, glacial melting, biodiversity, and forestry); and mutual consultation in international negotiation process as priority areas.

On the other hand, the prime responsibility for implementation of the SAARC Action Plan on Climate Change rests with member states (Climate Change Adaptation and Disaster, 2010: 1–44). The SAARC Action Plan on Climate Change (2008) is to be reviewed periodically by the appropriate institutional mechanism in SAARC at the technical level and national reports on its implementation are to be submitted to the SAARC Secretariat for subsequent consideration by the environment ministers.

VII Natural disasters

The pace and scale of the loss and damage caused by the *tsunami* of December 2004 and the earthquake of December 2005 in the region provided an immediate sense of urgency towards promoting regional cooperation in the area of disaster management. Following the *tsunami*, a Special Meeting of Ministers of Environment was held in Malé (2005). The outcome of this Special Meeting was the adoption of the Malé Declaration (2005) on a collective response to large-scale natural disasters, which is highlighted in **Annex XI** (SAARC, 2012: 121–124; SAARC, 2015: 7). This Special Meeting of Ministers of Environment (Malé on July 25, 2005) was held in between the Sixth Meeting of Ministers of Environment (Thimphu, June 12–13, 2004) and the Seventh Meeting of Ministers of Environment (Dhaka, May 24, 2006).

To pursue the Malé Declaration, a Comprehensive Framework on Disaster Management (2006–2015) was adopted in 2006 to address the specific needs of disaster risk reduction and management in South Asia (SAARC, 2015; White, 2015: 7; Shanahan, 2017: 21). The Framework is aligned with the Hyogo Framework of Action (2005–2015) (SAARC, 2005; White, 2015: 15). Member states are required to prepare their respective National Plans of Action for the implementation of the Regional Framework and thereafter, an Expert Group Meeting will be held to harmonize the national reports and articulate a Regional Plan of Action. The Comprehensive Framework is highlighted in **Annex XII** (SAARC, 2015: 7).

The SDMC was established in New Delhi in October 2006. The Centre has the mandate to serve the eight member countries of SAARC. Further, it provides policy advice and facilitates capacity building including strategic learning, research, training, system development, expertise promotion, and exchange of information for effective disaster risk reduction and management. The mandate of the SDMC was expanded to include the development of a Natural Disaster Rapid Response Mechanism by the Fifteenth SAARC Summit in Colombo (2008). According to Naseer Memon, the SDMC is a body of professionals who are working on numerous dimensions of disaster risk reduction and management in the region. It is communicating through the National Focal Points of the Member Countries with the various ministries, departments and scientific, technical, research and academic institutions within and outside the government working on different aspects of disaster risk reduction and management (Memon, 2012: 32).

To follow the directive of the Fifteenth SAARC Summit, the SDMC organized two Expert Group Meetings in New Delhi in February 2009 and July 2009 to discuss and recommend the modalities for setting up a Natural Disaster Rapid Response Mechanism. A draft SAARC Agreement on Rapid Response to Natural Disasters was also prepared by SDMC for the consideration of member states. The Thirty-seventh Session of the Standing Committee (Thimphu, April 25–26, 2010) noted that the text of the draft SAARC Agreement on Rapid Response to Natural Disasters required further consultations at the experts level and recommended that an Inter-Governmental Expert Group Meeting be convened at an early date to finalize the text of the draft agreement. Consequently, the Heads of State or Government at the Sixteenth SAARC Summit (2010) called for further negotiations and early finalization of the draft SAARC Agreement on Rapid Response to Natural Disasters (2011). The text of the draft agreement (SAARC, 2011) was finalized by an Inter-governmental Meeting (Malé, May, 2011) and the agreement was signed during the Seventeenth SAARC Summit (2011) and it is highlighted in **Annex XIII** (SAARC, 2015: 7–8; Radhakrishnan, 2011).

VIII Thimphu Statement on Climate Change (2010) and SAARC Convention on Cooperation on Environment (2010)

Climate change was the theme of the Sixteenth Summit (2010). Among others, this summit adopted the "Thimphu Statement on Climate Change", which outlines a number of important initiatives at the national and regional levels to strengthen and intensity regional cooperation to address the adverse effects of climate change in a focused manner (Sixteenth Summit, 2010). The Thimphu Statement on Climate Change (2010) called for a review of the implementation of the Dhaka Declaration and SAARC Action Plan on Climate Change and to ensure its timely implementation. There was an agreement to establish the IGEG.CC to extend a clear policy direction and guidance for regional cooperation as envisioned in the SAARC Action Plan on Climate Change. It was determined that the IGEG.CC shall meet at least twice a year to periodically monitor and review the implementation of this statement and make recommendations to facilitate its implementation and submit its report through the Senior Officials of SAARC to the SAARC Environment Ministers. The IGEG.CC is required to monitor, to review progress, and to make recommendations in order to facilitate the implementation of the Thimphu Statement on Climate Change (2010).

The SAARC Convention on Cooperation on Environment was also signed during the Sixteenth Summit. The Convention has been ratified by all member states and entered into force with effect from October 23, 2013. The Convention identifies 19 areas for cooperation in the field of environment and sustainable development through exchange of best practices and knowledge, capacity building, and transfer of eco-friendly technology in a wide range of areas related to the environment. The implementation of the Convention has been entrusted to a Governing Council, comprising the Environment Ministers of Member States.

IX Regional cooperation for food security

The SAARC Heads of State accorded the highest priority to the alleviation of poverty in all South Asian countries. To address the existing problem of food security in the region, an agreement to establish the SAARC Food Security Reserve was signed in Kathmandu in 1987 (Pant, 2014: 11–16; SAARC, 2016a; Haque, 2008). This agreement provides for a reserve of food grains to meet emergencies in member countries. It has been ratified by all the member countries and came into force on August 12, 1988 (Dawn, December 15, 2006; Lama, 2015; SAARC, 2016a). Recognizing that a great majority of these people remain below the poverty line, they showed keen interest in a *Daal-Bhaat*, or basic nutritional needs or assured nutritional standards approach towards the satisfaction of basic needs of the South Asian poor. They also underscored the importance of promoting sustainable agriculture in the region which could be met by the application of biotechnology (The Colombo Declaration, 1991; Pant, 2014: 9–34). Further, the SAARC Agriculture Ministers' Meeting was held in Islamabad from October 8–9, 1996. They adopted the Resolution and the Report of the Meeting which is highlighted in **Annex VIII**. This meeting considered the common SAARC position to be presented to the World Food Summit in Rome (SAARC Agriculture Ministers' Meeting, 1996). Another meeting of the SAARC Agriculture Ministers was held in May 15, 2002, to evolve a common SAARC position before the World Food Summit five years later to be held in Rome from June 10–13, 2002 (Meeting of the SAARC Agriculture Ministers, 2002). It is worthwhile to note that to meet food emergencies in the region, the Meeting of SAARC Agriculture/Food Ministers (2006) held in Islamabad agreed to create the SAARC Regional Food Bank (Dawn, December 15, 2006; Lama, 2015). It was renamed the SAARC Food Bank in 2007 (The Hindu, August 7, 2008). Through an agreement among the heads of the member states, at the Fourteenth SAARC Summit (2007) the SAARC Food Bank was established with the objective to collectively face the shocks of food supply following climate change and natural calamities. The main objective is, for food shortages and emergencies, to provide regional support to national food security efforts and to foster inter-country partnership and regional integration, and solve regional food shortages through collective action (Mukherji, 2012). Through this mechanism the countries can also obtain an early assessment of production of major food grains in the region as well as beyond the region (Sharma, 2011: 305–314). Another meeting of the SAARC agriculture ministers was held in Delhi (Extra-Ordinary Meeting of the SAARC Agriculture Ministers, 2008; SAARC Declaration on Food Security, 2008). Matters relating to cooperation of the SAARC Food Bank were taken up during the meeting. The Fifth Special Meeting of the SAARC Food Bank Board was held in Dhaka, on May 13, 2012. In this meeting, the member states came into an agreement to provide national efforts in ensuring food security in the region (Muniruzzaman, 2013: 46–73). To promote sustainable agriculture, the leaders agreed to eliminate the threshold criteria from the SAARC Food Bank Agreement so as to enable the member states to avail of food grains, during both emergency and normal time food difficulty. The leaders urged early ratification of the SAARC Seed Bank Agreement and directed to constitute the Seed Bank Board, pending completion of ratification by all member states. The leaders also directed the relevant SAARC bodies to finalize the establishment of the Regional Vaccine Bank and Regional Livestock Gene Bank (Eighteenth SAARC Summit, 2014; Jabed, 2016; Mitra, 2016). The Fourth Meeting of the SAARC Agriculture Ministers held in Thimphu, Bhutan, on June 27, 2019, reiterated their commitments to further strengthen regional cooperation by promoting cooperation in agriculture and rural development for enhancing food, nutrition, and livelihood security of South Asia (SAARC, 2019).

X Cooperation with inter-governmental organizations

In order to collaborate with many regional and international organizations in the area of environment, SAARC signed a Memorandum of Understanding (MOU) with the South Asia Cooperative Environment Programme (SACEP) on July 8, 2004. The Memorandum constituted a framework of cooperation between SAARC and SACEP in the area of environment protection (Daily News, July 20, 2004). The MOU between SAARC and the Asian Disaster Preparedness Center (ADPC) was signed on December 19, 2006. The MOUs are renewed whenever they may be required. These agencies extend technical and financial support in implementing SAARC programmes. The MOU was signed with the United Nations Environment Programme (UNEP) on June 13, 2007, and the United Nations International Strategy on Disaster Reduction (UNISDR) on September 3, 2008 (Shrivastava *et al.*, 2008). Moreover, the SAARC was appointed as an Observer with the UNFCCC at COP 16 in Cancun, Mexico (COP 16/CMP 6, 2010; SAARC, 2015).

XI Migration

Migration has been a long-standing feature in South Asia, especially between countries sharing common borders (Wickramasekara, 2011: 1; Sood, 2017: 35–36; Lal, 2016; Saidi, 2017; Garrote-Sanchez, 2017: 3). This trend has become a most sensitive issue when it comes to inter-state relations in South Asia. Climate change is further acting as a catalyst to the already pertinent problem of migration (Ratna, 2015; Bhattacharyya and Werz, 2012) because extreme events and deteriorating environmental conditions are forcing people to leave their homes temporarily or even permanently. In conflict-prone South Asia, climate change has added a new dimension to environmental security in the region (Ratna, 2015; Bhattacharyya and Werz, 2012). In terms of references to migration, SAARC has the Social Charter (2004, 6) and the Kathmandu Declaration (2014) and has been engaging in a consultative process to protect the interests of international and internal migrants (Srivastava and Pandey, 2017: 9; Rabbani, Shafeeqa and Sharma, 2016: 23). Another step taken by this regional cooperation group in this direction is the adoption of the 2016 SAARC Plan of Action on Labour Migration (ILO, 2018: xiv). This Plan of Action recognizes the importance of labour migration statistics, by committing to strengthening the capacity of government authorities and improving the policy impact of migration through measures such as "the creation of a shared database or web portal with information about migration trends and patterns" (SAARC, 2016b: ILO, 2018: xiv).

XII Conclusion

For the member states of SAARC, protection and preservation of the environment including disaster risk reduction and management remains a high priority on the agenda of their regional cooperation. At the regional level, they are pursuing it regularly. Several directives which are issued by successive SAARC Summits and Meetings of the SAARC Environment Ministers show their continued momentum for strengthening and intensifying regional cooperation in the areas of environment, climate change, and natural disasters. The meetings of the SAARC Environment Ministers and the Technical Committee represent the important mechanisms to guide and facilitate the agenda of cooperation. The existence of regional centres which are dealing with climate change effects also show that the members of SAARC are mitigating it actively at the regional level. For instance, regional centres such as the SAARC Coastal Zone Management Centre in the Maldives, the SAARC Forestry Centre in Bhutan, the SAARC Disaster Management Centre in India, and the SAARC Meteorological Research Centre in Bangladesh constitute a framework of SAARC institutions which address diverse aspects of environment, climate change, and natural disasters.

The SAARC Environment Action Plan (1997), Dhaka Declaration on Climate Change and SAARC Action Plan on Climate Change (2008), the Comprehensive Framework on Disaster Management (2006–2015), the

Thimphu Statement on Climate Change (2010), the SAARC Convention on Cooperation on Environment (2010), and the SAARC Agreement on Rapid Response to Natural Disasters (2011) are some of the most noteworthy policies/initiatives ever taken by SAARC to strengthen regional cooperation in the areas of environment, climate change, and natural disasters.

Though SAARC countries may have been comparatively poor, yet through the common position taken by the members it can be seen that SAARC is keen to contribute to adapt to climate change–related disasters. From the discussion, it is seen that the numerous directives issued by the successive SAARC Summits and Meetings of its Environment Ministers highlight the thrust of this regional cooperation group to strengthen and intensify regional cooperation in the areas of climate change, environment, and other natural disasters in the region. The next chapter will discuss the response of SAARC members to climate change.

5 Responses of the SAARC member countries

I Introduction

The vulnerability of climate change has been highlighted in the various texts and reports of the IPCC, the UNFCCC, and the Kyoto Protocol. As a scientific body, the IPCC provides a systematic assessment of climate change risks to policymakers (Watson, Zinyowera and Moss, 1998; Hare, 2003; Miller and Edward, 2001). The UNFCCC is an international environmental treaty negotiated at the UNCED. The signatories to this treaty officially recognized the reality of climate change and its associated problems and the need of engaging with the issue (UNFCCC, 1992; Couzens and Honkonen, 2011: viii). The Kyoto Protocol shares the objective and institutions of the UNFCCC (UNFCCC, 2002: 12, 2008). The Kyoto Protocol is considered as one of the most far-reaching climate change agreements ever adopted. Another international climate policy is the Paris Agreement which sets the frame for scaling up ambitions for climate protection (Burck, Marten and Bals, 2017: 7). Altogether, these bodies have different aims and functions. Each plays an important role in shaping the discourse of climate change at the global level.

The IPCC produces the scientific Assessment Reports on climate change and mentioned its liability since its Second Assessment Report (Houghton *et al.*, 1996). The Second Assessment Report stated that a highly vulnerable region would be the one that is highly sensitive to modest changes in climate. Based on this conceptualization, several regions and countries (including SAARC members) have been identified and categorized as one of the most vulnerable to climate change effects. The word *vulnerability* here carries a specific meaning within the Kyoto Protocol and the UNFCCC. Article 12, paragraph 8, of the Kyoto Protocol says developing countries are particularly susceptible to the adverse effects of climate change (Oberthür and Ott, 1999; Depledge, 2000). Within the context of the UNFCCC, vulnerability is used to describe the least developed countries that are in need of financial, technological, and other forms of assistance from advanced countries so that they will be in a better position to adapt to the grave threats of climate change (UN, 2007; UNFCCC, 2011). This perception is widely acknowledged as an important notion for the welfare of many developing countries (Yohe,

Jacobsen and Gapotchenko, 1999: 233–250; Smith *et al.*, 1996). Thus, the term *vulnerability* is understood to refer to the degree in which a system is susceptible to, or unable to cope with, the adverse effects of climate change, including climate variability and extremes. But a number of researchers have criticized the vagueness and limited value of this definition. Tim Forsyth (Forsyth, 2000) has pointed to the problem of using the word *system* as it is unclear whether this system refers only to an ecological system, or includes human society. Tim Forsyth (Forsyth, 2000) observes that the IPCC report used system to sometimes refer to an ecological system, and at other times include human society. Others have pointed out that the IPCC's understanding of vulnerability is predominantly concerned with the extent of changes in biophysical conditions (Kelly and Adger, 2000: 325–352). Understanding the vulnerability of humanity to the impacts of climate change has put pressure on the public and the scientific community to deal with it.

II Response of the SAARC member countries to climate change

All SAARC countries are identified as vulnerable areas to climate change owing to the scale of the ecological crisis, poverty, density of population, and their inability to deal with it. It has been discussed in Chapter 3 that climate change is a large-scale issue for all the SAARC members who will suffer from its numerous effects. Due to such problems all SAARC members have been implementing adaptive measures to attend to it. To implement any adaptive strategies each country requires strong and dedicated policy decision-makers, and the lessons for adaptations should build on previous experience (Burton, 1996; Kleinand Tol, 1997). Familiar with the threats caused by climate change, SAARC members have initiated a number of policies, programs, and projects to reduce the vulnerability of their people and communities in the region (Zafarullah and Huque, 2018: 27–28). With their own initiatives these countries are coming forward to prepare for and to respond to the foreseeable effects of climate change. This chapter examines the responses of the member states of SAARC with regard to the policy initiatives for cooperation.

Afghanistan

In view of its adaptive capacity to the problem, Afghanistan's ability to cope with and minimize its exposure to climate change effects is characterized by several factors (Mihran, 2011: 65). The factors include economic wealth, social networks, infrastructure, social institutions, social capital, and the range of technological adaptations available and the equity to access resources within the community (Kelly and Adger, 1999: 253–266; Smith, Klein and Huq, 2003: 356). In view of these factors, Afghanistan ranks poorly on adaptation and mitigation to climate change. The Government of Afghanistan took an active part to tone down climate change effects by signing the UNFCCC in Rio de Janeiro on June 12, 1992 (UNEP, May 6, 2007: 1; Mackenzie, 1993; Azimi and McCauley, 2003), though as a Non-Annex Party to the Convention. This signing had taken place less than a month after the overthrow of the former president, Mohammad Najibullah. Afghanistan has taken another step towards implementing its national climate strategy by ratifying the Kyoto Protocol in 2013 (Hazem, 2017: 455–484). Several general measures for environmental management have been undertaken. Such measures are done with the elaboration of overarching national policies, strategies, and legislation that address natural resource and land management issues (Azimi and McCauley, 2003; Mackenzie, 1993). Afghanistan is among the member states of the UN that adopted the Hyogo Framework for Action 2005–2015 (HFA) in January 2005 (GIRA, 2011: 82). The HFA aimed to achieve a substantial reduction of disaster losses by 2015 globally by building the resilience of nations and communities to disasters. This means reducing loss of lives and social, economic, and environmental assets when hazards strike.

Afghanistan's National Adaptation Program of Action for Climate Change (NAPA), which is jointly developed with National Capacity Needs Self Assessment for Global Environmental Management (NCSA), has noted the difficulty in projecting the future effects of climate change in the country. This is due to the inaccessibility of baseline data, scarce resources, and lack of capacity (GIRA, 2009; Afghanistan Statement, 2010). Nonetheless, the NAPA has identified the country's most vulnerable areas to climate change like agriculture, biodiversity and ecosystems, energy, forests and rangelands, periodic drought, floods due to untimely and heavy rainfall, flooding due to thawing of snow and ice and rise of temperature etc. Owing to continuous conflicts and wars, Afghanistan is not able to keep pace with other South Asian countries in terms of preparing for climate change and executing essential measures. The joint NAPA and NCSA, which was completed recently, is the main policy document of Afghanistan to deal with climate change. The three high-level objectives in Afghanistan's NAPA/NCSA are to:

- Identify priority projects and activities that can help communities adapt to the adverse effects of climate change;
- Seek synergies with existing multi-lateral environmental agreements (MEAs) and development activities with an emphasis on both mitigating and adapting to the adverse effects of climate change; and
- Integrate climate change considerations into the national planning processes.
  (GIRA, 2009: 71)

Until 2003, the environment in Afghanistan could not be independently recognized as a government mandate. It is only after the Constitutional *Loya Jirga*, or Grand Council, that environment was added to the portfolio of the former Ministry of Irrigation and Water Resources and the institution was renamed as the Ministry of Irrigation, Water Resources and Environment

(GIRA, 2009: 29; UNEP, 2008). After the presidential elections in late 2004, the Cabinet got reshuffled and the environment mandates were carved out from its previous institutional home. Known during the interim period as the Independent Department of Environment, in May 2005 the fledgling institution was renamed the National Environmental Protection Agency (NEPA). This department was established by a presidential decree (GIRA, 2005b). The Environment Law (Official Gazette No. 912 dated January 25, 2007) clarified the mandates, powers, responsibilities, and functions of NEPA (GIRA, 2005b). The position of NEPA within Afghanistan's government structure is as a largely environmental regulatory, policymaking, coordination, monitoring, and enforcement institution, with the line ministries responsible for actual management of environmental resources. These are:

- Promote the sustainable use of natural resources, conservation and rehabilitation of the environment.
- Coordinate environmental affairs in Afghanistan.
- Develop national environmental policies, strategies and legislation.
- Regulate and permit activities having an adverse impact on the environment, in the fields of environmental impact assessment, air and water quality management, waste management, pollution control.
- Improve environmental awareness and outreach.
- Gather and monitor environmental information.
- Implement the international environmental conventions that Afghanistan is a Party to.
- Enforce the provisions of the Environment Law.

(GIRA, 2009: 30)

The goal of NEPA is to protect the environmental integrity of Afghanistan and maintain sustainable development of its natural resources (Salter, 2009: 17). All these have to be initiated through the provision of effective environmental policies, regulatory frameworks, and management services that are also in line with the Afghanistan's MDGs (Salter, 2009; GIRA, 2009). The MDGs, adopted by all members of the UN, set measurable targets for enabling more human beings to enjoy the minimum requirements of a dignified life by 2015 (GIRA, 2005a: xvii). Through the MDGs, Afghanistan takes a step towards fulfilling some of the goals enumerated in the preamble to its 2004 constitution: "ensuring a prosperous life and sound environment for all those residing in this land" and "regaining Afghanistan's rightful place in the international community" (The Constitution of the Islamic Republic of Afghanistan, 2004: 1–2). There exist adaptation projects and programmes active in this country. The main priority sectors being addressed through those projects are the water sector, agriculture, forestry, land, and infrastructure as well as energy and food security. Funders which have been helping Afghanistan include the Asian Development Bank (ADB), the Food and Agriculture Organization (FAO), the Federal Ministry for Economic

Cooperation and Development (BMZ), the Least Developed Countries Fund (LDCF) etc. (Islam, Hove and Parry, 2011: 43).

In order to generate hydropower energy, expand agriculture, improve the economy, provide flood and drought control, and control its transboundary waters, Afghanistan has started to put forward some efforts for the management of its water resources. The Afghanistan National Development Strategy (ANDS, 2008: 18) emphasizes the need to "manage and develop water resources so as to reduce poverty, increase sustainable economic and social development, and improve quality of life for all Afghans and ensure an adequate supply of water for future generations". The National Environment Strategy (NES) was finalized in 2008 as part of the ANDS. The NES focuses on the two priority objectives of environmental governance and environmental management, and lays out thematic strategies for the following six areas: (i) forestry and rangelands; (ii) protected areas and biodiversity; (iii) water and wetlands; (iv) air quality; (v) urban and industrial environmental management; and (vi) environmental education and awareness (Van Epp *et al.*, 2017: 18).

In 2008, the Water Sector Strategy (WSS) highlighted the need for development of irrigation and dam infrastructure. The Water Sector Strategy (2008: 18) explains its strategic vision as to "manage the Nation's water resources so as to reduce poverty, increase sustainable economic and social development, and improve the quality of life for all Afghans and to ensure an adequate supply of water for future generations". The WSS (2008) insists on the storage and control of water so that the country can be prepared for the periods of drought due to climate change. The government feels that managing water resources will significantly help poverty reduction in the country as well, which is the biggest challenge for Afghanistan. The WSS also details Afghanistan's priorities for the development of its water resources: (i) better access to safe drinking water; (ii) enhanced household food security; (iii) protection from the negative effects of droughts and floods; (iv) sustainable development and management of water resources; (v) establish mechanisms for facilitating more effective user participation; (vi) support to poverty reduction and private sector development; (vii) effective services for efficient water use to facilitate economic growth and social development (King and Sturtewagen, 2010: 3; Van Epp *et al.*, 2017: 20). Further, the Water Law was enacted in 2009 to enforce the protections afforded by Constitution Article 9 through regulations aimed at promoting "conservation, equitable distribution, and the efficient and sustainable use of water resources" (Water Law, art. 1, 2009). This Law anticipates that local water user associations will play a key role in protecting and managing water resources (UNAMA, 2016: 6–9).

From 2010 the UNDP National Disaster Management Project (UNDP/NDMP) has been the major project of disaster risk management supported by the UN in Afghanistan. To strengthen Afghanistan's ability to withstand climate change (Mall and Kumar, 2014: 52), the UN has announced a US$6 million initiative (2012). The Islamic Republic of Afghanistan, Ministry of Justice has promulgated the law on Disaster Management on October 1, 2012, for management of natural and unnatural disasters in the country (Disaster Management Law, 2012, article 4). In addition, Afghanistan has other significant policy documents guiding its direction in disaster management and risk reduction. The Strategic National Action Plan (SNAP) aims to create a safer and more resilient Afghanistan by lowering the risk of future catastrophes and climate change impacts in an organized way (GIRA, 2011: 8). On seeing such development, Michael Keating, the former UN Afghanistan Resident and Humanitarian Coordinator, said that the Government of Afghanistan has shown a remarkable commitment to work with communities for a landscape approach to deal with climate change (Oskarsson, 2012: 6). By looking at its policies, it can be stated that the Government of Afghanistan has paid considerable attention to the pressing climate change–related issues.

In October 2015 the Afghan president approved the Extended Policy on Trans-boundary Waters. In 2016, the Afghan government announced the commencement of 21 large-scale infrastructure projects for water resources management of the country, which is a significant figure for the country (Hayat and Elci, 2017: 43). Presently, Afghanistan is finalizing its national Climate Change Strategy and Action Plan (ACCSAP) with the aim to (i) integrate and mainstream climate change into the national development framework; (ii) support the creation of a national framework for action on climate change adaptation; (iii) identify low emission development strategies; (iv) improve coordination and partnerships between government institutions, civil society, the international donor community, and the private sector; and (v) increase availability and access to additional financial resources for effectively addressing climate change (NEPA and UNEP, 2015: 17; Van Epp *et al.*, 2017: 19). The Ministry of Agriculture, Irrigation and Livestock (MAIL) is an important institution in relation to environmental management as it is responsible for executing natural resource exploitation, management, and conservation activities in Afghanistan. It is primarily the role of MAIL to manage or oversee projects on watershed management, rangeland protection, and national parks, forests, and species protection (Van Epp *et al.*, 2017: 63).

There are several programme interventions which are directly and indirectly contributing towards addressing chronic nutrition and food security in the country. Some of the programmes are the National Agriculture Development Framework, Health and Nutrition Sector Strategy 2008–2013, and the National Nutrition Action Framework (NNAF). The National Health and Nutrition Policy 2012–2020 of the Ministry of Public Health (MoPH) is committed to improving the nutrition status of the people. The mission of the MoPH is "improve the health and nutritional status of the people of Afghanistan in an equitable and sustainable manner through provision of quality health services and ensuring universal coverage". The Afghanistan

Food Security and Nutrition Agenda (AFSANA) with the goal of "ensuring that no Afghan suffers from hunger and every Afghan is well-nourished at all times" is a high-level policy and strategic document. These programmes have been able to boost food production and accessibility of food to children and pregnant women (SDGF, 2017: 4; MAIL, 2012; MoPH, 2010).

The US government, the United Nations Children's Fund (UNICEF), and the MoPH aim to reduce malnutrition in children between the ages of 6 and 59 months and decrease the prevalence of malnutrition in pregnant and breastfeeding women (UNICEF Nutrition Program in Afghanistan, 2008–2009). The World Food Programme (WFP) supports the Government of Afghanistan in building Strategic Grain Reserves (SGRs) to hold emergency food supplies of 200,000 Mt to assist approximately 2 million people for up to 6 months (WFP, 2016a). In order to help some 7.6 million people who are food-insecure the government has devised a SGR programme with a budget of US$14 million to (i) provide emergency food assistance to transitory food insecure households; (ii) support communities and farmers with storage facilities; (iii) assist producers to secure fair prices at harvest times; and (4) contribute to domestic food price stabilization (Special Operation: SO 200635, 2014–2018; WFP, 2017: 5). The programme aims at establishing a food reserve of some 200,000 Mt capacity by 2015, to assist about 2 million people for 6 months, and 300,000 Mt by 2024 (Special Operation: SO 200635, 2014–2018). The FAO, WB, and the WFP are supporting the government's initiative. In 2017, the Government of Afghanistan launched a new multi-million-dollar project aimed at protecting vulnerable Afghan communities from the worst impacts of climate change. This project will help reduce losses and damages resulting from climate-driven disasters, facilitate recovery from climate-shocks, and underpin progress towards the SDGs in Afghanistan (UNDP, 2017a).

Bangladesh

In contrast to other LDCs, Bangladesh is highly proactive in attending to the adversities of climate change (Ayers and Huq, 2008b: 753–764). To mainstream climate change issues in its development activities, Bangladesh had instituted a Climate Change Cells (CCCs) in 2004 in climate-relevant ministries and line agencies (Pokharel, 2011: 11). The Government of Bangladesh (GoB) has designated the Ministry of Environment and Forest (MoEF) as the "focal ministry" for providing coordination and the technical lead on all climate change–related matters (GoB, 2012a). The MoEF led to the drafting of the Bangladesh Initial National Communication in 2002, the National Adaptation Programme of Action (NAPA) in 2005 (MoEF, 2005), and various other position papers as well as the Bangladesh Climate Change Strategy and Action Plan (BCCSAP) in 2009 (MoEF, 2009). The BCCSAP identifies six pillars under which climate change mitigation and adaptation will take place. It properly takes disaster management as one of the pillars of the strategy:

- Food security, social protection and health;
- Comprehensive disaster management;
- Infrastructures;
- Research and knowledge management;
- Mitigation and low carbon development; and
- Capacity building and institutional strengthening.

(MoEF, 2009: 3; Sterrett, 2011: 28)

The BCCSAP has incorporated 44 listed programmes of action (MoEF, 2009). Its objective is to increase the country's resilience to climate change, and reduce or eliminate the risks that it poses to national development. Subsequently, the BCCSAP is reflected largely in Bangladesh's Sixth Five Year Plan (2011–1015) with the government focus on significant improvement of science and technology based on increasing productivity and efficiency across the country (MoEF, 2012b: 63). The Plan also seeks to integrate "poverty, environment and climate change into the process of planning and budgeting" (MoEF, 2013a, 2013c). So, climate change is not just an ecological concern but in reality a developmental concern for Bangladesh.

Besides demonstrating exemplary commitment to manage climate change effects, Bangladesh is also committed to follow a low carbon development path provided its economic growth and poverty reduction goals and prospects are not compromised. In the international arena, it is a leading voice on behalf of LDCs and other climate change vulnerable countries (Poudel, 2013: 39). With its own limited resources, it has been taking concrete actions even when it does not receive expected support from the international community (Khan *et al.*, 2013: 61). On its own it possesses a number of early-warning systems in place to alert villages of advancing cyclones and floods (Bhalla, 2009; Khan *et al.*, 2013: 62; Haque *et al.*, 2012: 151–152). The community-based early warning system of Bangladesh has become a model for other countries throughout the world (MoEF, 2012b: 23). It has invested time and money to further boost these systems. Over the last decades, Bangladesh has invested more than US$10 billion to make the country climate resilient and less vulnerable to any climatic disasters (Groom, 2012: 34; Huq and Khan, 2017). It has implemented 106 projects to address climate change including better adaptation and mitigation (Rashid, 2012; World Bank, 2012a).

The National Water Policy (NWP) is the first comprehensive strategy to look at short-, medium-, and long-term perspectives for water resources in Bangladesh (GoB, 1999). After the NWP, there came the National Water Management Plan (NWMP). The NWMP supervises the implementation and investment responses to address the critical priorities identified in the NWP. Bangladesh has ratified the Kyoto Protocol in October 2001 though it is also assessing the implementation of Agenda 21 (Philander, 2008: 113; WARPO, 2001; Agrawala *et al.*, 2003a: 32; Worldwatch Institute, 2006). It has established the Bangladesh Wetlands Network to conduct the case study

"Sanitary and Phyto-Sanitary Barriers to Trade and its Impact on the Environment of Shrimp Farming in Bangladesh". Further, Bangladesh carries out another regional project titled "Sustainable Livelihood, Environmental Security and Conflict Management" (Philander, 2008: 113).

To control and prevent the low-lying areas from the increasing frequency of high-intensity flooding, Bangladesh has gone ahead in preparing flood disaster mitigation; such efforts include the building of embankments in coastal areas (Ahmad *et al.*, 2000: 142; Choudhury *et al.*, 2003: 21–60). This country has established a Comprehensive Disaster Management Programme in 2003 with donor support which aimed to refocus the government's hard work towards disaster preparedness and risk reduction (Huq and Ayers, 2007: 10). To ease the frequent threats from climate change, certain funds have been generated. These include a Climate Change Trust Fund (CCTF) with an allocation of US$100 million for fiscal year 2009–2010; another US$100 million for fiscal year 2010–2011; the Bangladesh Climate Change Resilience Fund (BCCRF) with commitments of £60 million from the UK; €8.5 million from the EU; and DKK10 million from Denmark (Groom, 2012: 34; Huq and Khan, 2017; ADB, 2010). With the support from the ADB, WB, and the International Finance Corporation, Bangladesh prepared a Strategic Programme for Climate Resilience under the Pilot Programme for Climate Resilience of the Climate Investment Funds which is in line with the BCCSAP. Its objectives are to fortify embankments; raise coastal greenbelts; improve drainage, connectivity, and water resource management; promote climate resilient agriculture and food security; and ensure that drinking water is safe in vulnerable coastal communities (ADB, 2010). So, Bangladesh is very active in terms of planning and action in this regard.

In order to remove hunger and undernutrition in the country, Article 18–1 of the Bangladesh Constitution says, "the State shall regard raising the level of nutrition and improvement of public health as among its primary duties" (Shahan and Jahan, 2017: 4). This country is firmly committed to ensure food and nutrition security for its people, especially the most vulnerable groups of mothers and young children – through a comprehensive approach to food availability, access, and utilization for nutrition. In 1974, the Institute of Public Health Nutrition was set up to assist the government in "formulating policy and strategy for nutrition related activities and programmes" (GoB, 2011). This was followed by the establishment of the Bangladesh National Nutrition Council (BNNC) in 1975. The country adopted its National Food and Nutrition Policy (NFNP) in 1997 (GoB, 1997; Shahan and Jahan, 2017: 4) and the National Plan of Action for Nutrition (NPAN) was created in order to implement the NFNP. This policy identified undernutrition as a major developmental problem and according to the policy document: malnutrition is endemic in the country, with high infant, under-five, and maternal morbidity and mortality (GoB, 1997). Another approach called Bangladesh Integrated National Plan (BINP) came into existence in 1996 (Shahan and Jahan, 2017: 6; GoB, 2011; Taylor, 2012). The BINP was managed by the Ministry of Health and Family Welfare (MoHFW) which continued until 2002. In 2006, Bangladesh initiated the National Nutrition Project which was integrated into the Health, Population, Nutrition, Sector Development Programme (HPNSDP). It aims to "ensure quality and equitable health care for all citizens by improving access to and utilization of health, population and nutrition services and the development objective is to improve both access and utilization of such services, particularly for the poor". Key strategies to improve nutrition include (i) mainstreaming nutrition services; (ii) micronutrient supplementation; (iii) treatment of severe acute malnutrition; (iv) behaviour change communication to promote good nutritional practices; (v) coordination of activities across different sectors; and (vi) mainstreaming gender into nutrition programming (Compact2025, 2016: 9). Other policies such as the Livestock Policy of 2007, the Education Policy of 2010, National Food Policy (2006), National Food Policy Plan of Action (2008–2015), the Women Development Policy of 2011, and the Agriculture Policy of 2013 are concentrating on improving the nutritional status of the people (Shahan and Jahan, 2017: 22). In 2011, Bangladesh decided to "Mainstream Nutrition", an initiative which actually started as a three-year project funded by the WB from 2006 to 2009. The main objective of this initiative was to move "nutrition more into the mainstream of national policies and programmes, especially in the health sector" (Pelletier *et al.*, 2012: 19–31). One of the strategies of the Sixth Five Year Plan (2011–2016) is for improving nutrition, which includes "improving maternal and infant nutrition, strengthening institutional capacity, improving overall nutrition status, treating severe acute malnutrition, behaviour change communication to promote good nutritional practice and mainstreaming gender into nutrition programming" (Shahan and Jahan, 2017: 22; FPMU, 2015). The national development framework for Bangladesh in its Vision 2021 sets the policy objective of transforming Bangladesh into a middle-income country by 2021. Its goal includes goals regarding food security and nutrition (GoB, 2012b). Bangladesh's Vision 2021 aims to eliminate food deficiency and attain self-sufficiency in food production to meet the nutritional requirements of the population (Compact2025, 2016: 8). The Seventh Five Year Plan proposes ambitious social, economic, and environmental targets to be achieved by 2021. This Plan centers around three themes: (i) economic growth and poverty reduction; (ii) inclusion of all citizens in the development process; and (iii) sustainable development that is resilient to disasters, climate change, and urbanization (Compact2025, 2016: 8). The Post 2015 Development Agenda includes other government policy and action plans that prioritize food security and nutrition along with the major specific objectives (Compact2025, 2016: 8).

Despite the challenges, Bangladesh has made great strides in improving food security and reducing hunger. From 1990–1992 to 2014–2016, the prevalence of undernourishment fell by half, from 32.8 per cent to 16.4 per cent – a sign of strong, yet incomplete progress. The total number of undernourished people in Bangladesh has fallen as well, from 36 million to 26 million from 1990–1992 to 2014–2016 (FAO, 2016b; NIPORT, 2013, 2016).

While hunger and undernutrition are still serious problems in Bangladesh, the country has made strong progress to reduce both (Headey *et al.*, 2015: 749–761). There are some active USAID programmes in Bangladesh which focus on nutrition. In Fiscal Year (FY) of 2018, Food for Peace (FFP), a programme designed to reduce poverty and vulnerability of the poor in this country announced the provision of US$18.4 million to the WFP to support vulnerable people in Bangladesh's Cox's Bazar District with emergency food and nutrition assistance (USAID, 2018).

Public Interest Litigation (PIL) has its own method for the extensive interpretation of the procedural rule of *locus standi* in Bangladesh. A remarkable sign of this evolution is the growing role of the judiciary in environmental citizenship (Reid, 2000: 177–194). In the realm of environmental protection, judicial activism in Bangladesh has gone beyond the mere enforcement of the statutory provision of the environmental laws and embraces restitutionary as well as injunctive relief on the basis of constitutional jurisprudence and weaving in environmental law buzzwords such as "sustainable development", "polluter pays", "precautionary principle", and doctrine of trust and intergenerational equity (Salve, 2004: 360–380; Faruque, 2010: 57). In Bangladesh, PIL gets instituted before the Supreme Court through writ petition to challenge the action of the public bodies or individuals and to prohibit the government from violating environmental laws. To date, Bangladesh has about 180 laws which deal with or have relevance to the environment (Hasan, 2005: 85–108). Apart from statutes Bangladesh possesses a well-developed environmental policy and many other sectoral policies which have relevance on the preservation and conservation of the environment (Faruque, 2010: 58).

There is also the Environment Conservation Act, 1995, Environment Conservation Rules, 1997, National Environmental Policy, 1992, and National Environmental Management Action Plan, 1995 (Parvaiz and Neera, 2000: 20). In 2003, the Ministry of Food and Disaster Management (MoFDM) launched the Comprehensive Disaster Management Programme (CDMP). This shows that Bangladesh's judiciary assumed a greater role in the protection of the environment in Bangladesh. The Reducing Vulnerability to Climate Change (RVCC) project operating in the six districts of southwest Bangladesh, the area that is already affected by climate change, has had some fruitful results (Nambi, 2014: 13). The country's Disaster Management Act (2012) mentions the rehabilitation of people affected by climate change to resume the natural way of living. The background document on climate change for the Seventh Five Year Plan of Bangladesh identifies displacement issues (Rabbani, Shafeeqa and Sharma, 2016: 30).

Bhutan

The Royal Government of Bhutan (RGB, 2015b: 1) signed the Kyoto Protocol in 1997 and submitted its Instrument of Acceptance to the UNFCCC on August 26, 2002, which came into effect on February 16, 2005. Through its Planning Commission, Bhutan has developed a Vision Statement with a visionary strategy known as Bhutan Vision 2020. The policy guidelines of Vision 2020 emphasizes the ramping up of climate change mitigation by developing hydropower and solar energy resources (RGB, 2000a; Namgyel, 2003). This is a Vision for Peace, Prosperity and Happiness. The document states "although our natural heritage is still largely intact . . . the conservation of natural environment must be added to the challenges in the years ahead" (RGB, 1999). It is a longer-term strategy document which gives the overall framework for sustainable development which sets the future directions for charting out a distinctive development. Such a path is based on the sustainability principles in certain main areas of environment, culture and heritage, governance, and human development (Duba, Gurung and Ghimiray, 2006: 13). It also emphasizes to maintain forest areas, developing environmentally friendly options and balancing economic development by means of environmental conservation (Alam and Tshering, 2004: 21–22).

Due to its unique flora, fauna, and general biodiversity, Bhutan gives special attention to the protection and preservation of the environment. This country is committed to keeping 60 per cent of its territory forested (Lama, 2018). The Bhutanese authority is very consistent in preserving its environment, which it makes one of the top policy priorities. Bhutan's restrictive tourism policy is an expression of the government's concern for the environment whereas cultural preservation may also play its part. Another policy document of Bhutan, the National Environment Strategy (NES), directs the country's development paradigm with regard to environmental safeguards and development (RGB, 1998). This is a guide for environmental conservation. Bhutan's Environmental Assessment Act, approved by its National Assembly, established the procedures for the assessment of potential effects of strategic plans, policies, programs, and projects on the environment. It is also meant for the determination of policies and measures to reduce potential adverse effects and to promote environmental preparation (RGB, 2000b: 2). Formulation of the National Environmental Action Plan and the National Environmental Protection Act (NEPA) approved by Bhutan's national assembly in June 2007 asserts that the people and the government should "strive to consider and adopt its development policies and plans in harmony with the various environment principles" (Padma, 2007). Bhutan has championed a new approach to development, which measures prosperity through formal principles of gross national happiness (GNH) and the spiritual, physical, social, and environmental health of its citizens and natural environment. For Bhutan, "the idea of being green does not just mean the environment; it is a philosophy for life". In line with the philosophy of sustainable development, there is a maxim in Bhutan which says "it is better to have milk and cheese many times, than beef just once" (RGB, 1998: 17; Kelly, 2012: 1).

Environmental conservation constitutes an important part of the special planning strategic framework. It always enjoys a high priority in Bhutan's development agenda (Singha, 2010: 52; RGB, 2009b). Bhutan's Tenth Five Year Plan (10th FYP) highlights environment as an issue that is closely intertwined with poverty reduction. So, it calls for all sectors, agencies, like the *Dzongkhags* (mountainous terrain) and *Gewogs* (a group of villages or administrative units) to mainstream environmental issues in all their policies, plans, programmes, and projects (Chophel and Dorji, 2015). Similarly, Bhutan has a policy of middle path so as to stress equal prospects for economic development and environmental conservation (Singha, 2010: 53). The National Adaptation Programme of Action (NAPA) identifies priority activities which respond to their urgent and immediate needs to adapt to climate change (RGB, 2006a).

The National Environment Commission (NEC) (RGB, 2013b) is a high-level autonomous agency of the RGB. Its mandate is to look after all issues which are related to environment. The aims of the NEC are to put in place the necessary controls, regulations, and incentives to the private/public sectors to achieve sustainable development through the sensible use of natural resources. The NEC is a focal point for climate change activities in the country whose high-level commission members form the National Climate Change Committee (NCCC), a technical-level task force who may convene for project implementation (RGB, 2009c: 6–7, 2004).

The efforts put forth by Bhutan's Clean Development Mechanism (CDM) have yielded the Green Power Development Project which became the first cross-border CDM project to be registered and recognized by the UNFCCC in April 2010 (ADB, 2010; RGB, 2009c). There is a Bhutan Transport 2040 Integrated Strategic Vision under the Development Partnership Program for South Asia (ADB, 2006). This transport vision gets financial assistance from the Government of Australia through the Australian Agency for International Development. The transport vision integrates all existing transport-related plans, policies, initiatives, and actions to create a long-term comprehensive strategy for the country. It is building on earlier studies to create one integrated and sustainable plan (ADB, 2013: 1). According to Article 5 of the Constitution of Bhutan (adopted in 2008), every Bhutanese should be a trustee of the kingdom's natural resources and environment for the benefit of the present and future generations (Ura *et al.*, 2012: 164). So, it must be a fundamental duty of every citizen in Bhutan to contribute to the protection of the natural environment, conservation of its rich biodiversity, and prevention of all forms of ecological degradation (Thinley and Chopel, 2012: 51). Bhutan has also launched a US$13.9 million GEF project aimed at enhancing the resilience of communities and protecting the country's unique and rich biodiversity in the face of a changing climate (UNDP, 2017b).

This country has been getting ecological help from several donor countries including the EU members (European Commission, 2003). The Climate Change Screening of Danish Development Assistance (DANIDA) approved a project titled "Support to Enhancing Adaptive Capacity to Climate Change" in Bhutan (RGB, 2009c: 5). The Austrian Development Cooperation (ADC) too implements a project on GLOFs mitigation measures (Alam and Tshering, 2004: 20–24). The governments of the Netherlands, Germany, Japan, Switzerland, and the UK have been funding projects which focus in the areas of risk reduction, water, policy formulation, and health (Islam, Hove and Parry, 2011: 78). Finally, at the Twenty-third Board Meeting the Green Climate Fund (GCF) approved US$25.3 million in support of Bhutan's efforts to prepare and adapt to climate change and ensuring that Bhutan is heading towards low-carbon and climate-resilient developments (UNDP, 2019).

In Bhutan, chronic malnutrition among young mothers and children under the age of 5 is steadily increasing (RGB, 2011). To achieve food and nutrition security, Bhutan is working with the Japan International Cooperation Agency (JICA) and International Financial Institutions (IFIs) to invest in irrigation and post-harvest infrastructure (UNDAF, 2012: 20). The UN is helping Bhutan's agriculture by focusing on the following areas: (i) Strengthening food and nutritional security; (ii) Fostering agricultural production and rural development; (iii) Enhancing equitable, productive and sustainable natural resource management and utilization; (iv) Improved capacity to respond to threats and disasters; and (v) Climate change (and its impact on agriculture and food security and nutrition) (UNDAF, 2012: 20). The WFP has been working with the Government of Bhutan to develop and implement its school meals programme. Currently, the WFP's assistance is focusing on feeding schoolchildren. The WFP in-charges at schools are provided with training on school feeding management, nutrition and reporting, personal hygiene, nutrition and food preparation (WFP, 2015). This humanitarian organization (WFP) also provided equipment to the Ministry of Education (MoE) and Food Corporation of Bhutan (FCB) (MoE, 2010: 12). In 2015, the government's school meals programme supported around 30,000 children. The WFP is simultaneously providing food assistance directly to another 25,000 children in 196 schools. All these good works have been to meet the objective of WFP to help the government achieve self-reliance in the management of the school meals programme across the country by 2018 (WFP, 2016b).

To check any unnecessary effects of chemicals and fertilizers Bhutan has banned the sales of pesticides and herbicides. This country relies on its own animals and farm waste for fertilizers as it is planning to become the first country in the world to turn its agriculture completely organic (Vidal and Kelly, 2013). The Royal Government of Bhutan is seriously taking the challenge to become a carbon-neutral country (NEC, 2011; Flagg, 2015: 202–212; Lama, 2018), by preserving forest cover as a strategy to achieve the goal. The Renewable Energy Policy 2011 aims to reduce heavy dependence on hydropower, reduce the use of fossil fuels in the transport sector, and introduce environment-friendly and locally available fuels/energy resources to meet demand for energy, especially in remote and dispersed locations

(BMCI, 2016: 80). It is also becoming an example of a developing country that puts environmental conservation and sustainability at the heart of its political agenda (Kelly, 2012). In addition, Bhutan has in place a Forest Act 1995, Environmental Impact Assessment Act 2000, Social Forestry Rules 2000, and Forest and Nature Conservation rules of Bhutan 2000, Biodiversity Act 2003, Waste Prevention and Management Act 2009, and Water Act of Bhutan 2011 (Singha, 2010). Due to its rich harmonious tradition with its environment, Bhutan enjoys impressive achievement in developing the national policy and regulatory framework for the environment and achieves international recognition (Singha, 2010: 55). This country ranked 40th out of 163 countries in the 2010 Environmental Performance Index (EPI) produced by a team of environmental experts at Yale University and Columbia University (Ura *et al.*, 2012: 13–73). To pay tribute to its efforts, Bhutan was given the UNEP's Champions of the Earth Award in 2005(Singha, 2010: 56; European Commission, 2003). As the tall leader of environmental protection in the country, its fourth king got the honour of the J. Paul Getty for Conservation Leadership Award in 2006 by the World Wildlife Fund (Singha, 2010: 54).

India

Cautious to make public commitments to deal with climate change's effects, several steps have been taken by India – be it through mitigation or adaptation (Mehra, 2009). India is committing to prevent its per capita emissions from ever exceeding those of developed countries. It is actively involved in the UNFCCC negotiations too (MoEF, 2004). The National Disaster Management Act, 2005, which came into the statutes on December 26, 2005, provided for an institutional mechanism at the national, state, and local levels for comprehensive disaster management. The National Disaster Management Authority (NDMA) is an apex organization in the country to deal with natural and man-made disaster (Chandran, 2012: 81–82). The NDMA focuses on prevention, mitigation, preparedness, rehabilitation, and reconstruction. It formulates suitable policies and guidelines for effective and synergized national disaster response and relief (NDMA, 2007: 2). The Government of India has approved an advanced Tsunami Warning System under the Ministry of Earth Sciences (Bandyopadhyay, 2007: 13–40; MHA, 2011: 55–79).

The Prime Minister's Council on Climate Change (PMCCC) has in June 2007 committed to develop India's first-ever national action plan on climate change. This plan intends for assessment, adaptation, and mitigation (The Hindu, June 6, 2007). Under former Prime Minister Manmohan Singh, India launched the National Action Plan on Climate Change (NAPCC) on June 30, 2008. This Action Plan outlines India's domestic response to the climate crisis (GoI, 2008; Sangal, 2008; Bidwai, 2012: 129). It includes priority areas such as energy efficiency, solar and forestry initiatives, as well as the provisions for a sustainable habitat. It focuses attention on eight priority National Missions. These are: (i) Solar Energy, (ii) Enhanced Energy Efficiency, (iii) Sustainable Habitat, (iv) Conserving Water, (v) Sustaining the Himalayan Ecosystem, (vi) A Green India, (vii) Sustainable Agriculture, and (viii) Strategic Knowledge Platform for Climate Change (Prasad and Kochher, 2009: 25; GoI, 2008). Through the NAPCC, India is ready to make its own contribution in this matter.

On June 24, 2010, India's cabinet approved the National Mission on Enhanced Energy Efficiency (NMEEE) (MoEF, 2010d: 2). The Mission includes several new initiatives – the most important being the Perform, Achieve and Trade (PAT) Mechanism. This will cover facilities which account for more than 50 per cent of the fossil fuel used in India. It intends to fuel savings of around 23 million tonnes per year and also motivates to help reduce CO_2 emissions by 25 million tonnes per year by 2014–2015 (MoEF, 2010d: 2). Launched in 2013, the National Mission for Sustainable Agriculture of the NAPCC aims at making India's agriculture more resilient to climate change. This goal can be achieved by identifying varieties of crops especially thermal resistant ones and alternative cropping patterns. This mission put special emphasis on soil and water conservation, water use efficiency, soil health management, and rain-fed area development. The Mission requires a budgetary support of approximately US$17.4 billion up to the end of the Twelfth Five Year Plan (FYP) in 2017 (MoEFCC, 2014: 8–10). India's Ministry of Water Resources released a revised comprehensive mission document (2009) which detailed the strategy of the National Water Mission under the NAPCC (MoWR, 2009, 2008). In pursuance to the announcement made by the Finance Minister (while presenting the Union Budget in 2007–2008), the Government of India set up the "Expert Committee on Impacts of Climate Change" (on May 7, 2007) under the chairmanship of Dr. R. Chidambaram, Principal Scientific Advisor to the Government of India (Agnihotri, 2008: 22). The Committee intends to study the effects of anthropogenic climate change. It also identifies the measures which India may take in the future in relation to addressing vulnerability to the anthropogenic climate change effects (Prasad and Kochher, 2009: 24). Besides, a Council has been set up since June 6, 2007, under the Chairmanship of the Prime Minister of India which consisted of eminent persons to evolve a coordinated response to issues relating to climate change at the national level. This Council is expected to provide the oversight for formulation of action plans in the area of assessment (Prasad and Kochher, 2009: 24). To ensure integrated water resource management, conserve water, minimize wastage, and ensure equitable distribution of water within states, the National Water Mission (NWM) was approved by the Cabinet in 2011. The Mission is run by the Ministry of Water Resources, River Development and Ganga Rejuvenation (Rattani, 2018: 15). The main goals of the mission are (i) creating a comprehensive water database in the public domain and assessing the impact of climate change on water resource; (ii) promoting citizen and state action for water conservation, augmentation and preservation; (iii) increasing water-use efficiency by 20 per cent, and (iv) promoting basin-level integrated water resources management (Rattani, 2018: 15). India has built concrete rainwater reservoirs in Rajasthan (Chadburn, 2007: 6–7) which is an important adaptation measure as parts of this state are increasingly drought-prone.

The Government of India initiated a Green India programme to undertake massive afforestation of degraded forest land in the country (F.No.8–14/2004-FP, 2012: 1–8; MoEF, 2010b). The Green India Mission (GIM) is one of the eight Missions under the NAPCC. This Mission aims at addressing climate change by enhancing carbon sinks in the state's forests and simultaneously attempts to increase resilience capacity of the forest ecosystem while it enables forest-dependent communities to adapt in the face of climatic vulnerability (MoEF, 2010b; MoEF, 2010c). The other aims of the GIM are both at increasing the forest and tree cover as well as improving the quality of the existing forest covers. The scheme was proposed for a period of ten years. Its main objective is to increase forest cover to the extent of 5 million hectares (mha) and improve the quality of forest cover on another 5 mha of forest and non-forestlands (GoI, 2012a). Several other initiatives have been undertaken by the Government of India under the National Climate Resilient Agriculture Programme (NICRA) as part of the National Agriculture Mission of the NAPCC (Nambi, 2014: 15).

On February 28, 2014, the National Mission for Sustaining the Himalayan Ecosystem was approved by the Indian government. This Mission aims at evolving conservation measures for sustaining and safeguarding the Himalayan glaciers and mountains through establishment of a monitoring network, promotion of community-based management, human resource development, and strengthening regional cooperation (Rattani, 2018: 20). The National Mission on Sustainable Habitat (NMSH) (NMoSH, 2010) was approved with the aims to integrate mitigation and adaptation into the urban planning process with a view to make cities sustainable through improvements in energy efficiency of buildings, management of solid waste, and shift to public transport. The Planning Commission of India visualized that many of the activities expected under NMSH would be taken under Atal Mission for Rejuvenation and Urban Transformation (AMRUT). The thrust areas of AMRUT – water supply; sewerage facilities and septage management; storm-water drains to reduce flooding; pedestrian, non-motorized, and public transport facilities; parking spaces; creating and upgrading green spaces, parks, and recreation centres in cities – align closely with that of the Mission (GoI, 2012b; MoUD, 2014: MoHUA, 2018). The Strategic Knowledge Platform for Climate Change came into existence in 2010. Some of its targets are creation of regional climate models, 50 chair professorships, 200 specially trained climate change research professionals, public-private partnerships, collaborations with other countries, outreach and public awareness. The Mission as well envisions asignificant role of states in their active participation (MoST, 2010; Rattani, 2018: 21). The National Solar Mission

(NSM) was launched in 2010 with the primary aim of achieving grid parity by 2022 and with coal-based thermal power by 2030. On June 17, 2015, India revised the cumulative targets for grid-connected solar-power projects initially envisioned from 20,000 MW by 2021–2022 to an ambitious 100,000 MW by 2021–2022 (MoNRE, 2011; Rattani, 2018: 11). In October 2014, the Government of India launched the *Swachh Bharat Abhiyaan* (Clean India Mission) towards achieving a "clean and open-defecation-free India" by 2019. This mission is concentrating on the areas of sanitation and solid waste management. Under this mission, India plans to build 1.52 million toilets in rural areas along the Ganga River and 1.45 million toilets (private and public) in cities adjoining the river banks (MoDWS, 2014; Rattani, 2018: 25). By 2019, this mission has achieved significant mile stones in both the rural and urban front (Dutta, 2019). It is still working to achieve and sustain cleanliness all over the country.

While preparing the Approach Paper (of the Twelfth Five Year Plan), the Planning Commission of India consulted more experts as people currently realized that India's economic development would be sustainable only if it could pursue that in a manner which protects the environment (GoI, 2011). The Twelfth Plan therefore aspires to transition the environmental governance system towards managing the Environment, Forests, Wildlife and other challenges of climate change for faster and equitable growth, where ecological security for sustainability and inclusiveness is restored (MoEF, 2012c: 5; MoEF, 2013b). For the first time, India integrates low-carbon growth into its Twelfth FYP (2012 to 2017). If India can implement its current climate change initiatives, by 2020 the country will reduce its emissions intensity by 25 per cent over 2005 levels with a "determined-effort" scenario, and by 33 to 35 per cent with an "aggressive effort" scenario (Parikh, 2011). Though the NAPCC is closely in line with the government's adopted policy of shared but differentiated responsibility, India reaffirmed it would pursue the principle of sustainable development as its per capita GHG emissions must not exceed the per capita GHG emissions of developed countries, despite its developmental imperatives (Prasad and Kochher, 2009: 24–25). The Twelfth FYP has set clear goals and targets for monitoring and these cover environmental protection, climate change, forests, ecosystems and biodiversity and propose organizational, regulatory, investment, and capacity-building strategies. On regional cooperation, the Twelfth FYP proposed an institutional mechanism "for developing and implementing policies, laws and action plans" (Mehra, 2009: 214). But for implementing adaptation actions in agriculture, forestry, fisheries infrastructure, water resources, ecosystems etc. (Lama, 2018), India will require US$206 billion (at 2014–2015 prices) between 2015 and 2030. Thus, the challenges of arresting the pace of degradation of environment will be difficult for India.

The National Green Tribunal Act (NGT) 2010, which is an Act of the Parliament, enabled the creation of a special tribunal to handle the expeditious disposal of the cases pertaining to environmental issues. It gets enacted under India's constitutional provision of Article 21, with the assurance that the citizens of India must have the right to a healthy environment (Balaji, 2010). According to Mr Jairam Ramesh, the former Union Minister for Environment and Forests of India, the NGT will give Indian citizens a first-time judicial remedy as far as environmental damages are concerned (MoEF, 2010a). In line with this, the Supreme Court of India had temporarily banned the registration of diesel-powered Sport Utility Vehicles (SUVs) and cars with engines over 2,000cc in Delhi (Hindustan Times, December 16, 2015). In addition, after it faced criticism from the Delhi High Court over the Delhi's mounting pollution problem, the Delhi Government announced that private vehicles would be allowed to run on the streets of Delhi on alternate days depending on whether their registered licence plates end in even or odd numbers. But this decision will not apply to compressed natural gas (CNG)-driven buses, taxis, auto-rickshaws, and emergency vehicles but will cover automobiles entering Delhi from other states (Hindustan Times, December 29, 2015). Further, the Sundarbans Programme in West Bengal (initiated in 2005) is working in three major, albeit non-exclusive thematic areas: biodiversity conservation, climate change and energy, and sustainable livelihoods (Das, 2009).

India has been pressing the UNFCCC and other international conferences for collaborative development of clean technologies and immediate transfer of the existing environment friendly equipment to developing countries. To address climate change–related issues in the developing countries, India called for early operationalization of the Adaptation Fund and Special Climate Change Fund under the UNFCCC (Persson, 2011: 3). Such funds are meant to finance projects and programs which intend to help developing countries to adapt to the harmful effects of climate change. It also meant to support adaptation and technology transfer in all developing countries who are parties to the UNFCCC (Ayers and Huq, 2008a; Harmeling and Bals, 2008: 6–7). India is a partner to the new Asia Pacific Partnership on Clean Development and Climate which consists of key developed and developing countries – Australia, China, Japan, South Korea, and the United States besides India. This joint venture focuses on development, diffusion, and transfers of clean and more efficient technologies. It is consistent with the principles of the UNFCCC and complements the efforts under the UNFCCC and will not replace the Kyoto Protocol (US Department of State Archive, 2007; Oxley, 2005).

For India, ensuring food security ought to be an issue of great importance where more than one-third of its population is estimated to be absolutely poor, and as many as one-half of the children have suffered from malnourishment over the last three decades (Ittyerah, 2013: 1). Government schemes related to food security such as the Targeted Public Distribution System (TPDS), the Integrated Child Development Services (ICDS), Public Distribution System (PDS), Mid-Day Meal Scheme (MDMS) of India, and the National Food Policy (NFP) have been working successfully (Dev and

Sharma, 2010: 33–35; Saxena, 2011; Ittyerah, 2013: 1–34). India passed the National Rural Employment Guarantee Act, 2005 (NREGA), which aims at enhancing livelihood security in rural areas by providing at least 100 days of guaranteed wage employment in a financial year to every household (Ittyerah, 2013: 32). The National Food Security Act (2013) is a milestone in the history of India's fight against hunger and malnutrition, as it empowers more than 800 million Indians (75 per cent of the rural and 50 per cent of the urban population living below and just above the national poverty line) to legally claim their right to highly subsidised staple foods (Puri, 2017; Satpathy, 2016; Saini and Gulati, 2015; Dev and Sharma, 2010: 34–35). Through the Global Strategic Objectives, the WFP has been supporting India to meet international goals and standards regarding food security and nutrition. The WFP's engagement in India is guided by the commitment towards: (i) ensuring access to safe, nutritious and sufficient food for all people all year round and (ii) ending malnutrition according to internationally agreed targets, with a focus on stunting and wasting in children under 5 years of age, and addressing the nutritional needs of adolescent girls, pregnant and lactating women, and older persons (WFP, 2015: 24).

The Maldives

Non-action to the crisis of climate change is never an option for the Maldives as it is one of the most liable countries to the adverse effects of climate change. Although it is not possible to accurately predict its adverse effects, the first Climate Risk Profile and the Disaster Risk Profile authorize the need to take preventive and adaptive actions (MoEEW, 2006a; Ghina, 2003; UNDP, 2006; MoEEW, 2006b: 4–5). Anxious to climate change vulnerabilities, attention is given for adaptation. To reduce the emission of GHG, the Maldives started the pilot projects on alternate sources of energy. Solar power has been used to power telecommunication sets, navigational aids, government office buildings, and in mosques. A variety of programmes have been designed and implemented in areas such as freshwater management, coastal and coral reef protection. The Maldivian government is taking up important measures to protect the coral reefs. Reduction of import duty on construction materials, prohibition on the use of coral for government buildings and tourist resorts and banning of coral mining from house reefs are identified as key measures (MoHTE, 2010a: 8; MoHAHE, 2001a: 134).

Through its representative such as Maumoon Abdul Gayoom, the Maldives' has been playing a crucial role in bringing the crisis of climate change to the attention of the international community. Alert about the hazardous effects of climate change, speeches by its government representatives have been given at various important international gatherings – the Commonwealth Heads of Government meetings, the Earth Summit in Rio in 1992, the UN Millennium Summit in 2001 (Barnett and Campbell, 2010: 86; UNFCCC, 2005; World Bank, 2012a), and the SAARC Summit meetings.

The Maldives has also participated in the Second World Climate Conference in 1990 along with other small island states (MoHAHE, 2001b; Hein, 2004: 9). This Conference demands the best environmentally sound technologies be transferred swiftly to small island states (MoHAE, 1994). In this way, the Maldives is playing a leading role to encourage all the small island states to group together to devise a unified position on the problems. With the help it receives from the Commonwealth Secretariat, in the year 1989 the Maldives hosted the Small States Conference on Sea Level Rise at Ministerial Level. The outcome of this Conference was the Malé Declaration on Global Warming and Sea Level Rise (Pernetta, 1992: 19–31; Psaila, 2010: 16; UNICEF, 2012: 183), one of the first ministerial declarations on the impacts of climate change. It is the Malé Declaration which paved the way for the establishment of the Alliance of Small Island States (AOSIS) and contributed to the establishment of the UNFCCC (UNICEF, 2012: 183). Due to its modest political influence, however, the ability of the Maldives to bring about such help is limited, though it is largely involved in and instrumental to the formation of the AOSIS (Philander, 2008: 617–618). This alliance is a lobbying and negotiating body representing the needs of small island developing states at the level of the UN.

The Maldives has signed the UNFCCC on June 12, 1992, and it ratified the same on November 9, 1992 (Inaz *et al.*, 2004: 21). Similarly, it signed the Kyoto Protocol on March 16, 1998, and ratified it on December 30, 1998 (MoHAHE, 2001a). The submission of the first National Communication of Maldives to UNFCCC was at the Seventh Session of the COPs to UNFCCC that occurred in Marrakesh in 2001 (MoHAHE, 2001a). Its National Greenhouse Gas Inventory, National Mitigation Plan, Vulnerability Assessment and Adaptation Options were included in this National Communication (MoHAHE, 2001a: 35). Due to its small size, it began to invest for the implementation of various augmentation and adaptation strategies to offset current water shortages at great financial cost (Parakoti and Scott, 2002: 9).

Without much ado after the *tsunami*, a temporary National Disaster Management Center (NDMC, 2010: 28; RoM, 2010) was established with the mandate of coordinating all emergency relief work. With the assistance and cooperation from various international donors and non-governmental organizations (NGOs), reconstruction efforts have been made (NDMC, 2010). Efforts to mainstream environmental problems got boosted by the adoption of the National Adaptation Programme of Action (NAPA) (UNICEF, 2012: 181; MoEEW, 2007), which was adopted on January 2007 to identify the most urgent and immediate adaptation needs of the country with regard to predicted climate change effects. In its NAPA, the Maldives identify eight critical sectors which need immediate adaptation action. The agricultural sector is identified as one of the key areas (MoEEW, 2007). Adaptation actions in the areas of risk reduction, coastal zones, infrastructure, water, and tourism are viewed as top priorities for the government. There are ongoing actions that touch on the areas of risk reduction, costal zones, and water

(MoEEW, 2007). Before its acceptance it was expected that the adaptation project of NAPA would focus on coastal and coral reef protection, adaptation in agriculture, freshwater and fishery sectors, food security and health (MoEEW, 2006c). The goal was to present a coherent framework to climate change adaptation to enhance the buoyancy of the natural, human, and social systems and ensure their sustainability in the face of predicted climate risks (MoEEW, 2006c: 5). In this way, NAPA identifies the urgent and immediate actions for climate change adaptation. The cost of adaptation needs as identified in the NAPA of the Maldives was estimated around US$100 million (MoEEW, 2007).

The implementation of this NAPA got bolstered by the signing of a 4-year initiative in December 2009. This initiative is called the "Integrating Climate Change Risks into Resilient Island Planning in Maldives" (MoHTE, 2010b: 1). The project is funded by the Least Development Country Fund and UNDP and Co-financed by Government of Maldives and the GEF (MoHTE, 2010b: 1; Sovacool, 2012; Elrick-Barr, Glavonic and Kay, 2015; GEF, 2009). In addition, the Maldives has the National Safer Island Strategy and the Safer Island Development Program. This strategy arose from a preference to make islands more resilient to external threats. It is based on the assumption "that any island could be made safer using appropriate technology" (RoM, 2010: 69; Stojanov *et al.*, 2017a: 9). This strategy focuses on hard adaptation and structural engineering solutions, such as land reclamation and rising islands (Elrick-Barr, Glavonic and Kay, 2015).

The Maldives has a Strategic Plan of Action known as the National Framework for Development and National Adaptation Programme of Action. A solid policy foundation is given to environmental sustainability, climate change adaptation, and low-carbon development (MoEEW, 2007). The National Sustainable Development Strategy, the Third National Environmental Action Plan (2009), and the Strategic National Action Plan (2009–2013) stress the importance of strengthening resilience to climate change and reducing disaster risk (UNICEF, 2012: 184).

There is a Presidential Advisory Council on Climate Change which will offer expert advice on how the Maldives can cut its GHG emissions (Haveeru Daily, April 2, 2009). The reason is because the Maldives aims to become the first country in the world to attain carbon neutrality in ten years i.e., by 2020 (Haveeru Daily, December 4, 2010). To meet this objective, fossil fuels will be virtually eliminated from the country. This can be done when generation of electricity will be carried out through renewable means (Bansal and Datta, 2012: 255). This project not only covers homes and business, but even cars and boats, where diesel engines will be replaced by electrical engines run by batteries. Environmental sustainability is accepted as basic rights in the Maldives (MoEEW, 2006d: 12–13). With all its efforts the Maldives has been trying to minimize climatic hazards by building central and local government capacity to undertake long-term and integrated planning (Clark, 2009). Moreover, the Maldives has organized the first-ever underwater cabinet meeting in 2009 in a symbolic gesture to draw global attention to the peril of climate change (Muricken, 2010: 69). This country has also proposed to buy land abroad to create artificial islands for their people (Ramesh, 2008).

The Maldives recognize the importance of environment in human health. The Ministry of Health (MoH, 2016, 2014) in the Maldives has plans to strengthen the resilience of its health system to cope with and adapt to climate change. For instance, the National Public Health Protection Act (7/2012) of 2012 gives importance to addressing environmental and climate change–related diseases. Through the National Environmental Health Action Plan (NEHAP), the Maldives gives a special focus on addressing the detriments of climate change on human health. The Maldives' Health Master Plan (MHMP) for 2016 to 2025 identifies climate change as a key issue and intends to address it via "monitoring health impacts of climate change and developing strategies for reorienting programmes to address the emerging health issues" (MEE, 2016: 102–103; MoH, 2016). The "Tourism Adaptation Project" (TAP) which was initiated in 2012, facilitated and provided support to bring about the required amendments to the existing laws and regulations that govern the tourism sector (MEE, 2016: 105). Tourism sector projects identify and maintain community-based adaptation projects in tourism-associated communities. It teaches how tourism operators and tourism-dependent communities can cooperate on joint initiatives to reduce common vulnerabilities. One of the successful community-based adaptation was through "Sustainable Water Management and Community Awareness in Maalhos Island, Baa Atoll" (MoEE, 2016: 106).

Reducing poverty is one of the stated goals of the Maldives' Seventh National Development Plan (2006–2010) (MoPND, 2010). The Agricultural Development Master Plan 2006–2020 (ADMP), which had been developed with assistance from FAO, envisions a rapid transformation that will make the sector a driving force in the economy, after tourism and fisheries. At the request of the government, WFP extended its recovery operation until the end of 2005 to provide food supplies for 14,000 people displaced by the *tsunami*. In the Maldives, WFP successfully concluded a School Feeding programme which provided food to 24,000 school children (WFP, 2005; Morris, 2005: 21–23).

Nepal

Increase in temperature extremes, sporadic rainfall, and amplified, unpredictability in weather patterns are regular phenomena of Nepal (Oxfam, 2009: 1–5). This makes it imperative that Nepal adapts to these challenges. National policies of Nepal are committed to meet sustainable development targets. Some of the policies and legislations are National Conservation Strategy (NCS) (1988), Master Plan for the Forestry Sector (1989), National Environmental Policy and Action Plan (1993), Industrial Policy (1992)

(Khadka *et al.*, 2012: 53), the Environmental Protection Act, 1996, Environmental Protection Regulation, 1997, and the National Environmental Impact Assessment Guidelines, 1992 (Bhatt and Khanal, 2010: 586–594). The Ministry of Environment, Science and Technology (MoEST) is the focal department for climate change issues, where there are many challenges to meet (Rai and Gurung, 2005: 316–320). In 2002, Nepal developed the National Water Resources Strategy which recognizes climate variability and its potential impacts on the country's water resources (Water and Energy Commission Secretariat, 2002).

Being a poor country, Nepal's 2001 MDG initiatives and its 2003 Sustainable Development Agenda both included addressing climate change issues as key to achieving its goals. The Sustainable Development Agenda was based on longer-term development goals of the country's Ninth (1997–2002) and Tenth Plans (2002–2007), the Poverty Reduction Strategy Paper (PRSP), the MDGs, and Agenda 21 (Patra and Terton, 2017: 14; NPC and MoPE, 2003). The PRSP in particular indicates that the government will focus and dispense all available resources on fulfilling an agenda of sustainable development by reducing poverty. The objective of the Thirteenth Plan for 2013–2014 to 2015–2016 is to bring about a positive change in the living standards of its citizens by reducing the economic and human poverty prevalent in the country (NPC, 2013: 10; Patra and Terton, 2017: 14). To achieve this objective, this Plan identifies a set of seven strategies, one of which is to "implement development programmes which support climate change adaptation" (NPC, 2013: 11). This strategy spells out the common understanding among the country's policymaking and planning communities about the larger dividends, both economic and environmental, which could be accrued through integrated adaptation-centric development planning (Patra and Terton, 2017: 15).

This country ratified the three Rio Conventions: the UNFCCC, the Convention to Combat Desertification (UNCCD), and the Convention on Biological Diversity (UNCBD) (MoEST, 2008). Its most recent national report to the United Nations and Commonwealth Department (UNCD) was prepared for the Fourth COP to the UNCD in 2000 (Agrawala *et al.*, 2003b: 21). Nepal possesses a National Strategy for Sustainable Development (NSSD) under the name of the Sustainable Development Agenda for Nepal (SDAN) (Agrawala *et al.*, 2003b: 21–22). With the approval of His Majesty's Government, this plan (SDAN) aims to guide and influence national-level planning and policies up to 2017. Its agenda is drawn in conformity with the long-term goals envisioned in its Tenth FYP (2002–2007) (Durbar, 2002). With the support it receives from the Department for International Development (DFID), Danish International Development Agency (DANIDA), and other multilateral organizations, a regional conference, "From Kathmandu to Copenhagen" was organized in 2009 (Uprety, 2013). Nepal prepares its NAPA with the support of various international agencies like the GEF, DFID, and DANIDA. The UNDP gives fee-based technical assistance (MoE,

2010a). The NAPA set out the country's strategy and action plan to respond to the challenges posed by climate change, and the government of Nepal finalized it in 2010. It identifies six major areas that are affected by climate change. These are:

> *Agriculture and food security*: Because Nepal's perpetual farming economy is in jeopardy to changes in precipitation, rise of temperatures, flooding, and irregular monsoon rainfall;
>
> *Water resources and energy*: Because climate-related water pressure directly affects agricultural production, human health and sanitation, and changes in river flow may directly affect micro-hydro projects in the hills and mountains;
>
> *Climate-induced disasters*: Because Nepal is now exposed to various hydro-meteorological disasters, and this may be exacerbated in the future with climate change;
>
> *Forests and biodiversity*: Because rise of temperatures and rainfall inconsistency may lead to shifts in agro-ecological zones;
>
> *Public health*: Because of possible increase in vector-borne and water-borne infectious diseases, including an increased risk of malaria and other climate change–related diseases;
>
> *Urban settlements and infrastructure*: Because climate change is anticipated to hit infrastructure like roads, bridges, community and public buildings, and schools. Those effects are expected to be concentrated around urban water and energy resources, and may also affect human health.
>
> (MoE, 2010a; Islam, Hove and Parry, 2011: 120–121;
> Tiwari *et al.*, 2012)

The NAPA is Nepal's first wide-ranging and climate change–dedicated government document which was released to the public audience in September 2010 (Helvates Nepal, 2012; MoE, 2010). This document was submitted to UNFCCC in 2010 and it has since become a pioneer in climate change adaptation planning in the country (Bahadur *et al.*, 2017: 4). In 2011, Nepal became the first LDC to issue a national framework on Local Adaptation Plans for Action (LAPAs) to strengthen and implement the NAPA prioritized adaptation actions (GoN, 2011a). The rationale of the LAPAs is to implement the NAPA efficiently by means of influencing public participation to identify and execute local adaptation action and to integrate climate change adaptation into sectoral plans and policies. The LAPAs also ensure that the process of integrating climate change resilience from local to national planning is governed by the four guiding principles of being bottom-up, inclusive, responsive, and flexible (Patra and Terton, 2017: 16; MoE, 2011). The Government of Nepal (GoN, 2011b) endorsed the National Climate Change Policy in 2011 which supports NAPA and LAPA implementation (Maharjan and Maharjan, 2017: 1–14). The Policy specifies to "allocate at least 80 per cent of available funds for field level climate change activities" (GoN, 2011b). This Policy outlines seven principal policy objectives, of which three are specifically related to adaptation-oriented activities:

- Implement climate adaptation–related programs;
- Improve local communities' adaptative capacity and resilience to enable better use and more efficient management of natural resources;
- Improve capacity to identify, quantify, and adapt to climate risks and climate change impacts.

(GoN, 2011b; Patra and Terton, 2017: 15–16)

The Ministry of Environment serves as a focal ministry to climate change–related activities (MoE, 2010; Workshop Report, 2010). To foster a unified and coordinated climate change response, a multi-stakeholder Climate Change Initiatives Coordination Committee has started since April 2009. The three-year interim sustainable development plan of Nepal expired in June 2013 (Bhushal, 2013). This plan focused on improving environmental management and sustainable natural resources with the aim of maintaining a 39.6 per cent forest cover. The ADB is assisting in strengthening Nepal's capacity to manage climate change and the environment. The activities are expected to provide necessary information for awareness raising, training activities, and planning future projects (Rahman, 2014). To adapt to the changing seasons and flash flooding, projects have taken place over a period of three years (2004–2007), in one watershed in the southwestern area of the lesser Himalayan foothills (Nambi, 2014: 16).

The Government of Nepal is highly committed to improve the nutritional status of children and women as a foundation for future social economic growth and development (MoHP, 2004, 2006, 2010). The National Agricultural Policy of Nepal sees the need to ensure food security and alleviate poverty along with the conservation, promotion, and use of natural resources and the environment (Nepal Law Commission, 2004: 2). The Government of Nepal has developed a National Multi-Sector Nutrition Plan (MSNP) for improving maternal and child nutrition and reducing chronic malnutrition for the Nepalese people (GoN, 2012a: 12; GoN, 2012b: 3–4). In 2011, the Climate Change Adaptation and Disaster Risk Management in Agriculture: Priority Framework for Action 2011–2020 was developed, with a ten-year vision ending in 2020 (MoAC, 2011). This document was written with the technical assistance of the FAO. Its purpose is to provide a road map for the Ministry of Agriculture and Cooperatives (MoAC) to shift its approach from reactive emergency response to proactive climate adaptation and climate risk management in the agricultural sector. In May 2013 the National Nutrition and Food Security Secretariat (NNFSS) was established so as to provide technical support to the National Nutrition and Food Security Coordination Committee (NNFSC) (SUN Movement, 2014: 86). The WFP has been trying to provide food assistance in Nepal through three sub-offices with three major projects namely: the Country Programme (DEV 200319), Social Protection Livelihoods and Assets Creation Programme (PRRO 200152), and the Refugee Operation (PRRO 200136) (WFP, 2010).

Additionally, Nepal's National Urban Development Strategy (2015) identifies climate change as a major risk factor, mainly in the context of rising poverty trends in urban areas and the possibility of increased numbers of refugees moving from rural regions to urban areas due to disasters. In response, Nepal commits to improving the urban surroundings by promoting multi-hazard approaches to deal with disaster and climate change (MoUD, 2015). Nepal's Strategic Programme for Climate Resilience (SPCR) was drafted in 2011 under the leadership of the government and in coordination with ADB, members of the WB Group (IBRD, IFC), and key national stakeholders (Trabacchi and Stadelmann, 2013: 43; ADB, 2015: 36). The SPCR supports projects that will improve the resilience of watersheds in mountainous regions and reduce the vulnerability of communities and ecosystems to climate-related hazards, particularly through initiatives funded by the private sector. It has these components: Building Climate Resilience of Watersheds in Mountain Eco-Regions (BCRWME); Building Resilience to Climate-Related Hazards (BRCH); Mainstreaming Climate Change Risk Management in Development (MCCRMD); Building Climate-Resilient Communities through Private Sector Participation (BCCPSP); and Enhancing the Climate Resilience of Endangered Species (ECRES). Besides administering the BCRWME and MCCRMD, the ADB is supporting the ongoing Energy Access and Efficiency Improvement Project in lighting the way for clean and efficient energy (ADB, 2015: 38). In a response to the inadequate supply of energy, Nepal established the National Rural and Renewable Energy Program which gives priority to the development of electricity-generation facilities powered by renewable-energy sources, both on- and off-grid (ADB, 2015: 38). It is also funding the Kathmandu Sustainable Urban Transport Project to help reduce congestion and pollution in Nepal's city (Rahman, 2014). Being a signatory to the UNFCCC, Nepal agreed to take climate change considerations into account in its national development agenda (Rai and Gurung, 2005: 316–320; NPC, 2003).

An attempt to attract worldwide attention on the threat posed to Himalayan glaciers by global warming, a meeting of 21 ministers was held at Mt. Everest base camp. This meeting passed a resolution on climate change to highlight the threats especially to the glaciers and mountains of the Himalayas (Parashar, 2009; Lang, 2009). This event was a message to the world to minimize the negative effects of climate change on Mt. Everest and other Himalayan mountains. But in comparison to many developing nations who have signed the Kyoto Accords, Nepal is still far behind in developing a national adaptation plan. This result is due to its intense poverty though it receives huge amounts of funds from donors (Oxfam, 2009: 6; Bartlett *et al.*, 2010: 35). That is why Oxfam says more work needs to be done (Oxfam International, 2009). On seeing the efforts put up by Nepal, an important point to note is that the lack of adaptive measures does not necessarily mean a lack of attention to climate change in this country.

Pakistan

In 2003, Pakistan submitted the National Communication to the UNFCCC. To deal with climate change–related issues, the National Environmental Policy was set up in 2005. This Policy provides an overarching framework for addressing the environmental issues facing the country, mainly pollution of freshwater bodies and coastal waters, air pollution, lack of proper waste management, deforestation, loss of biodiversity, desertification, natural disasters, and climate change (GoP, 2005: 9). It also focuses on adaptation and mitigation policies to reduce the effects of floods, to safeguard economic growth and food security. Besides, afforestation has been given priority in the agricultural and environmental policy (Saddiqui, 2010: 74–78).

In 1995, the Government of Pakistan formed the Cabinet Committee on Climate Change to provide a policy coordination forum to deal with climate change (Iqbal *et al.*, 2014: 4). In 2004, this Committee was converted into the Prime Minister's Committee on Climate Change as a high-level Inter-Ministerial platform to forge linkages and coherence between climatic change challenges and the risks that climate change poses to national development and planning (GoP, 2017a: 14; Khan and Munawar, 2011: 4; Salman, 2011: 4). Speaking at the UN, Pakistan's former environment minister voiced concern for climate change effects on its environment, economy, and people. He indicated Pakistan's major goals to lessen climate change effects include water management, improving energy production, changing agricultural practices, tree planting for carbon sequestration, establishing a committee for research, advising, and action (Daily Times, September 26, 2007). Research training sessions have been organized to increase understanding of climate change's effects like drought and flood in the coastal areas (Ensor and Berger, 2009: 72–86: Rasul, 2012).

There is the National Conservation Strategy, 1992, whose objectives are conservation of natural resources and sustainable development (Banuri, 1993; GoP, 1993). There is also the Federal Flood Commission (FFC) in Pakistan which was established in January 1997. Its mandate includes the preparation of National Flood Protection Plans on a nation-wide basis, approval of flood control schemes and arrangement of funding for the approved flood protection works and measures, reviewing the flood damages, improving the flood forecasting and warning system (Habib and Nawaz, 2010: 131–132; MoWP, 2010). The Pakistan Environmental Protection Act (PEPA), 1997, is an act to provide for the protection, conservation, rehabilitation, and improvement of the environment, for the prevention and control of pollution and promotion of sustainable development (PEPA, 1997a, 1997b). There are also various sectoral strategies in Pakistan which mentioned the potential effects of climate change. These are the National Forest Policy,

National Energy Conservation Policy (2006), National Renewable Energy Policy (2006) and Policy for Development of Renewable Energy for Power Generation (2006), National Environmental Policy, National Water Policy (Khan, 2011: 15–17). A research study on Mainstreaming Community Based Climate Change Adaptation was launched to examine the vulnerability to climate change–induced hazards (Salman, 2014). There is the Planning Commission too. This body is responsible to prepare the National Plans for the country's main economic sectors who had established a Task Force on Climate Change that was given the job for preparing the country's climate change policy. The Task Force released its Final Report in 2010, which outlined the country's present approach to address climate change from both a mitigation and adaptation perspective (GoP, 2010b). Over the years, therefore, Pakistan has made some efforts related to climate change mitigation and adaptation.

There is a programme to integrate disaster risk reduction into the educational curricula and supporting awareness building in educational institutions. From this point of view, Pakistan is aware that all such efforts should graduate from small-scale pilot-projects to large-scale institutionalization in the country. The University of Peshawar uses geographic information systems for planning and natural resources management performs research and informs government policymakers in sustainable industrial growth, resource conservation, and environmental preservation (Habib and Nawaz, 2010: 131–132). To encourage local communities to conserve, biodiversity efforts were made. This was done with the help of the Global Environment Facility/United Nations Development Fund (GEF/UNDP) and the Pakistan chapter of International Union for Conservation of Nature (IUNC-Pakistan) (Saddiqui, 2010: 77).

Already a resource-poor country with a very high and fast-growing population, Pakistan is being permanently harmed by the crisis of climate change (Oxfam, 2009; GoP, 2010a, 2012; Park, 2013: 1–13). Certain about its dismal security environment, Pakistan approved a National Climate Change Policy (NCCP) in March 2012 (GoP, 2012; Naeem, 2013; Khan, 2013; Femia and Werrell, 2012). It has ten policy objectives and one of them is to sustain economic growth by addressing the challenges of climate change (GoP, 2012: 1). According to the Minister for Climate Change, Rana Farooq Saeed Khan, this policy aspires to ensure water security, food security, and energy security (Business Recorder, October 19, 2012). It will also minimize the risks arising from the expected increase in frequency and intensity of extreme events such as floods, droughts, and tropical storms. Further, it will strengthen inter-ministerial and inter-provincial decision-making and coordination mechanisms on climate change. This national policy is expected to advance the development of appropriate economic incentives and to encourage public and private sector investment in both adaptation and mitigation measures (Femia and Werrell, 2012). This Policy includes the term *climate-induced migration* (GoP, 2012). To Dr Qamaruz Zaman Chaudhry, who is the lead author of this policy, its major focus is the adaptation to climate change by different sectors of the national economy which comprise water resources, agriculture, livestock, human health, biodiversity, and disaster preparedness (Aftab, 2012; Khan, 2013). According to him Pakistan is in fact among the few developing countries which have prepared such a comprehensive national policy on a subject which is on the top of the global priority agenda, maybe after the war on terror (Yusuf, 2011). Jawed Ali Khan, Director General (Environment) at the Ministry of Environment, says this policy is a multi-sector approach in which the long-term project will come under the National Climate Change Action Plan – a road map for adaptation and mitigation of serious problems (Yusuf, 2011).

Besides the NCCP, there is a Framework for Implementation of NCCP 2013, and completion of a Climate Public Expenditure and Institutional Review (CPEIR) 2015, the 2017 CPEIR update, and the recently enacted 2017 Pakistan Climate Change Act (GoP, 2017a: 13). The Pakistan Climate Change Act promises to fast-track measures needed to implement actions on the ground in a country that has so far lagged on climate action. This law establishes a policymaking Climate Change Council, along with a Climate Change Authority to prepare and supervise the implementation of projects to help Pakistan adapt to climate impacts and hold the line on climate-changing emissions (Khan, 2017). Cognizant of the threats of climate change Pakistan has taken significant steps to address the issues like establishment of the Global Change Impact Studies Centre (GCISC). The Vision 2025 highlights resource scarcity issues among others, but the linkage with consequences of climate change and its negative impact needs amplification (MoPDR, 2015a; Kiran and Ain, 2017: 55). To help the government in implementing its policies, there is the Climate Change Financing Framework (CCFF) which allows the government to augment the Public Financial Management (PFM) system capabilities for responding to climate change challenges, promoting adaptive measures and minimizing negative fiscal impacts. It is expected that the integration of CCFF in the national PFM system will ensure public investment is hedged against future climate-related losses. The CCFF will provide the Climate Change Authority with a head start for devising uniform institutional arrangements and introducing fiscal rules for climate change finance (GoP, 2017a: 13).

The Ministry of National Food Security and Research (MNFSR) was created on October 26, 2011 (GoP, 2017b). This is a government department which is responsible for implementing, enforcing, developing, and executing the policy on agriculture, rice, livestock, fishing, and farming. There are other policies which are related to nutrition such as the Protection of Breastfeeding and Child Nutrition Act, the Food Fortification Act, and the Early Marriage Restraint Act (WFP, 2018). FAO is a long-term partner for the Government of Pakistan on agriculture and food security issues. Through various programmes, it has been helping over 8 million people in Pakistan. It addresses the hardships caused by high food prices in the country, launched a

US$71 million project to provide 86,295 tonnes of food and it also shows its strong focus on school food programs to assist over 450,000 girls and boys in 5,400 primary schools (Bengali and Jury, 2010: 130–131; FAO, 2017a: 20–29). The Pakistan Agricultural Research Council (PARC) and FAO are collaborating on several projects, including work on soil conservation and seed sector development (FAO, 2017b: 12).

Pakistan's Vision 2025 seeks to improve social capital, food security, and nutrition (WFP, 2018: 2). The government has three social assistance programmes to address the issues of hunger, poverty, and malnutrition: Benazir Income Support Programme (BISP), Pakistan Bait-ul-Mal (PBM), and Zakat (Haq, 2015: 3). The UN Sustainable Development Framework for Pakistan for 2018–2022 focuses on economic growth, food security, nutrition, resilience, education, productive livelihoods as well as social protection (WFP, 2018: 2). The International Fund for Agricultural Development (IFAD) supports the government in its efforts at helping people emerge from poverty, and to build resilience for sustainable food security and nutrition. The World Bank is supporting Pakistan's efforts to reduce poverty (WFP, 2018: 8). The UNICEF's draft country programme for 2018–2022 focuses on neonatal and child survival, nutrition for girls and boys, with a special focus on treating severely acutely malnourished children. The United Nations Development Programme (UNDP) works on crisis prevention and recovery, the environment, and climate change (WFP, 2018: 8).

Sri Lanka

Over 13 per cent of Sri Lanka is protected land. The government of Sri Lanka has been taking action to conserve wildlife. The Sinharaja Forest Reserve which protects rainforest in the country was declared a World Heritage Site in 1988 (Philander, 2008: 928). Though this country has a long history of conservation in the world, decades of perpetual conflict have contributed to the environmental vulnerability (Harrison, 2012: 1–53). This forces the government to identify its vulnerability to climate change (Yamane, 2003; 1). Being a small island nation, it falls under serious threat from various climate change effects, such as sea level rise and other water-related problems (MoE, 2010; McCarthy *et al.*, 2001). Thus, to administer the uncertain effects of climate change is no longer a choice for Sri Lanka. In reality, it is essential.

Sri Lanka is a signatory to the UNFCCC and the Kyoto Protocol when it ratified the two treaties in 1993 and 2002 respectively (Daily Mirror, April 28, 2017). As a signatory, it is obliged to prepare a national communication report to the UNFCCC. Following the guidelines provide by the UNFCCC, the report was produced, the Initial National Communication (INC), and submitted to the UNFCCC in October 2000 (Yamane, 2003: 5).

Owing to its vulnerability to climate change particularly to the expected increases in floods, droughts, landslides as well as sea level rise, this National Communication identifies these various areas as the most important (MoFE,

2000; Baba, 2010: 12). As the country is frequently facing these disasters, the National Council for Disaster Management (NCDM) was also established under Disaster Management Act No. 13 of 2005 (SLDM Act, 2005; MoDM, 2005). This is a high-level inter-ministerial body which provides direction to disaster risk management work in the country (Hettiarachchi, 2012: 139). In the same year, the Disaster Management Centre (DMC) was established under the NCDM to be functional under the president of Sri Lanka (USAID, 2007: 4). In February 2006, the Ministry for Disaster Management and Human Rights (M/DM&HR) was created with the subject of Human Rights listed under its purview (Hettiarachchi, 2012: 139–140). The National Disaster Management Policy (updated in 2013) elaborated the act, using an integrated approach, the "Total Disaster Risk Management System" (MoDM, 2013).

Sri Lanka tries to contain the escalating rise of the sea level when it builds new coastal structures by promoting alternative timber such as rubber as well as ensuring conservation of natural forests and banning deforestation for commercial practices (Baba, 2010: 4–16). Another policy where this country responds to climate change is the National Climate Change Adaptation Strategy for Sri Lanka (NCCAS) (MoE, 2010b). This policy puts in place a prioritized framework for action and investment for the 2011–2016 period. It aims at systematically moving Sri Lanka and its people towards a climate change resilient future (MoE, 2010b: 27). The document provides a thorough overview of the country's adaptation priorities and suggested actions. It identifies agriculture and fisheries, water, health, urban development, human settlements, economic infrastructure, and biodiversity and ecosystem services as the most vulnerable sectors to the effects of climate change (MoE, 2010b). The NCCAS developed through a three-phase procedure. These are: (i) Preparing Sector Vulnerability Profiles (SVPs) for key sectors, which outlined the current status of the sectors and the main potential climate change risks facing Sri Lanka; (ii) Adopting a participatory process for the above through working groups comprising a range of stakeholders, as well as individual consultations with key people, to refine content of the SVPs and to identify and prioritize areas for future investment; (iii) Synthesizing these sector-based analyses into one cohesive national adaptation strategy, which includes a clear programme for priority action and investment based on clearly defined strategic priorities (MoE, 2010b: 3–4). The NCCAS supports Sri Lanka's national development strategy as articulated in the Mahinda Chintana (Mahinda Vision). The NCCAS is organized around five strategic areas, under which priority adaptation measures are identified as follows:

i) Mainstreaming climate change adaptation into national planning and development;
ii) Enabling climate resilient and healthy human settlements;
iii) Minimizing climate change impacts on food security;

iv) Improving climate resilience of key economic drivers; and
v) Safeguarding natural resources and biodiversity from climate change impacts.

(MoE, 2010b: 7)

Another document which is in the same direction with the NCCAS is the National Climate Change Policy of Sri Lanka in 2012 (MoMDE, 2016a). Together they elaborate the national vision and strategic priorities with regard to facing the threat of climate change in Sri Lanka. The Climate Change Secretariat (CCS) which is headed by the Director of the Climate Change Division has taken up a broad national approach to address climate change challenges that are also a development issue of Sri Lanka. The Ministry of Environment and Natural Resources (MENR) is responsible for all national policy-level activities on climate change. The CCS which is headed by the Director of the Climate Change Division, MENR, adopts a comprehensive national approach to contribute towards local, regional, and global efforts in combating climate change scenarios into national sustainable development plans. To promote better decisions by means of wider inputs, to integrate diverse viewpoints, and to bring together the principal actors, the National Advisory Committee on Climate Change (NACCC) was created in 2008 under the MENR and all activities pertaining to climate change will be coordinated by NACCC (Mall and Kumar, 2014: 124–125).

The Second National Communication (SNC) was submitted to the UNFCCC Secretariat in 2011 with GEF and UNDP support, and expands on the strategies and priorities initially outlined in the INC. The National Climate Change Policy of Sri Lanka (2011) was developed to provide high-level guidance and direction on climate change response for all stakeholders, with a particular policy focus on environmentally friendly economic development (MoERE, 2011).

The National Adaptation Plan for Climate Change Impacts in Sri Lanka (NAPCCI) was prepared in line with the broad set of guidelines set forth by the UNFCCC for the development of national adaptation plans (MoMDE, 2016b). The NAPCCI covers adaptation needs at two levels, namely: adaptation needs of key vulnerable sectors and cross-cutting national needs of adaptation. Nine vulnerable sectors are identified in the consultative process are (i) food security; (ii) water resources; (iii) coastal and marine sector; (iv) health; (v) human settlements and infrastructure; (vi) eco-system and biodiversity; (vii) tourism and recreation; (vii) export agriculture sector and industry; and (ix) energy and transportation (MoMDE, 2016b: 48–51). Broader stakeholder consultation adopted in the preparation of the NAPCCI has helped to identify adaptation needs of each vulnerable sector based on logical criteria involving projections, vulnerabilities, impacts, and socio-economic outcomes (MoMDE, 2016b). The National Climate Change Adaptation Plan of Sri Lanka 2015–2024 which incorporates health sector adaptation was developed with all relevant stakeholders by the CCS (WHO, 2015: 6).

There are other national-level plans and strategies in place in Sri Lanka, including INDC and SDG Action Plans, as well as roadmaps to implement them (MoMDE, 2016c; MoSDWRD, 2018). Under the INDCs, for example, Sri Lanka expects to reduce GHG emissions from the energy sector by 20 per cent against business-as-usual projections. It has already developed and implemented certain actions towards this. One of these is the Battle for Solar Energy project, and Sri Lanka has already exceeded its solar energy–related INDC target (Pallawala, Muller and Woods, 2018).

Furthermore, there is the National Environment Act (NEA), 1980 (amended in 1988), Coastal Conservation Act, 1981, Fauna and Flora Protection Ordinance (amended 1993), and the National Policy on Industry and Environment, 1996 (REAP-IUCN Asia, 1999: 20). There are several environmental policies, legal enactments, and plans which contain provisions that will contribute in reducing or mitigating the effects of climate change. These are the National Forestry Policy (NFP) of 1995, the National Policy for Wildlife Conservation (NPWC), the Agriculture Policy (AP) of 1996, the Energy Policy of Sri Lanka (EPSL) of 1997, the National Transport Policy (NTP), and the National Policy on Air Quality Management (REAP-IUCN Asia, 1999: 20). Sri Lanka has proposed relocation of coastal communities due to environmental events (Rabbani, Shafeeqa and Sharma, 2016: 30). Such are the existing policies that have a direct bearing on climate change and other associated confrontations in this country.

The Government of Sri Lanka is committed to ensure optimal nutrition for its citizens. In order to achieve optimum nutrition for all, the National Nutrition Policy (NNP) was proposed (MoHN, 2010: 1). The policy was meant to provide overall guidance for the development of national strategic plans of action for nutrition activities. The NNP upon adoption will serve as the base document in which the strategic approaches will be developed, leading to the phase of implementation (MoHN, 2010: 1). The NNP was adopted by the Government of Sri Lanka and put under the responsibility of the Ministry of Health. In 2012, Sri Lanka joined the Scaling Up Nutrition (SUN) Movement, a movement which tries to eliminate all forms of malnutrition, based on the principle that everyone has a right to food and good nutrition. Ever since, it has been able to make considerable effort and progress to mainstream nutrition in a multi-sectoral approach and in partnership with the donor community. In December 2013, the president launched a Multi-sector Action Plan for Nutrition (MsAPN) seeking to obtain specific objectives in terms of nutrition between 2014 and 2016 (IFAD, 2017: 40; NNC, 2013). This is because its target is to reduce the prevalence of under-nutrition, anaemia, and stunting amongst key population groups, as well as to improve food security and provide access to safe water, sanitation, and hygiene to households-at-risk.

In Sri Lanka, FAO is helping in providing technical expertise in the agriculture, fisheries, and livestock sectors. It is also lending its help to the Sinhalese farmers who are struggling with lack of knowledge in cattle production and feeding techniques (Sriskantharajah, 2014). The WFP is supportive to the National School Meals Programme (SMP), which is one of the main social-protection programmes in the most war-affected districts and has been helping to restore the education system (Lister *et al.*, 2017: vii). Due to chronic food insecurity and recent food security shocks, a few districts like Jaffna, Vavuniya, Mannar, Killinochchi, and Mullaitivu are the priority operational areas for WFP (Vhurumuku *et al.*, 2012: 1).

Considering its capacity, the approach which Sri Lanka has pursued is indeed encouraging, though many other measures pertaining to climate change and its related effects are still necessary to be taken up. Besides the efforts being put up at the national level, the effective adaptation strategies required more coherent, coordinated policies and cooperation among governments, civil society, and the private sector. It has been acknowledged that poor governments need support from other international agencies to provide appropriate capacities and resources (Lagos and Wirth, 2009: 22). So, being a developing country Sri Lanka still needs international assistance to support adaptation in the context of national planning, though the bulk of responsibility it will have to do on its own.

III Conclusion

As climate change–related effects are occurring across the SAARC countries and across many sectors of their economies, these developing countries are preparing to lessen the effects that are expected to occur. Recognizing the danger, they have been actively involved in building national policies to respond to such threats. There are a growing number of policies, programs, and projects which have been initiated to lessen the exposure of people and communities in South Asia to climate change–related threats. These countries are coming forward with their own policies and initiatives to prepare for and respond to the inevitable effects of climate change. This include formulating policies and strategies, supporting research on climate change impacts and adaptation, capacity building and setting up dedicated funds to tackle climate change. Some governments are allocating sizable amounts of their own scarce resources to set up necessary institutions and integrate climate change into sector interventions. Looking at the efforts that have been developed within these countries, it shows that its members are committed to move ahead to do their part in conserving nature and warding off the adverse effects of climate change. No doubt the arduous work lies ahead; perhaps, their attempts are an important first step towards the right direction. A number of climate change adaptation strategies have been raised in the region. As discussed previously, at the national level, all of these South Asian LDCs have prepared their own NAPAs. Bangladesh, considered to be one of the most active countries in addressing climate change, released a NAPA in 2005 (updated in 2009) and a BCCSAP in 2008. By 2010, Bhutan, the Maldives, Afghanistan, and Nepal released their NAPAs as well. Nepal even developed a LAPA as a bottom-up approach. With the approval of Pakistan's cabinet in March 2012, the NCCP was accepted in principle, ratified in September 2012, and officially launched in February 2013. Sri Lanka had developed the NCCAS for 2011–2016 and the NAPCC of India is in operation. These plans are the cornerstones for climate change adaptation in this disaster-prone region facing the unpredictable effects of climate change. By showing their willpower to battle the common challenge, together they may be able to soften their past differences as well. The next chapter will discuss the problems faced by SAARC as a regional body in addressing climate change and the future prospects.

6 Climate change and regional cooperation in SAARC

Problems and prospects

I Introduction

To address scientific and policy issues on the matter of climate change, the policymakers, climate scientists, and environmentalists have been meeting throughout the 1980s and 1990s and continued till today (Hague Declaration on the Environment, 1989a: 1308–1310; UNEP, 2007; Olivier *et al.*, 2016; Brack and Gray, 2003: 18–33; Reusswig and Lass, 2010: 155–81; Gupta, 2010a: 636–653). Owing to the awareness of its negative effects they have sincerely stressed the need of a world-wide action (Schenck, 2008: 323; Cronin, 1995). So the needs for a fundamental duty that every state has to carry out to preserve the ecosystem have been recognized. The right to live in a viable global environment has also been discussed. Decisions have been made to take a collective tangible action on climate change (Bodansky, 1999: 596–604). The nation-states have realized the unfeasibility of solving climate change–related problems by an individual action alone. The policymakers, climate scientists, and environmentalists have understood that an international institution would be a suitable catalyst to promote cooperation. To gather in an international institution it was assumed even sovereign states having divergent interests must cooperate because none can attend to it alone (Desai, 2010: 17). Even if collective action may be difficult to realize, the world leaders have been convinced that such an idea will be possible for the global community to achieve (Esty and Mendelsohn, 1998). Therefore, the international community has been actively engaging to find solutions and promote global cooperation on climate change. So far, however, cooperation on climate change has revealed a compromise is hard to reach and maintain because misunderstanding exists amongst the states (Ohl, 2003; FitzRoy and Papyrakis, 2010). Confusion exists even for those states who consider climate change a priority. Misunderstanding goes on as countries see environmental policies as purely domestic matters (Porter and Brown, 1996). Thus, the existing agreed-upon global climate change objectives are yet to be realized. In other words, a global response to the threat of climate change has been less effective so far.

Predictions have projected that the negative implications of climate change would be far-reaching and largely irreversible (Brunner, 2001: 1–33;

UNEP, 1997: 141–142). To deal with it, scholars say the approach must start from the farms where food is produced to the homes where it is consumed (Parry *et al.*, 2009: 55). So the states need to develop cooperation efforts to study and collect unswerving data on the effects of climate change to facilitate adaptation. On realizing the need for a regional cooperation to deal with climate change in the region, SAARC, which is a platform for the people of South Asia, has undertaken thorough efforts to tackle its menaces (Majaw, 2012: 73). The main concern is the increasing frequency and strength of climate change–related disasters in South Asia, which has stimulated the SAARC countries to enter into regional cooperation to deal with it. Its member countries also have shown their fervour in dealing with it in their capacities. In Chapter 4, it is seen that there are numerous policies/initiatives of SAARC which are supposed to help and complement each other's advantages and address their disadvantages in the process of adaptation to and mitigation of climate change in South Asia. In Chapter 5, there are various national efforts which have been approved and implemented by the individual SAARC members in mitigating the risks. Through these national strategies the SAARC countries have been taking path-breaking adaptation actions on the ground, while the questions remain.

II Problems faced by SAARC as a regional body

Although SAARC was formed for political and economic cooperation, its scope has been expanded to include, *inter alia*, the preservation, protection, and management of fragile ecosystems. Through its various policies/initiatives, SAARC tries to have a role to address the crises of climate change. Climate change adaptation within SAARC has taken centre stage through the guidelines written down in its policies/initiatives. The projects under SAARC are identified under various technical committees and working groups. But there are many problems the SAARC countries have to face which ultimately result in lack of progress in their implementation. As a regional body the SAARC has its own problems to deal with. The subsequent paragraphs will portray those policies/initiatives that SAARC as a regional cooperation of South Asia is working to promote and the difficulties it faces to make these policies/initiatives promising.

The SAARC has the Dhaka Declaration on Climate Change (2008) and the SAARC Action Plan on Climate Change (2008) which was adopted by a Ministerial Meeting on Climate Change in Dhaka on July 3, 2008. But the Dhaka Declaration on Climate Change and the SAARC Action Plan on Climate Change did not portray this regional cooperation as a single entity at the world climate summit in Copenhagen (Prakash and Kalita, 2010). The summit at Copenhagen could have been an opportunity for the South Asian countries to come up with a common agenda to address the challenges regarding climate change as a regional group. While studying the SAARC's policies/initiatives, it is seen this regional cooperation group takes climate change as one of its regional agenda, but it is difficult to understand why SAARC as a group of South Asian countries did not adequately represent in international environmental negotiations such as in Copenhagen (Roul, 2014). The SAARC Action Plan on Climate Change (2008) and the Thimphu Statement on Climate Change (2010), which was adopted during the Sixteenth SAARC Summit, provide the necessary strategies for mitigating the effects of climate change in the region. So far, very little progress has been made in implementing these action plans on the ground (White, 2015: 1–3; Khan and Arora, 2014). Both these plans are not flourishing as expected. The Ninth Meeting of the SAARC Environment Ministers (2011) extended the timeline for the SAARC Action Plan on Climate Change by three years i.e. from 2011 to 2014. Its accomplishment is yet to be seen.

The most significant step taken by SAARC was at the Sixteenth Summit at Bhutan (2010) when it adopted the Thimphu Statement on Climate Change (2010) with the theme "Towards a Green and Happy South Asia", through which SAARC would undertake a number of initiatives in a focused manner (The Hindu, April 28, 2010). It is worth mentioning here that SAARC approved the Thimphu Statement on Climate Change in order to further toughen and build up a regional cooperation to address the effects of climate change to which the region was susceptible. The Thimphu Statement on Climate Change agreed to review the implementation of the Dhaka Declaration on Climate Change and SAARC Action Plan on Climate Change and ensure its timely implementation. The main intentions of plans like the Dhaka Declaration on Climate Change (2008) and SAARC Action Plan on Climate Change (2008) as well as the other commitments made in Thimphu during 2010 SAARC Summit are meant for combating climate change (SAARC, 2015). In their national capacities, all South Asian countries have adopted adaptation measures to reduce the effects of climate change. But there is a serious shortage of strategies at the national level (Ahmed and Suphachalasai, 2014: 103). To realize its ambitions, SAARC countries have not yet agreed on how to finance the SAARC Action Plan on Climate Change and other commitments made in Thimphu during the 2010 SAARC Summit as they cannot generate enough funds on their own. The poverty of its members makes it difficult for them to generate enough funds at their own expense. The existing initiatives are unlikely to see quick action although the Kathmandu Declaration signed at the Eighteenth Summit (2014) underscored the urgency for effectively recommendations of the Thimphu Statement on Climate Change (2010). On the climate crisis, reliable data is lacking. In SAARC Member States, much of the climate change information available lacks socio-economic data. Without this required information, understanding of the varying susceptibilities of the communities will be considerably limited. Thus, it will be difficult to translate climate scenario information into adaptation strategies. Some of the SAARC countries do not have sufficient trained manpower and lack technologies required to realize its policies/initiatives.

The idea of establishing the SAARC Agreement on Rapid Response to Natural Disasters had been voiced since the Fifteenth SAARC Summit in Colombo (2008). The SAARC "Agreement on Rapid Response to Natural Disasters" became a reality at the Seventeenth Summit (2011). It is indeed a major development for the region because the agreement is not just a statement but also a charter of great importance. Its main objective is cooperation of member states in developing and implementing measures to reduce the consequences of disasters and to identify disaster risks, develop monitoring, assessment, and early warning systems and making standby arrangements for disaster relief and emergency response, exchanging information and providing mutual assistance (Zafarullah and Huque, 2015). Moreover, the agreement focuses on disaster preparedness, emergency response, and institutional arrangements, stipulates that an affected country will be primarily responsible for responding to disasters and external assistance will only be extended when requested. Though it was signed at the Seventeenth Summit, it is still in the process of being ratified by all the SAARC members. The Kathmandu Declaration or Eighteenth Summit (2014: 3) put emphasis on "the relevant bodies/mechanisms for effective implementation of SAARC Agreement on Rapid Response to Natural Disasters, SAARC Convention on Cooperation on Environment and Thimpu Statement on Climate Change, including taking into account the existential threats posed by climate change to some SAARC member states".

Established by the Thimphu Statement on Climate Change, the 2010 IGEG.CC is supposed to meet twice a year with the mandate to monitor and review its implementation and to make recommendations to facilitate its implementation. Nevertheless, meetings of the IGEG.CC have never taken place regularly as intended to be from its creation in 2010. The First Meeting of the IGEG.CC was held in Colombo on June 29–30, 2011 (SAARC, 2015: 3). It reached a broad consensus on an agreement that aimed to put in place an effective mechanism for rapid response to natural disasters (Radhakrishnan, 2011; SAARC, 2011). A number of recommendations have been made to ensure timely implementation of the Thimphu Statement. The second meeting took place on April 16–17, 2012. It was intended that the Report of the Meeting would be submitted to the Fifth Meeting of Technical Committee on Environment and Forestry and Tenth Meeting of the SAARC Environment Ministers.

The SAARC Action Plan on Climate Change (2008), which covers from 2009 to 2011, focuses on seven thematic areas, from adaptation of climate change to regional stance for international negotiations. It emphasizes policies and action for climate change mitigation, technology transfer, financing and investment mechanisms, education, training and awareness, monitoring, assessment, and management of impact and risks due to climate change. It has yet to reach any consensus on mobilizing funds for the implementation of such a plan. The SAARC ministerial meeting on climate change which was held in Dhaka on July 1–2, 2008, had suggested diverting funds from the SAARC Development Fund, apart from seeking funds from donors such as the Asian Development Bank (Habib, 2008). There are also members of SAARC who have committed themselves to promote programmes for advocacy, awareness, and mitigation of climate change, including incorporation of science-based educational material in educational curricula. It is seen that the SAARC Environment Action Plan directed all countries to formulate their own national environment action plans and submit reports to the SAARC Secretariat for assimilation and developing a coherent regional plan for South Asia (Goel, 2004: 41–42) because the need for concerted shared efforts at meeting challenges has become critical. But SAARC has been in a difficult situation to find meaningful regional cooperation on environmental issues (Swain, 2002: 61–85). It is unfortunate to note therefore that many of the SAARC policies/initiatives have not been entirely implemented (Wijenayake, 2014).

It is worthwhile to note the four centres of SAARC, i.e. the SAARC Forestry Centre (2007) in Bhutan, Disaster Management Centre (2006) in New Delhi, Coastal Zone Management Centre (2005) in the Maldives, and Meteorological Research Centre (1995) in Dhaka, each have a respective role to play with regard to climate change mitigation (White, 2015: 9). But the Forty-ninth Meeting of the Programming Committee comprising joint secretaries of the SAARC countries decided to merge them into one centre. The reason for merging was the four centres could not produce the required input. So, a decision was unanimously taken (White, 2015: 9; The Kathmandu Post, November 22, 2014) to merge these four centres to form the SAARC Environment and Disaster Management Centre (SEDMC). These centres were created to provide credible institutional support for taking up climate change and disaster risk reduction issues in the region in their relevant authority. But as they lack credibility in their efforts, the decision was taken to dissolve them (White, 2015: 9) with the intention of having a better mechanism in dealing with climate change. Also, there seems to be an expectation that when the SEDMC is established, the new centre will be more effective as it will be a more focused and concentrated centre to respond to climate change–related disasters. Another reason is that the sharing of reliable data could be easier in a centralized centre. That is why it is desirable for SAARC to establish the SEDMC. However, the SEDMC has yet to be operationalized. Therefore, when the devastating earthquake hit Nepal in April 2015, it posed a major challenge not only to Nepal but also raised questions about the existence of SEDMC and the preparedness of the SAARC countries to meet such challenging disasters in the future. Since the SEDMC was not yet established, the response to deal with the immediate crisis in Nepal was left to the SDMC, which was officially dissolved in November 2014. In spite of that, the already bungled SDMC was ineffective to help Nepal. According to Nihar R. Nayak (Nayak, 2015), the ineffectiveness of SDMC is that its activities currently appear to be more that of a consulting institute whose authority is only to coordinate with National Focal Points

(NFPs) in the member countries in the normal course. The SDMC did not have any operations wing or field activities during crises. When the deadliest disasters like the Nepal earthquake happened, despite having spent large sums of money on research and documentation, the SDMC only circulated a brief composed of information from media sources and details of Nepalese agencies involved in disaster management (Nayak, 2015). It did not have any visible role in coordinating between the NFPs of member countries and others for smooth rescue and rehabilitation operations. SAARC's problem to respond to the Nepal tragedy exposed the absence of its collective response machinery to mitigate the common and immediate threats to the region. Again, the 2018 flood in Kerala, India, has demonstrated the ineptness of SAARC when it comes to a large-scale disaster. Only the National Disaster Response Force (NDRF) of India, Indian Navy, Air Force, Army, and Central Armed Police Forces (CAPFs) were engaged in taking on the gigantic task of evacuating flood victims stranded on rooftops, highlands etc. of the floods that devastated regions of Kerala (*India Today*, 2019b; Gupta, 2018; Kumar, 2018). Similarly, SAARC failed to help the flood victims of 2019 in South Asia (Yeung, Pokharel and Gupta, 2019; *India Today*, 2019a). Over the years, SAARC has produced regional guidelines, conducted technical trainings, and developed a mechanism for collective emergency response for ratification by states through SDMC. Despite these efforts, there is little sense that SAARC has been able to make use of them in any meaningful way.

The leaders of SAARC have been apprehensive of the fact that millions of South Asians are living in extreme poverty and many of those poor people are malnourished and become more susceptible to the diseases that are caused due to hunger. They have realized droughts, floods, storms, and other disasters triggered by climate change are rising in frequency and severity over the last three decades, increasing the damage caused to the agricultural sectors of SAARC countries and putting them at risk of growing food insecurity. Therefore, the Meetings of SAARC Agriculture Ministers acknowledge the scale of the challenges related to food insecurity and poverty eradication in South Asia. To minimize climate change's effects upon food security, the SAARC Food Bank is a mechanism which has been undertaken by SAARC members to enhance food security in the region. The SAARC Food Bank offers a device for SAARC members to sustain each other both in times of emergencies due to natural disasters and even during normal conditions (Bhattarai *et al.*, 2015: 42–45; Sharma, 2011: 305–314). Despite being ratified several years ago, this machinery continues to remain only notional, mainly due to the lack of political will and other differences within the regional cooperation (Chatterjee and Khadka, 2011: 111; Dawn, December 26, 2007). To date, no one knows where this bank is physically positioned (Pant, 2014: 23). Maybe no one has a clear idea about the institutions involved in its distribution and the transportation mechanisms.

The devastating earthquake of April 2015 in Nepal led to a food crisis. During the crisis, a decisive regional intervention was needed to check the food crisis and hunger faced by the Nepalese. But SAARC failed to invoke the operationalization of the SAARC Food Bank which is supposed to provide emergency supplies to a nation facing a crisis resulting from a natural calamity such as floods, draught, or earthquake. Being the Chair of SAARC during the crisis, it is seen Nepal could not make use of the provision of the SAARC Food Bank in dealing with its unprecedented food insecurity and devastating crisis after the earthquake in April 2015. Thus, the SAARC Food Bank is fraught with problems and inefficient (Ahmad, Iqbal and Farooq, 2015: 30; Pant, 2014: 15–18). In fact, a few months before the earthquake in Nepal, i.e. at the Eighteenth SAARC Summit (November, 2014), the leaders directed "to eliminate the threshold criteria from the SAARC Food Bank Agreement so as to enable the member states to avail food grains, during both emergency and normal time food difficulty" (Lama, 2015; Eighteenth SAARC Summit, 2014: 3; Rahman, Bari and Farin, 2017: 14). This shows that SAARC has no food reserve available to its members even during a time of national crisis as the member countries are unwilling to meet their commitments in providing food to Nepal. Similarly, when Kerala was facing the worst flooding in 2018, various organizations across India have come forward to provide food to help the flood victims in the hour of need (Nambudiri, August 20, 2018; Business Standard, August 20, 2018), but the SAARC Food Bank remains silent. So, it can be stated that SAARC has very much prioritized climate change and other natural disasters in its action plans, but on paper only. While lecturing on "The Challenge of Food and Nutrition Security in SAARC Region", Venkatesh B. Atreya (advisor on food security to MS Swaminathan Foundation, Chennai) said that while most of the SAARC members have experienced relatively rapid GDP growth, this has not always translated into significant improvements in food security (The New Indian Express, April 21, 2015; Bhattarai *et al.*, 2015: 54–56). The vision of its members in promoting sustainable agriculture in the region, which may well be possible by the application of bio-technology, is still an illusion. The lack of food will create severe challenges to the livelihood of the people in the region, particularly the poor who will bear the brunt of this security threat.

A large number of cross-border migration in this region is "illegal", "undocumented", or "clandestine", ranging from simple border crossings to organized trafficking and smuggling of people (Wickramasekara, 2011; Srivastava and Pandey, 2017: 45), thereby leading to exploitation (including sexual exploitation) and the violation of migrants' human rights. To curb this problem, SAARC has framed its strategies like the SAARC Social Charter (2004, 6) and Kathmandu Declaration 2014 (Eighteenth SAARC Summit, 2014: 5). These are the policy documents which confirm the SAARC recognizes climate change, environment and disaster issues as the causes of migration in the region. However, none of the SAARC policy documents provide comprehensive action plans or the guidance to address migration due to climate change and environmental disasters (Srivastava and Pandey, 2017: 45; Rabbani, Shafeeqa and Sharma, 2016: 30).

Given the SAARC's track record for regional cooperation, a principal challenge facing the region is the ill-tempered relation between nations with a negative bearing on all policy areas, including climate change. Some of its member countries have a tendency to take a bi-partisan approach in dealing with environmental problems, bypassing the regional body (Zafarullah and Huque, 2015; Sharma, 2011: 305; Karim, 2014: 302; Sidhu and Sandhu, 2014: 6). This misunderstanding at the regional level overshadows the implementation of the existing plans. Although various environment-related policies or initiatives have existed since the 1990s, the governments of SAARC have not revised or updated them despite so much serious thinking and discussions on climate change. As a result, Ashok Swain says due to the lack of unity and commitment to a stronger regional arrangement among its member states, SAARC has to face the obstacle of finding meaningful regional cooperation on climate change (Swain, 2002: 61–85).

III Global inequality sways SAARC's climate change adaptation

It is seen that inequality of economic development makes it harder for developed and developing countries to trust each other in establishing mutually acceptable agreements (Parks and Roberts, 2008: 621–648). Such global climate change disagreements between the developed and developing countries seem to have their role to play in South Asia. The SAARC countries are perpetually emphasizing the overriding importance of socio-economic development and poverty eradication in the region (Oli *et al.*, 2014). Perhaps their slow performance in this regard need not be blamed; as developing countries, they lack the means to deal with it. Indeed, all these countries are poor and weak. Even India, which is the highest polluter of GHG emissions in the region, still needs help such as the transfer of technology and financial assistance from developed countries for adaptation and mitigation to climate change (Gopalakrishnan and Gopal, 2018). Likewise, there is simply no proof to offer that SAARC countries will suddenly become comfortable and be in a position to install more environmentally friendly technologies any time soon as most of them are very poor. The nonexistence of an effective global agreement to bring both the rich and poor countries together to protect them all from climate change effects seems to be the determinant for the ineffectiveness of SAARC.

Though SAARC has several policies/initiatives, meanwhile it is also closely following the events taking place at the COPs, particularly the question over the Kyoto Protocol. This is the reason which makes its members vehemently welcome the Kyoto Protocol at the Tenth SAARC Summit in 1998. As a regional body of a poor region, SAARC never stops urging the rich countries to ratify the Kyoto Protocol and to take swift action in implementing commitments laid down in the treaty (Khatun and Hossain, 2012: 29). Its members keep demanding an equitable distribution of entitlements for emissions as proposed in the Kyoto Protocol. Being poor, its member countries stress the importance of the principles of equity and differentiated responsibilities (Liberatore, 1997: 119; Guérin and Wemaere, 2009). For these countries, any effort at addressing climate change must take into account the historical responsibility and must be in accordance with the principles of the UNFCCC, the Kyoto Protocol and the Bali Action Plan which, centred on the four main building blocks – mitigation, adaptation, technology, and financing (Allan *et al.*, 2017: 2; Majaw, 2012: 76; McGovern, 2006; Averchenkova, 2010: 2–5). This shows the SAARC or any of its members will not be able to sincerely deal with climate change until and unless the developed countries also pursue it eagerly (Gopalakrishnan and Gopal, 2018; Das and Bandyopadhyay, 2015: 40–54; Common SAARC Position, 2010). The SAARC expected tangible action to materialize at the global annual talks when parties shared a vision for long-term cooperative action (UNFCCC, 2010). Some experts still put forth suggestions where developed countries should commit to global partnerships that will help less developed nations to build the capacity to better manage climate change effects (Bialos *et al.*, 2008: 293; Wijaya, 2014: 5). Therefore, despite the presence of various policies/initiatives, there is still a lack of evidence to show the adaptation programmes to address climate change in South Asia will be successful.

IV Problems faced by SAARC countries to deal with climate change

It is clear that the developing countries of SAARC are hard-hit, and the policymakers of these countries have realized climate change is real and one of the greatest threats faced by the people of the region. These countries have begun deliberations on climate change and also evolved national strategies to adapt to the changing climate and emerging risks. So, it is better to take a look at the workings of these national strategies in the SAARC countries. Adaptation measures are there in Nepal, both at local and national levels. The latter is usually working within the framework of NAPA (Ojha *et al.*, 2015: 427; Vivekananda, 2010: 8). But there are gaps in current adaptation actions in this country. The ongoing activities do not appear to address the issue of vulnerability to climate change in the energy sector, which in Nepal is inextricably linked to the water sector (Islam, Hove and Parry, 2011: 135). At the local level, the process of formulating LAPA has not been easy (Ojha *et al.*, 2015: 427). Local communities do not fully understand the problems their areas are facing from climate change (Kumar, 2015). The country is blessed with an abundant water supply but the number of communities that are in the grip of severe water scarcity are increasing day by day (Baishyal, 2017). Owing to the lack of management and the pollution of rivers and aquifers, people living in Kathmandu also suffer the consequences. The burden of malnutrition and diarrhoeal diseases is expected to increase under climate change scenarios (WHO, 2016).

The Allachy Trust UK has been supportively working in the Chitwan district by identifying and developing adaptation and coping strategies since

2004 (Shrestha, Dhakal and Rai, 2007). Nepalese workers and their trade unions are committed to fighting climate change by protecting Nepal's beautiful forests, and concrete actions have been taken to reduce deforestation. For over 30 years, Nepalese communities have been working hard to manage their forests to assist in protecting their environment (Timilsina-Parajuli, Timilsina and Parajuli, 2014: 5; Webb and Prasad, 2001: 146–157; Basnet, 2009; Ojha, Persha and Chhatre, 2009). Even the government is doing the same. But it is found that the capacity to monitor and address the crisis appears to be inadequate. Like many developing countries who have signed the Kyoto Protocol, Nepal is far behind in developing a national adaptation plan, including its NAPA for the UNFCCC (Bartlett *et al.*, 2010: 17). There is a lack of political contestation around climate policy (Ojha *et al.*, 2015: 427) because climate policy processes have not really been a matter of concern in the national political arena. Allegedly, there is a dearth of capacity at all levels of the bureaucracy, from extremely isolated local communities to leaders in Kathmandu (Nightingale, 2017: 11–20). Very little understanding exists as to how potentially detrimental the effects of climate change will be for Nepal's economy. With regard to climate change–induced migration, there is no official document mentioning it, not even the NAPA. Certainly Nepal has the Foreign Employment Act, 2007, Foreign Employment Rules, 2008, and Foreign Employment Policy, 2012 (Rabbani, Shafeeqa and Sharma, 2016: 176). All of these policies are focused on the protection of migrant workers working abroad. These policies do not say anything about the climate change–induced migration.

Afghanistan is a signatory to the three Rio Conventions – the UNFCCC, the UNCBD and the UNCCD (World Bank, 2005:114). Despite climate change presenting a significant threat to its cross-sectoral development, the 2008 ANDS (ANDS, 2008) and Agricultural Sector strategies do not clearly recognize climate change as a real threat to the country. Its Water Sector Strategy (GIRA, 2008) does not make explicit reference to climate change, though drought and decades of conflict are key drivers behind the reduction of irrigated land. Perhaps the number of existing measures contained within Afghanistan's strategies can be classified as adaptive, but they are without clear assessments of climate thresholds. Nowhere are such effects analyzed in the context of a larger process (GIRA, 2008). So, it can be asserted that climate change is yet to be appropriately mainstreamed in the country's policy making (Savage *et al.*, 2009b: 23). Numerous barriers to adaptations such as conflict, weak policy and legal frameworks, and low capacity and awareness are likely to remain challenging in the future (Islam, Hove and Parry, 2011: 43). With the approval of the Afghan Ministers Cabinet on October 7, 2015, the Ministry of Economy (MoEc) has been designated as the lead ministry and focal point to take the lead in coordinating, planning, monitoring, and reporting on SDGs (Najafizada, 2017: 13). In July 2017, the Government of Afghanistan reported its progress towards the SDGs at the UN High Level Political Forum. But, the official SDG implementation plan of Afghanistan does not cover SDG 16 (Nanayakkara, 2018). Afghanistan still has to pay extensive attention to environmental issues in the coming years. Afghanistan needs more than 17 billion Afghanis (US$235 million) in its annual budget to combat climate change. Being poor, the Afghanistan government currently has only 100 million Afghani (US$1.4 million) in its budget annually to cope with climate change (Ikram, 2018). This country has yet to receive funds from the GCF- financial mechanism created under the UNFCCC in 2011 to promote low-emission and climate-resilient development pathways by providing financial resources to developing countries to limit or reduce GHG emissions and adapt to the impacts of climate change (NEPA and UNEP, 2015: 15).

Bhutan's preparation for NAPA is considered as an opportunity to look at the country's climate change–related vulnerabilities (RGB, 2006b). Though this NAPA noted about the rural-urban migration in Bhutan, it doesn't mention the migration issue relating to climate change and environment. As per expectation the National Sustainable Development Strategies (NSDS) of Bhutan must consider issues such as climate change, disaster management, and energy sustainability in order to guide sectors in the formation of their environment chapters for the Tenth FYP (RGB, 2009a: 2). Surprisingly, the main document (volume 1) of the draft of the Tenth FYP (2008–2013) did not make specific reference to climate change. Even its NEC does not have staff assigned to climate change issues on a full-time basis (RGB, 2009a: 6–7). Thus, it is suggested that national instruments designed to respond to immediate and urgent climate change adaptation need to be operationalized through targeted projects and activities that will contribute towards both mitigating the threats and enhancing the national capacity to adapt to climate changes (Tshering, 2007: 76). There is also a lack of proper waste management which jeopardized Bhutan's reputation as a clean and green country (RGB, 1999).

In Bhutan, climate change mitigation and adaptation policies are directly and indirectly related to climate change through the programmes of sustainable forest management, biodiversity conservation, watershed management and water source protection. But evidenced-based climate change policies are fewer, and these policies and strategies need further strengthening to be more effective (Suberi et al., 2018: 97–98; MoAF, 2016). At a national level, Bhutan is considered a "reasonably equitable and sustainable society" (Ura, 2015) with largely happy people despite the low per capita income. It has achieved the status of a carbon-neutral country (NEC, 2011; RGB, 2015a). But it now faces the test of maintaining its carbon neutrality status (Yangka et al., 2018: 6) when about 60 per cent of the population depends on forest resources. Further, most of the farmers are not aware of the proper utilization of forest resources, nor are they familiar with the impacts of farming on forest biomass that ultimately affect carbon sequestration capacity of forests (Suberi et al., 2018: 97; GoB, 2009; Locatelli, 2010). The reason is common people in general lack awareness regarding the future climate change–related threats. It is reported that the newspapers and televisions in Bhutan never present reports of climate change effects as their most important items (Chhetri, 2010). So, the role of media to create awareness may be necessary to enable the general public to become familiar with this crisis. These programmes can be very helpful for the Bhutanese to understand the intensity of the crisis. The media and NGOs are expected to play a vital role in promoting better understanding of climate change particularly at the grassroots level.

The former Minister of Disaster Management of Sri Lanka, Rishard Bathiudeen, said he did not want his people to become climate refugees as happened in other parts of the world (IRIN, 2011). That's why adaptation should be considered at the earliest. There are several ongoing projects in Sri Lanka which covered certain priority sectors such as coastal zones, water, forestry, and agriculture (Mawilmada, 2010). The NCCAS, which sets out a prioritized framework for action and investment for 2011–2016, intends to move the country and its people towards a climate change–resilient future (SLMOE, 2010: 27; Sterrett, 2011). Its post-*tsunami* coastal zone project or Sri Lanka's participation in the regional-level Mangroves for the Future Program is so far quite successful (Islam, Hove and Parry, 2011: 160). There are many national-level plans and strategies in place in Sri Lanka, including a NDC, National Adaptation Plan for Climate Change (NAP), and SDG Action Plans, as well as roadmaps to implement them (MoMDE, 2016b, 2016c; MoSDWRD, 2018). These ambitious climate change plans and strategies (NDC, NAP, and SDG Action Plans) will be difficult to implement in isolation as "achieving these cannot be done as an effort of a single institution or ministry," says Dr Jayathunga, former Director of Climate Change Division. He also said, "We have realised the importance of bringing subnational governments – Provincial Councils and Local Government Authorities – on board to implement, as well as to review and update, these plans" (Pallawala, Muller and Woods, 2018: 1). While its low-lying areas are sensitive to sea level rise, the adaptation responses are concentrated mostly in the coastal zones (Baba, 2010: 4–16). The perpetual armed conflicts and uncertainty of the war put Sri Lanka in a difficult position to make regular development investments, particularly in its northern part. Adaptation efforts have so far been fragmented, lacking a strong link between national climate change strategies, plans, and existing disaster risk reduction, agricultural, and other relevant policies (Pathiraja, Balaraman and de Silva, 2014: 72). There is also a noticeable dearth of resources related to agriculture and water projects contributing towards making its economy more climate-resilient. There is no focal point responsible for ensuring climate resilience criteria considered in national level planning initiatives. For example, the Strategic Environmental Assessment (SEA) and Environmental Impact Assessment (EIA) guidelines do not specifically include any climate change considerations (SLMOE, 2010). Further, measures which intend to deal with climate change lack adequate funding (Pathiraja, Balaraman and de Silva, 2014).

Speaking at the launching of a report entitled "Climate Change Issues in Sri Lanka", the former Environment and Renewable Energy Minister, Susil Premajayantha, said, in order to mitigate climate change's negative effects "the country needs to mainstream climate change adaptation into the national planning and development process" (Nizam, 2013: 1). During the speech, he stressed that his ministry would not be able to address and manage any climate change–related issues without the support and help of the relevant stakeholders and scholarly efforts (Nizam, 2013). So, it is understood that he foresaw "Climate Change Issues in Sri Lanka" would attempt to begin as a national dialogue between stakeholders in order to encourage a better understanding in the academic and policy spheres that could lead to better informed mitigation policies and adaptation measures (Sunday Observer, February 22, 2013).

The Maldives has shown its commitment to adapt to climate change when former President Maumoon addressed at various crucial international gatherings, including the Commonwealth Heads of Government meetings, the UNGA, the Earth Summit in Rio in 1992, the UN Millennium Summit in 2001, as well as at the SAARC Summits and Meetings. The Maldives' NAPA aims to increase resilience of the vulnerable systems against climate hazards and risks to achieve sustainable development outcomes (MoEEW, 2007: 3). But the lack of land to accommodate citizens and to construct infrastructure is a hindrance to the Maldives (Khan, 1997: 471; Karthikheyan, 2010: 346). The Maldivians are living in fear of future ecological dangers. Knowing the danger his country was facing former President Mohamed Nasheed said, "We can do nothing to stop climate change on our own and so we have to buy land elsewhere" (Ramesh, 2008: 1). This is due to their past experiences like the great *tsunami* of 2004 (Hameed, 2007; Fox News, December 29, 2004). To show what the future will hold for the Maldives, the historical 30-minute cabinet meeting at 6 metres below sea-level was organized in 2009 (Omidi, 2009). Its population is so spread throughout the archipelago. Thus, it is difficult for the Maldivian government to provide services to its people, such as water and sanitation to everyone. For this reason, a small and weak country like the Maldives is still expecting to get financial assistance from the developed countries; otherwise the Maldivian government will be in dire crisis to fund any effective policy implementation and technology adaptation to save its fragile environment. This country does not have the required resources to shoulder the burden of climate change on its own. It needs help to find durable solutions since the burden of climate change on the lives of its people is real and noticeably apparent. But there is a feeling that a country which has recently slipped backwards into autocracy and corruption should be excluded from accessing the US$100 billion in climate finance promised to developing nations (Lynas, 2015).

Apparently, the need for a national policy on climate change in Bangladesh has been expressed time and again since the early 1990s (Ipiv and Reinhardt, 2010: 64). But it is seen in most developing countries the adaptation planning is only a marginal activity, particularly in building infrastructure to provide protection against extreme climate change events (Lagos, 2009: 22). There can be no exception in Bangladesh too. While it is significantly affected by climate inconsistency (Worldwatch Institute, 2006; MoEF, 2012a: xix; Mazumdar, 2012: 46), still there is no substantial national policy in place to comprehensively address the risks. In Bangladesh climate change is not mentioned in the context of planning vulnerability reduction measures (except for a proposal for further research on impacts) (GoB, 1997: 559). Outside of the section on natural hazards, the national development plan which is known as the Poverty Reduction Strategy Paper (PRSP) (Agrawala *et al.*, 2003: 32) does not contain any references to climate change. The BCCSAP, which is the basis for the government's efforts to combat climate change, fails to mention what will be the quantifiable indicators by which achievement of implementing interventions to reduce climate change effects is noted (The Daily Star, May 8, 2010). Although the BCCSAP states the need of mainstreaming women in all activities under it, it does not describe how this will be done or specify any specific instruments or strategies designed to implement gender mainstreaming (Sharif *et al.*, 2016: 12; MoEF, 2013a). Some elements of climate change adaptation are addressed through specific sectoral policies (Ahmed, 2004: 58; UO-Oxfam, 2008: 30; Rahman *et al.*, 2010: 24). With the exception of the Coastal Zone Policy (MoWR, 2004), and the recently renewed National Agriculture Policy (MoA, 2011), climate change issues have not been sufficiently highlighted in the national policy regime. Environmental disasters have been displacing people internally with a projection that millions of Bangladeshis have moved to India illegally (Homer-Dixon and Percival, 1996). Such illicit movement of people doesn't find any place in Bangladesh's policies. Even the BCCSAP fails to mention illegal migration in the context of climate change. Given the importance of climate change and its potential adverse implications on economic development and people's lives and livelihoods, revision of sectoral policies and explicit inclusion of climate change impacts and considerations in these policies were highlighted by survey respondents as priorities (Mazumdar, 2012: 25).

Bangladesh put up significant work in terms of disaster management by investing in flood protection and drainage schemes, coastal embankment projects etc. (Bansal and Datta, 2012: 227). The BCCSAP and NAPA emphasize the adverse effects of climate change and identify adaptation needs. The National Climate Change Fund, which focuses mainly on adaptation, was established in 2008 with the support of the UK. Under this plan Bangladesh intended to establish a Centre for Research and Knowledge Management on Climate Change to ensure it has access to the latest ideas and technologies from around the world. But it is now apparent the donors are not providing enough assistance to the projects which required large-scale investment (Harvey, 2012; Huq, 2016). To avail foreign funds, it has to demonstrate good practice in transparency and accountability of climate funding (Huq,

2016). Even if it shows that it can, the funding available is scanty compared to what is required to support the adaptation and mitigation activities in the developing countries (US$100 billion by 2020). It has been projected that US$5.5 billion and US$112 million in annual recurrent costs to Bangladesh will be needed by 2050 to protect against storm surge risk (Huq, 2016). Presently, funds are given to Bangladesh for technical studies and planning only (Khan, 2010; Bansal and Datta, 2012: 227). With almost half of its population living below the UN-designated poverty line, Bangladesh neither will be able to finance any fundamental adaptation and mitigation programmes, nor will it think it should have to in future (Yasmin, 2018: 36–44; Pathmarajah, 2012: 109; Matthews, 2009). Although the worry of Bangladesh about climate change risks started in the early 1990s, it is still the most vulnerable nation to its effects. Perhaps the resources it has spent are too little to undertake the massive cost of adaptation infrastructures that it needs for protection against sea level rise, floods, and storm surges. Hence, like many developing countries, Bangladesh must have shared the view that the industrialized nations who are historically responsible for climate change should be obligated to take the lead to halt its effects.

India is cautious about making commitments in dealing with climate change – be it through mitigation or adaptation. But it is poorly equipped to cope effectively with the adversities due to low capabilities, weak institutional mechanisms, and lack of access to adequate resources (Kumar, 2008). Though it is recognized that climate change will affect India tremendously, the government affirmed "the process of adaptation to climate change must have priority" and "the most important adaptation measure is development itself"(Mehra, 2009: 81). Its negotiating position relies heavily on the principles of historical responsibility, enshrined in UNFCCC (Brechin, 2003: 106–135; De Leo *et al.*, 2001: 478–479). The NAPCC (GoI, 2008) outlines India's domestic response to the climate crisis. To critics like Praful Bidwai (2012: 129), the NAPCC lacks a long-term vision, a coherent strategy, and an overarching policy thrust. It offers no baseline data on current GHG emissions from different sources or sectors. It therefore, remains only a wish list, a statement of pious intentions and a basket of half measures (Bidwai, 2012: 132–133). It also falls short of detailed proposals and contained no specific targets (Giddens, 2011: 224). Nevertheless, this is no surprise when India always maintained its stand that the right to pollute the atmosphere should be assigned to all the countries based on their historical emissions. It has repeatedly said that it would not compromise on growth by committing to emission reduction goals set by developed nations, which it deemed bigger culprits when it comes to global pollution (Lakshmi, 2009). India's effort is to boost economic growth and development by rapidly industrializing and transforming itself into a manufacturing hub. India's overriding objectives, which include the eradication of poverty, enhancing economic well-being, improving public health, providing basic amenities, and improving infrastructure, are behind its stand (Sklias, 2010: 1–10). India's economic ambition is integral through the Twelfth FYP which recommended innovative and long-term planning to explore new approaches for sustainable development (GoI, 2012). From its stand at the world climate summits, India never sees it as important and it avoids taking a proactive step in order to mitigate the GHG emissions.

National efforts pertaining to climate change have been made in Pakistan over the years. Despite the presence of policies and legislation to protect and conserve forests, the country's forests are degrading. This country generates more than 50,000 tonnes of solid waste per day. Only 20–25 per cent of the waste is being collected but not properly disposed. As a result, it causes air and water pollution and creates serious health hazards too (Saddiqui, 2010: 74–78). Pakistan's Environmental Act 1997, part 11 states, "no person shall discharge or is allowed to discharge or emissions of any affluent or waste or air pollutant" in any amount (PEPA, 1997a, 1997b). Conversely, air pollution crosses the threshold levels in Pakistan, which is six times higher than the World Health Organization (WHO) guidelines (Saddiqui, 2010: 74–78). Though its former environment minister voiced concern about the burdens of climate change effects in Pakistan, a cohesive programme to combat the issue on the ground is absent (Saleh, 2007).

Being poor, Pakistan relies on donations to protect itself from the great risk of climate change (Javed, 2016: 10; Parsons, 2011: 1427). Thus, access to international finance is crucial. But a lack of international systems to provide necessary finance is creating hindrance. Several of Pakistan's top climate change scientists have complained about the insufficient funding allocation for the climate change ministry and the problems they faced to address such challenges. The key author of the NCCP of Pakistan, Dr Qamaruz Zaman Chaudhry, pointed to Bangladesh which received US$500 million as international assistance, an opportunity Pakistan was not getting (Singh, 2012). Lately, the total losses assessed in the index for the 15-year period for Pakistan stand at US$ million 3823.17 Purchasing Power Parity (PPP). The average cost for annual adaptation and mitigation to climate change for Pakistan is estimated to range annually from US$14 billion to US$32 billion leading to 2050 (GoP, 2017a: 1–19). On its own, Pakistan will be in a difficult position to fund such an amount for climate change mitigation. This country remains ineffective (to date) in cashing on both external and internal opportunities (Javed, 2016: 10). Its readiness to face climate change is also affected by a lack of reliable data on the extent of financial losses caused by climate change and its impact on GDP (GoP, 2017a: 19). The authorities responsible for disaster risk reduction have not made adequate use of recent developments in scientific methodologies and tools for cost-effective and sustainable interventions (Ullah *et al.*, 2018: 359–378).

Pakistan has the world's third-largest population of stunted children (UNICEF, 2015: 7). About 15 per cent of children under 5 are wasted and 30 per cent are underweight (WFP, 2018: 4). A framework to provide a mechanism to address food security called the National Food Security Policy which aims at promoting sustainable food production with goals of improving food availability, accessibility, and sustainability was released only in 2018 (Huq, 2018; Irfan, 2018; Aazim, 2018). In 2016, Pakistan's new poverty index reveals that four out of ten Pakistanis live in multidimensional poverty (UNDP, 2016). But poverty is one of the least-discussed issues in Pakistan. While other debates centre on different economic issues, poverty is mostly left out (Nelson, 2018).

Members of the UNCCCC (UN, 1992) are committed to launch national strategies for adapting to expected impacts, including the provision of financial and technological support to developing countries, and to cooperate in preparing for adaptation to the impacts of climate change. Article 4.9 of the Convention calls for special attention to the least developed countries "with regard to funding and transfer of technology" (UN, 1992: 9). Consideration has been given to the scientific and technical aspects of adaptation and technology transfer by the Convention's Subsidiary Body for Scientific and Technological Advice. This includes the Nairobi Work Programme on impacts, vulnerability, and adaptation to climate change. The Programme was adopted by the COP to the UNFCCC in 2005 and renamed in 2006 and its objective is twofold: to assist countries, in particular developing countries, including the least developed countries and small island developing states, to improve their understanding and assessment of impacts, vulnerability, and adaptation; and to assist countries to make informed decisions on practical adaptation actions and measures to respond to climate change on a sound, scientific, technical and socio-economic basis, taking into account current and future climate change and variability (UNFCCC, 2007b: 2, 2007a: 10). To mitigate and adapt to climate change, developing countries need an almost incomprehensible amount of money – likely trillions of dollars. To help them, the US and other wealthy nations, considered as the main polluters, have in 2009 agreed to share the financial burden to spend US$100 billion annually to help poorer countries (The New York Times, November 14, 2014; Gunter, 2016). The UN has created the GCF, which is the largest multinational cash pool for financing climate action in developing countries. However, investment from the richest nations totalled just US$10 billion in 2016 (Ryan, 2017). This amount is only a pittance as compared to the US$100 billion a year promised by rich nations collectively to their poor counterparts in 2009. It is seen here that the rich countries are very far from raising billions of dollars they have promised to help the poor countries in dealing with climate change. Thus, poor countries don't appear to be impressed with the announcement of the rich countries. Indeed the promise made by the rich countries to the poor countries remains extensively only on paper.

Climate change has threatened the key natural resources of South Asia, affecting water and food security, health impacts, and environmental stresses. It creates tension in certain parts due to mass migrations and may also raise national security issues. That is why the member countries of SAARC have realized the value of a regional cooperation, especially on matters of climate change. At the regional level, SAARC leaders have been trying to collectively attend to climate change when this issue becomes its regular agenda. Its member countries have done their part too to address the challenge in their respective territories. But the lack of knowledge, expertise, and financial help from the rich countries as well as their own willingness to concentrate on adaptation are the serious constraints. As a result, the transfer of technologies and financial help from the rich countries is required for the SAARC countries, which is not easy to obtain. If developed countries continue to deny supporting the poor in their mitigation plans, there is the possibility that climate change may threaten the mitigation progress which may lead to an uncertain future.

V Genesis of the problems to mitigate climate change

At the world discourse, managing climate change risks is vital because of the irreparable threats it poses to sustainable development. But "responding to climate change involves an iterative risk management process which includes both adaptation and mitigation and takes into account climate change damages, co-benefits, sustainability, equity, and attitudes toward risk" (Pachauri and Reisinger, 2007b: 22). The distinctive fact about managing climate change risk is that this problem drives changes in earth's life support systems which are vital for sustainable development (Yohe, 2012: 503–510; Arndt *et al.*, 2012: 369–377). So, climate change mitigation generally involves reductions of anthropogenic emissions of GHGs. Efforts to cut emissions – mitigation – must therefore be universal. Without international cooperation and coordination, some states may free ride on others' efforts, or even exploit uneven emissions controls to gain competitive advantage (Council on Foreign Relations, 2013: 5).

The UNFCCC (1992) was the first international treaty to acknowledge and address human-driven climate change (UN, 1992). Rich countries including the US ratified the treaty in 1992 (US Treaty Document 102–38, 1992). Among the obligations outlined in Article 4, the higher-income parties or developed countries have committed to pay for the cost incurred by developing countries to mitigate and adapt to climate change. The Kyoto Protocol, which emerged from the UNFCCC, is another global climate change agreement. It has "heightened the expectations for large scale collective action with stringent mitigation measures" to combat climate change. The 160 nations signing on to this agreement include six out of eight of the highly industrialized countries (Kaur, 2017: 80; Laub, 2014). This Protocol was adopted in such a way as to impose more specific mitigation commitments on UNFCCC parties, because UNFCCC lacks commitments to address the climate change–related problems. According to Article 3 of the Kyoto Protocol, the developed countries and economies in transition listed in Annex B to the Protocol agreed to reduce their GHG emissions to a specified percentage of their 1990 emissions by 2012 (The Kyoto Protocol, 1998). The emissions reductions required under the Protocol are approximately 30 per cent and 24 per cent for the US and Japan, respectively (Goulder and Nadreau, 2001: 5). Article 3(2) says those parties commit themselves (The Kyoto Protocol, 1998) to make demonstrable progress in achieving their targets by 2005. No targets for the reduction of emissions were set for the developing countries, but they are required to report on their emissions. This is because their priorities are economic growth and poverty reduction, whereas industrialized countries consume far more energy, and thus produce far more GHGs (Darragh, 1998).

The meeting in Copenhagen witnessed the adoption of the Copenhagen Accord which is another important milestone in international efforts to address climate change. For the first time, all the major emitters of GHGs have agreed under the Copenhagen Accord that global average temperature increase should be kept below 2°C. The most significant part of the succinct three-page 12-paragraph CHA (1) is the following: "We underline that climate change is one of the greatest challenges of our time" in its opening paragraph, followed by the second paragraph, which begins with "We agree that deep cuts in global emissions are required according to science, and as documented by the IPCC Fourth Assessment Report with a view to reduce global emissions so as to hold the increase in global temperature below 2 degrees Celsius, and take action to meet this objective consistent with science and on the basis of equity" (Stavins, 2009; Ramanathan and Xu, 2010: 8055). However, developed countries who have agreed to cut GHG emissions on the basis of their historically high emissions have not put any significant efforts to do the same. The poor developing countries too are apparently lacking commitment as they do not have legally binding emission reduction commitments under the Kyoto Protocol. These countries are expecting to get necessary donations from the rich countries to help them deal with climate change. In this way, efforts to adapt to climate change which need to be universal lack international cooperation.

It has been discussed in Chapter 2 that global cooperation on climate change met serious differences among the highly industrialized countries on the one hand and the developing countries on the other. Both the highly industrialized countries as well as the poor countries perceive dealing with climate change means sacrificing economic growth and development. When these countries exercise that excuse, none has attended to it in a strict approach. As such, the world leaders, policymakers, climate scientists, and environmentalists find it tricky to discover a way out to address scientific measures to mitigate it (WMO, 2005). The existing environmental negotiations seem unlikely to come to an end soon (Park, Conca and Finger, 2008; Agarwal, Narain and Sharma, 1999: 7). A serious flaw exists with the UNFCCC convention itself as it does not mention clearly that every huge emitter has to undertake strong emission cuts, irrespective of whether it qualifies as a developing or developed country (Saha and Talwar, 2010: 174). The AR5 of the IPCC is not an exception from such flaws as it is grounded in speculative conjecture based on well-documented climate model analytical limitations instead of solid science (Hamlin, 2014). The existence of these peculiarities contributes to the ineffectiveness of the global response to the threats of climate change.

The current global political character of climate change may be understood as problems of collective action (Paterson, 2009: 261–262). In this volume, collective action may be referred to the joint actions of the number of states which aim to achieve and distribute some gain through coordination or cooperation (Holzinger, 2003: 2). The degree of climate change crisis is understood as the ideal global-scale collective action problem when its implications require careful policy management or coordination along with multilevel governance (Esty and Moffa, 2012: 777–791). To adapt and mitigate the crisis, governments take an international institution as a key instrument because universal action is required. With the existence of an international institution, countries show their willingness to participate in global climate change negotiations. Principles essential to adapt and mitigate it have been discussed and agreed upon. But when it comes to implementing these principles, the international community faces difficulty since many countries have their own stands to disrespect these principles which have been agreed upon by all member states (Rafferty, 2011: 296–299). Due to such differences, the existing climate change policies become weak. Therefore, it becomes a crucial dilemma for climate negotiators (Wiegandt, 2001: 127–150). The reason is there are too many countries vying against each other at the global level. Thus, at the most basic level, countries disagree over climate monitoring and financing stipulations in the Kyoto Protocol and other legally binding accords. Climate frameworks also struggle to effectively monitor GHG outputs, especially in developing countries. For example, India is often a force that keeps opposing mitigation commitments at the global level (Haidar, 2010). Though, it may have made the NAPCC years ago, it has yet to be fully realized. The reason is being a developing country, India does not have legally binding emission reduction commitments under the Kyoto Protocol. Until recently, it has repeatedly rejected calls to quantify its targets for reducing GHG emissions on the grounds that this would jeopardize national poverty alleviation goals (Costa, 2018).

At the global level climate change conferences and meetings take place on a regular basis with the presence of a large number of actors. Large numbers of climate negotiators have diverse opinions. So there exists no easy set of solutions (Depledge, 2006: 6; Sandbrook, 1993: 29–30; Doyle and McEachern, 2001: 174–175; Victor, 2001; Grubb, Vrolijk and Brack, 1999). Strong disagreement exists among these actors on the causes and severity of global climate change and the avenues for its mitigation (Balling Jr., 1992). In the next century (and beyond), climate change could be extremely disruptive, spreading disease and sparking wars. It could also be beneficial for some people, businesses, and nations. For instance, in some countries of colder regions the human race will benefit from the increasing global temperature level (Dolsak, 2001: 414–415). Given that the negative effects of increasing global temperature will not be uniform across the globe, in return, it creates massive organizational challenges. So, the mess of climate change remains basically unsolved (Victor, 2001; Grubb, Vrolijk and Brack, 1999). The reason is countries differ in their willingness to mitigate global climate change, to enact the policies to stabilize or reduce emissions. Some countries prefer the status quo and continue with their current energy use patterns (Dolsak, 2001: 415).

Looking at the anthropogenic GHG emission which is assumed to be largely responsible for climate change, countries with large populations, large economies, or both tend to be the largest GHG emitters (Olivier *et al.*, 2013; Olivier, Schure and Peters, 2017: 46). As of 2017, the total group of the 20 largest economies (G20) accounted for 78 per cent of GHG emissions (Olivier and Peters, 2018: 19). In other words, the GHG emission trends of the largest countries and regions have continued to grow in 2017, although in the US and Japan only by 0.1 per cent and 0.3 per cent respectively. In absolute values, the largest emitters for CO_2 and total GHG emissions are China, the US, and the EU, followed by India, the Russian Federation, and Japan (Olivier and Peters, 2018: 20–21). Table 6.1 shows GHG emissions

Table 6.1 Total GHG emissions per capita per country/group, 2010–2017 (tonnes of CO_2 per person)

2010	2011	2013	2013	2014	2015	2016	2017	Country/Group
8.43	9.09	9.23	9.55	9.58	9.55	9.49	9.55	China
22.34	21.92	21.25	21.38	21.65	2105	20.61	20.47	US
9.72	9.38	9.29	9.12	8.78	8.88	8.87	8.95	European Union (28)
2.28	2.35	2.44	2.43	2.56	2.59	2.63	2.68	India
15.86	16.53	16.46	16.02	15.84	15.94	16.15	16.32	Russian Federation
10.51	10.93	11.20	11.50	12.16	11.75	11.67	11.73	Japan
9.43	9.78	9.81	9.52	9.52	9.51	9.55	9.65	Other OECD G20
27.25	30.66	30.32	26.87	26.88	26.69	26.02	25.63	Australia
22.00	22.12	22.00	22.45	22.78	22.29	22.22	22.57	Canada
6.05	6.07	6.21	6.28	6.45	6.40	6.30	6.29	Other G20 Countries
20.71	21.06	21.69	22.16	22.73	23.09	22.98	22.89	Saudi Arabia
7.93	8.13	8.15	8.20	8.25	8.19	8.14	8.17	Total Group of Twenty (G20)
6.08	6.12	6.12	6.11	5.96	5.74	5.77	5.83	Other Large Emitting Countries
3.16	3.14	3.18	3.18	3.17	3.16	3.18	3.17	Remaining Countries (186)

Source: Olivier and Peters (2018: 46).

per capita from different major emitting countries, the EU, the rest of the world, and for the world average. Except for India, all major emitters have per capita emission levels that are significantly higher than those for the rest of the world and the world average. Canada, Australia, and Saudi Arabia, the US, the Russian Federation, and Japan are the top GHG emitting countries, per capita (Olivier and Peters, 2018: 21).

The year 2017 was significant as it was globally the third-warmest year since records began in 1880, behind the record year 2016 and second-warmest year on record 2015 (NOAA, 2018). The year 2017 was also a remarkable year because the growth in total GHG emissions has resumed at the rate of 1.3 per cent (±1%) per year, reaching a new GHG emission record of about 50.9 gigatonnes of CO_2 equivalent (Gt CO_2 eq) (excluding those from land-use change), after two years of virtually no growth: 0.2 per cent in 2015 and 0.6 per cent in 2016 (which was a leap year and therefore 0.3 per cent longer). This ends speculation about peaking of global emissions in 2015 and 2016 (Olivier and Peters, 2018: 11). While looking at the per capita emissions of India (Table 6.1), it still remained much lower than those of developed countries even if it is expected to rise slightly, to about half of the world average, and one-fourteenth of that of the US (Rahman, 2007: 93).

Some countries try to impress upon the others that the uncertainty in their understanding of climate change is still large enough to postpone action or that technical solutions will be sufficient in the future to reverse trends and avoid future critical situations (Gautier and Fellous, 2008: 2–3). The step for the governments to take concrete action to prevent climate change still appears sceptical (Pauwelyn, Wessel and Wouters, 2014; Pearce, 1994). The commitments they are pursuing domestically are insufficient. It is happening this way as whatever climate consensus has been made is only considered as agreements which are not enforceable in a court of law. Thus, the countries are reluctant and have no obligation to it as global climate change agreements carry less force (Liftin, 1998). The crisis of climate change as of now presents a sizeable collective action problem, though it may be the greatest collective action problem the international community has ever faced (Chestnoy and Gershinkova, 2017: 215–223; Böhringer and Rutherford, 2017; Stern, 2006; Arrow, 2007).

On December 12, 2015, 196 parties to the UNFCCC adopted the Paris Agreement, a supposedly legally binding framework for an internationally coordinated effort to tackle climate change. The Paris Agreement constitutes a milestone in international climate policy as it sets the frame for scaling up ambition for climate protection (Burck, Marten and Bals, 2017: 7). It intends to set a goal of limiting global warming to less than 2°C compared to pre-industrial levels. The proviso of the Paris Agreement is similar to the Kyoto Protocol. For example, Article 9 of the Paris Agreement reiterates the obligation in the Convention for developed countries, which are Parties including the US, to seek to mobilize financial support to assist developing countries, which are Parties with climate change mitigation and adaptation efforts (Article 9.1). For the first time as well under the UNFCCC, the Paris

Agreement encourages all Parties to provide financial support voluntarily, regardless of their economic standing (Article 9.2). The Paris Agreement further states developed countries which Parties should take the lead in mobilizing climate finance and that the mobilized resources may come from a wide variety of sources. It also adds the mobilization of climate finance "should represent a progression beyond previous efforts" (Article 9.3) (Lattanzio, 2017: 2; CoP, 2015: 22–26; Abeysinghe and Prolo, 2016: 1). The Paris Agreement was opened for signatures on April 22, 2016, where countries would have to sign it in New York between April 22, 2016 (Earth Day) and April 21, 2017, and also adopt it within their own legal systems (through ratification, acceptance, approval or accession) (Hersher, 2016; Abeysinghe and Prolo, 2016: 4). The Paris Agreement will only become legally binding if joined by at least 55 countries which together represent at least 55 per cent of global GHG emissions. By mid-November 2017, 197 countries joined the Paris Agreement, while 170 of these countries with a combined emission share of 88 per cent had already ratified (Lippelt and Mayer, 2017: 43).

Unlike the earlier climate change accords, after signing the Paris Agreement emerging, developing, and industrialized countries are committed to climate protection measures in the form of INDC. In 2016 (Marrakech), a consensus was reached on a regular review of national action plans and the development of transparency plans. In essence, the contracting parties endorse to submit concrete rules to this end by 2018 in order to steadily tighten national climate contributions. The reason for this is the fact that the climate contributions submitted to date by states are not sufficient to reduce global warming to below 2°C, or even to 1.5°C compared to its pre-industrial levels (Hickmann, 2017). Under the Paris Agreement, the EU has set common targets where GHG emissions are to be reduced by 20 per cent compared to 1990, by 40 per cent by 2030, and by 80–95 per cent by 2050 (European Commission, 2017; Lippelt and Mayer, 2017: 43). China set several environmental targets to be achieved by 2030 where CO_2 emissions per unit of GDP, for example, are expected to fall 60–65 per cent compared to 2005 levels, to raise the share of non-fossil fuels in primary energy consumption to around 20 per cent by 2030 and to achieve the peaking of CO_2 emissions around 2030 and make best efforts to peak as early as possible (Zhidong, 2016: 66; Lippelt and Mayer, 2017: 43). Under the Paris Agreement, India is committed to cut its GHG emissions intensity by 33 to 35 per cent below 2005 levels by 2030, and to achieving 40 per cent of its electricity generation from non-fossil sources by the same year and 2.5–3 billion tonnes CO_2 will be additionally bound by larger forest areas (GoI, 2015; Issue Brief, 2017). India's economic plan gives priority to clean energy to fuel economic growth and includes ambitious targets of 100 gigawatts (GW) of solar power and 60 GW of wind power by 2022 (Rattani, 2018: 11–12; MoNRE, 2017; GoI, 2015). In just one year, from 2016 to 2017, India increased its renewable energy capacity by 11.3 GW (Bridge to India, 2017; Issue Brief, 2017: 3). Table 6.2 summarizes the pledges or legally binding commitments of a few countries as well as of the SAARC members for post-2020.

Table 6.2 Selected parties' pledges to abate greenhouse gas emissions (as of October 6, 2015)

Party	Post-2020 Pledge in INDCs
China	Announced that, by 2030, it would: Achieve peaking of CO_2 emissions around 2030 and make best efforts to peak earlier; Increase the share of non-fossil fuel energy sources to around 20 per cent of primary energy supply; Lower CO_2 emitted per unit of GDP by 60–65 per cent compared with 2005 levels; Expand forest stock volume by around 4.5 billion cubic meters (m^3) compared with 2005 levels; and "Proactively" adapt to climate change.
US	Reduce GHG emissions by 26–28 per cent below 2005 levels by 2025.
Canada	Reduce GHG emission by 30 per cent below 2005 levels by 2030
Australia	Reduce GHG emissions by 26–28 per cent below 2005 levels by 2030. Establish an emissions budget for 2021–2030. Targets National Energy Productivity Plan to achieve 40 per cent improved productivity by 2030.
EU (28)	Reduce GHG emissions by 20 per cent below 1990 levels by 2020, binding in 2nd commitment period of the Kyoto Protocol. Binding target to reduce GHG emissions by at least 40 per cent below 1990 levels by 2030
Russian Federation	"Long-term indicator" to limit GHG emissions to 25–30 per cent below 1990 levels by 2030, subject to the "maximum allowance" of credits for CO_2 removals by land use changes and forestry. Target conditioned on what "major emitters" pledge
Republic of Korea	Reduce its GHG emissions by 37 per cent below 2030 BAU level (reported as 851 million metric tons of CO2e). Intends to use international carbon credits to achieve this target in part. No decision yet on whether emissions and removals from the land sector will be included. States that it aims to reduce GHG emissions by 49–70 per cent from 2010 levels by 2050

INDC of SAARC Countries

Member	Post-2020 Pledge in INDCs
Afghanistan	In Afghanistan's INDC, the country offers a 13.6 per cent emissions reduction from business as usual levels by 2030 contingent upon receiving international support in the estimated amount of $17.4 billion. Afghanistan's INDC also includes a section on adaptation and reducing vulnerability to climate impacts.
Bangladesh	In Bangladesh's INDC, the country offers an unconditional 5 per cent reduction in GHG emissions by 2030, compared to business-as-usual levels, in the power, transport and industry sectors. To be accompanied by a further 15 per cent reduction, conditional upon international support. These three sectors will represent 69 per cent of total emissions in 2030.

(*Continued*)

(Continued)

Party	Post-2020 Pledge in INDCs
Bhutan	In Bhutan's INDC, the country which has already achieved carbon neutrality will remain carbon neutral, so that emissions of greenhouse gases do not exceed carbon sequestration by forests. Bhutan also commits to maintaining current levels of forest cover. Includes a selection of low-emissions policies. Includes a section on adaptation. Successful implementation will depend on level of support received. Repeats commitment to keep 60 per cent of territory forested.
India	India's official INDC is committing to: Reining in the emission intensity of per unit GDP by 33 to 35 per cent below 2005 levels by 2030; Increasing non-fossil fuels in its electrical mix to 40 per cent installed capacity by 2030 with the help of transfer of technology and low cost international finance including from the Green Climate Fund (GCF); Adding 175GW of new renewable energy generation by 2022 (of which 100GW will be solar); Adding forest and tree cover to create a carbon sink for 2.5 to 3 billion tons of $CO2$ by 2030; India intends to cover the $2.5 trillion cost of its pledge with both domestic and international funds. Includes information on adaptation.
Maldives	In the Maldives' INDC, the country offers an unconditional 10 per cent reduction in energy sector emissions by 2030, compared to business as usual, or a 24 per cent reduction conditional on international support. Business as usual 2030 emissions are projected to be triple those in 2010. Contains a section on adaptation.
Nepal	Aims to reduce dependency on fossil fuels by 50 per cent by 2050 and achieve 80 per cent electrification through renewable energy sources with appropriate energy mix. Plans to maintain 40 per cent of the total area of the country under forest cover.
Pakistan	In Pakistan's INDC, the country submitted a two page broad statement of support for climate action but did not lay out any specific targets. Pakistan said the specifics of its climate policies would be forthcoming, but specified that "Pakistan is committed to reduce its emissions after reaching peak levels to the extent possible subject to affordability, provision of international climate finance, transfer of technology and capacity building. As such Pakistan will only be able to make specific commitments once reliable data on our peak emission levels is available".
Sri Lanka	In Sri Lanka's INDC, the country offers an unconditional 7 per cent emissions reduction target from 2030 business as usual levels that could increase to 23 per cent conditional upon international support. Sri Lanka estimates that the costs of adaptation will amount to $420 million over the next decade. Includes a section on loss and damage.

Source: Country-reported INDCs available at www4.unfccc.int/submissions/indc/Submission%20Pages/submissions.aspx; accessed 5/2/2020; Also see at (Leggett, 2015: 7–9).

What kind of accord is the Paris Agreement? The mitigation commitments embodied in the INDC are considered voluntary and only certain provisions are considered legally binding. According to Richard Falk (Falk, 2016), the Paris Agreement is a "voluntary international law". For Daniel Bodansky (Bodansky, 2015: 55), the Paris Agreement will be a treaty within the definition of the Vienna Convention on the Law of Treaties. It does create legal obligations for its parties as compliance with the obligations is not voluntary (Bodansky, 2016: 142). But debate continues over which provisions of the agreement should be legally binding. The 195 signatories to the Paris Agreement created a non-binding system for raising their individual obligations every five years, starting in 2020 (Adoption of the Paris Agreement, 2015). The US fought very hard to ensure its interpretation of INDCs as voluntary, which was later accepted and allowed by the Obama administration to effectively argue that the Paris Agreement is not a "treaty" under Article II of the US Constitution, even if it is a legally binding agreement or treaty in the broader sense of the term (Busby, 2016). Though each new INDC must be more "ambitious", the UNFCCC required parties only to adopt national policies that would mitigate GHG emissions and did not require parties to submit specific "targets and timetables" in perpetuity (Groves, 2016). By traditional international legal standards, the Paris Agreement is basically a statement of good intentions, but not a treaty since it is a non-binding accord. Law scholar Anne-Marie Slaughter (Slaughter, 2015: 1), writes treaties must contain "enforceable rules" with "sanctions for non-compliance" and must be "ratified by domestic parliaments so that they become a part of domestic law". When the Paris Agreement is "none of these things", she concludes the agreement is "essentially a statement of good intentions" rather than a law. Like any international agreement, the Paris Agreement has flaws. The elements of the Paris Agreement are considered legally binding, but mitigation commitments embodied in the INDC are considered voluntary. Thus, signatories can renege on their commitments.

Initially, climate change sceptics had foreseen that the US Senate would reject or even shred the Paris Agreement into pieces. Disregarding its critics, Barack Obama bypassed (DeMint, 2016) the approval of the Senate. This makes the Paris Agreement politically weak in the US. When Donald J. Trump, a scornful critic of the concept of climate change (Matthews, 2017), won the presidential election in 2016, he asserted that meeting the terms of the Paris Agreement would pull down America's economy. This is because (i) under the Paris Agreement, the US is obligated to undertake "economy-wide absolute emission reduction targets" and (ii) provide an unspecified amount of taxpayer dollars "to assist developing country Parties with respect to both mitigation and adaptation". Developed countries like the US have the obligation to contribute to the GCF while developing nations are merely "encouraged" to make "voluntary" contributions. So, the amount the US is obligated to pay into the GCF beginning in 2020 is likely to be several billion dollars each year (Darwall, 2015a; Darwall, 2015b; UNFCCC, 2016; Rajamani, 2016: 19; Lee, 2016: 97; Thompson, 2016: 88). According to Donald J. Trump, the terms of the Paris Agreement will pull down America's economy because it states that each new INDC must be more "ambitious" than the party's previous INDC. In his opinion, Barack Obama agreed to a bad deal for Americans that would handcuff the economy and put the US at a disadvantageous position against its international competitors. He says the US will stop contributing to the GCF as it would cost the US "billions and billions and billions" of dollars (Popovich and Fountain, 2017). His grievance is that the Paris Agreement restricts America's coal production whereas India and China are the culprits who will benefit from it. By this announcement, many hailed his actions to "put America's interests first" (DeMint, 2016). He even appointed Scott Pruitt, a climate change sceptic, to run the Environmental Protection Agency (EPA). By Executive Order 13783, "Promoting Energy Independence and Economic Growth", in March 2017 Donald J. Trump directed a number of agencies and departments to review and then revise or rescind the Obama administration's regulations focused on climate and energy (Memorandum, 2017: 1–2; U.S. Environmental Protection Agency, 2017: 1–2). This includes the Clean Power Plan, which will limit power sector carbon dioxide emissions to one-third below their 2005 level by 2030, along with regulations affecting the oil and gas industry, as well as federal coal leasing policy (Aldy, 2017: 2). In June 2017, Donald J. Trump stated the "U.S. will cease all implementation of the non-binding Paris accord and the draconian financial and economic burdens the agreement imposes on our country" (Han, 2017: 337–349; Mooney, 2017; Roberts, 2017; Resnick, 2017). Now the Paris Agreement faces the same fate as the Kyoto Protocol, which also ended in not getting the support of the US, which is historically responsible for more emissions than any other country. As such, agreement for collective action in dealing with climate change as of now is still difficult to realize. The lack of harmony on essential facts can become disastrous for the globe in the near future. Hence, understanding the crisis of climate change is required when trying to reach agreements about mitigation and on how to share responsibilities between nations (Bolin, 1997: 139–149).

For years the Climate Change Performance Index (CCPI) has been tracking countries' efforts to combat climate change. The CCPI is an essential tool in gaining a clearer understanding of national and international climate policy. In 2017 the methodology of the CCPI was revised, to fully incorporate the Paris Agreement, which marked a milestone in the international climate negotiations (Burck *et al.*, 2019: 9). From 2018, the CCPI is monitoring the development of all GHG emissions of the 56 countries and the EU that are assessed in the index. The index now is even better suited to measure how well countries are on track to meet the global goals of the Paris Agreement. The 2019 CCPI measures countries' performances in three categories: GHG Emissions, Renewable Energy, and Energy Use. For each category, the index analyzes countries based on their INDCs under the Paris Agreement on climate change to determine if they are on track to a below 2°C pathway. Each category includes four indicators on recent developments, current levels, 2°C compatibility of the current performance, and an evaluation of the countries' 2030 targets in each category. Tables 6.3, 6.4, and 6.5 evaluate the countries' 2030 targets within the important categories – "GHG Emissions", "Renewable Energy", and "Energy Use" to determine if they are on track to a well-below-2°C pathway.

Table 6.3 GHG Emissions – Rating Table

Rank	Country	Score	Overall Rating	GHG per Capita – current level (incl. LULUCF)	GHG per Capita – current trend (excl. LULUCF)	GHG per Capita – compared to a well-below-2°C pathway	GHG 2030 target – compared to a well-below-2°C pathway
7	UK	75.9	High	Medium	High	Medium	High
12	India	71.8	High	Very High	Very Low	Very High	High
18	Italy	67.0	Medium	Medium	High	Medium	Medium
22	France	62.1	Medium	Medium	Medium	Medium	Medium
23	EU (28)	61.6	Medium	Low	Medium	Medium	Medium
25	Brazil	60.6	Medium	Medium	Low	Medium	Medium
28	Indonesia	58.8	Medium	Low	High	Very Low	Low
29	Mexico	58.7	Medium	Medium	Low	Low	Medium
34	Germany	55.5	Low	Low	Low	Low	Medium
37	Turkey	54.1	Low	Medium	Very Low	Medium	Low
39	South Africa	52.7	Low	Low	High	Low	Low
44	Russian Federation	49.1	Low	Very Low	Low	High	Low
46	Argentina	46.4	Low	Low	Low	Very Low	Low
47	Japan	46.1	Low	Low	Low	Very Low	Very Low
49	Australia	44.2	Very Low	Very Low	Medium	Low	Medium
51	China	43.6	Very Low	Low	Low	Low	Very Low
54	Canada	32.5	Very Low	Very Low	Medium	Very Low	Low
57	US	21.4	Very Low	Very Low	Medium	Very Low	Very Low
59	Republic of Korea	13.5	Very Low	Very Low	Low	Very Low	Very Low
60	Saudi Arabia	2.3	Very Low	Very Low	Very Low	Very Low	Very Low

Source: Burck *et al.*, *Climate Performance Index 2019*, 2019: 9.

Table 6.4 Renewable Energy (RE) – Rating Table for G20 Countries

Rank	Country	Score	Overall Ranking	Share of RE in Energy Use (TPES) – current level (incl. hydro)	Share of RE in Energy Use (TPES) – current trend (excl. hydro)	Share of RE in Energy Use (TPES) (excl. hydro) – compared to a well-below-2°C pathway	RE 2030 Target (incl. hydro) – compared to a well-below-2°C pathway
11	Brazil	54.15	High	Very High	Low	Medium	Medium
15	Turkey	47.24	High	Medium	Very High	Medium	Low
20	Italy	38.71	Medium	High	Medium	High	Medium
21	Germany	37.69	Medium	Medium	High	Medium	Medium
22	EU (28)	37.13	Medium	Medium	Medium	Medium	Medium
24	United Kingdom	35.90	Medium	Medium	High	Medium	Very Low
27	India	35.03	Medium	Medium	Medium	Medium	Medium
31	China	33.89	Medium	Low	Very High	Low	Very Low
34	Republic of Korea	30.34	Medium	Very Low	Very High	Very Low	Very Low
38	Indonesia	28.18	Low	Medium	Low	Low	Low
41	France	25.63	Low	Medium	High	Low	Low
44	Canada	23.40	Low	High	Low	Low	Low
47	US	19.20	Low	Low	Medium	Low	Very Low
48	Japan	18.30	Low	Low	Medium	Low	Very Low
49	Australia	17.93	Low	Low	Medium	Low	Very Low
50	Mexico	17.44	Low	Low	Medium	Very Low	Low
51	Argentina	15.39	Very Low	Medium	Low	Very Low	Very Low
53	South Africa	14.46	Very Low	Very Low	Low	Very Low	Very Low
59	Saudi Arabia	2.86	Very Low	Very Low	Very Low	Very Low	Very Low
60	Russian Federation	2.05	Very Low	Very Low	Very Low	Very Low	Very Low

Source: Burck *et al.*, *Climate Performance Index 2019*, 2019: 11.

From Tables 6.3, 6.4, and 6.5, it can be seen that no country performed well enough to reach the top ranking in the 2019 Climate Index. The UK leads the rankings with high ratings. France, Mexico, and Germany fall into the medium-performing countries. Indonesia and others rank as low performers. The bottom performers on the 2019 Index are Saudi Arabia, US, and the Republic of Korea (Eckstein, Hutfils and Winges, 2018; Wahlen, 2018). On GHG emissions, when considering emissions from land use, land use change and forestry (LULUCF), Sweden leads the ranking, followed by

Table 6.5 Energy Use – Rating Table

Rank	Country	Score	Overall Rating	Energy Use (TPES)** per Capita – currentlevel	Energy Use (TPES) per Capita – currenttrend	Energy Use (TPES) per Capita – compared to a well-below-2°C pathway	Energy Use 2030 Target – compared to a well-below-2°C pathway
10	India	72.3	High	Very High	Very Low	Very High	High
11	Mexico	71.7	High	High	High	High	High
18	Italy	65.0	High	Medium	High	Medium	Medium
19	Brazil	65.0	High	Very High	Low	Medium	Medium
20	UK	64.7	Medium	Medium	High	High	Medium
26	Indonesia	61.1	Medium	Very High	Very Low	High	Low
28	South Africa	59.5	Medium	Medium	High	Low	Low
30	EU (28)	57.6	Medium	Low	Medium	Medium	Low
31	Argentina	57.1	Medium	High	Low	Low	Low
32	France	55.6	Medium,	Low	High	Low	Medium
35	Germany	54.7	Low	Low	Medium	Medium	Medium
36	Japan	54.3	Low	Low	High	Low	Very Low
42	Russian Federation	51.2	Low	Very Low	Medium	Medium	Medium
48	China	42.2	Very Low	High	Very Low	Very Low	Very Low
49	Turkey	41.8	Very Low	High	Very low	Low	Very Low
52	Australia	38.1	Very Low	Very Low	High	Very Low	Very Low
55	US	32.0	Very Low	Very Low	Medium	Low	Very Low
58	Canada	25.2	Very Low	Very Low	Low	Very Low	Very Low
59	Republic of Korea	14.7	Very Low	Very Low	Low	Very Low	Very Low
60	Saudi Arabia	8.8	Very Low	Very Low	Very Low	Very Low	Very Low

Source: Burck *et al.*, *Climate Performance Index 2019*, 2019: 13.

others like Egypt and Malta. On renewable energy, Latvia performed the best, followed by Sweden and New Zealand. Morocco had the greatest improvement among countries in this category, moving to the medium-performing countries. On energy use, Ukraine, Malta, Morocco, and Romania ranked the highest, mostly due to their low current levels of energy use. The index observes that emerging economies tend to perform well on the energy use category but some countries, such as Thailand, Turkey, Algeria, India, and Indonesia, have increased their energy use over the past few years (Eckstein, Hutfils and Winges, 2018; Wahlen, 2018). The CCPI 2019 shows only a few countries have started to implement strategies to limit global warming below 2 or even 1.5°C. While there is continued growth and competitiveness of renewable energy, especially in countries that had low shares before, the CCPI shows a lack of political will of most governments to phase out fossil fuels with the necessary speed. Because of that, after consecutive years of stable CO_2 emissions, emissions are rising again (Olivier and Peters, 2018: 11).

On seeing the problem of climate cooperation at the global level, where in most countries the climate policy evaluation by national experts is significantly lower than in the last years, a good regional cooperation in South Asia is the only alternative to help in mitigating the impact of climate change in the countries of the region. The reason being that inaction in dealing with it will expose the populace of the region to more catastrophes. The successive summits of SAARC and meetings of the SAARC Environment Ministers have focused more directly on the climate change–related disasters of the region. Additionally, they have also reiterated the need to strengthen and intensify regional cooperation to preserve, protect, and manage the diverse and fragile ecosystems of the region including the need to address the challenges posed by it. That is why there are many policies/initiatives taken by SAARC. This regional cooperation has its own difficulties to implement these policies/initiatives, as it cannot afford to have well-equipped technologies and to create enough finances due to the extreme poverty of its members. But on seeing some of the significant steps taken by it and the progress being made to build up regional cooperation by SAARC in the areas of climate change and natural disasters, there are good prospects for regional cooperation in the region. The succeeding paragraphs will therefore discuss the future prospects of SAARC as a regional body.

VI Future prospects of SAARC

Despite many difficulties, the member countries of SAARC have displayed that broad progress is being made on climate change adaptation in the region. Precisely, the two regional studies – (i) Causes and Consequences of Natural Disasters and the Protection and Preservation of Environment and (ii) Greenhouse Effects and its Impact on the Region (SAARC, 1992a, 1992b) – which provide a comprehensive assessment of the state of the environment in member states have been completed since 1992. The completion of the two regional studies is a significant step forward in promoting regional cooperation in climate change. The Sixteenth SAARC Summit (2010) in Bhutan concluded with the Thimphu Declaration on Climate Change, which emphasizes renewable energy, cutting carbon emissions, and reducing poverty while strengthening resilience to climate change. Although not much progress has been made in the implementation of the SAARC Action Plan on Climate Change as well as Dhaka Declaration on Climate Change, the SDMC had commissioned a study in 2013 to understand the institutional and policy landscape for Disaster Risk Reduction-Climate Change

Adaptation (DRR-CCA) in the region based on which an action plan will be developed for implementation of the Thimphu Statement (SAARC, 2014b: 25; Mall and Kumar, 2014).

The SAARC Convention on Cooperation on Environment which was signed during the Sixteenth SAARC Summit (2010) was ratified by all member states and entered into force with effect from October 23, 2013. This enables the prospects for enhanced cooperation among the members, since the Convention provided for cooperation in the field of environment and transfer of eco-friendly technology in a wide range of areas related to the environment in the region. The SAARC Agreement on Rapid Response to Natural Disasters is in the process of being ratified. So far, it has been ratified by five countries. India was the first country to ratify it (Daily Outlook, August 23, 2012). It is expected that there will be no difficulty for the remaining three members to ratify the SAARC Agreement on Rapid Response to Natural Disasters as there is urgent need to protect their people from the effects of climate change. When it is ratified by all the members, it would institutionalize regional cooperation among member states in the critical area of response in the aftermath of natural disasters in the region. India is taking a leadership role in implementing an agreement signed by all SAARC states on Rapid Response to Natural Disasters (*Dhaka Tribune*, April 27, 2015). India is also helping in setting up a SAARC monitoring system that involves developing tools for an early warning system and risk mitigation in member states (*Dhaka Tribune*, April 27, 2015). After an earthquake that struck Nepal, India promised to take a leadership role to implement an agreement signed by all SAARC states on Rapid Response to Natural Disasters (The Times of India, April 27, 2015).

The SAARC Plan of Action on Environment (1997), the Dhaka Declaration and SAARC Action Plan on Climate Change (2008), and the Comprehensive Framework on Disaster Management (2006–2015) present the common priorities of member states (SAARC, 2015: 1). To demonstrate SAARC's emerging united voice at the global arena, common positions have been taken by its members at COP 15 and COP 16 to the UNFCCC.

The SAARC may not be so successful with all its policies/initiatives in adapting climate change–related effects, be it the Dhaka Declaration on Climate Change (2008) or its other plans. Poverty of the SAARC members has hindered the progress in their implementation. Though the SAARC Development Fund (SDF) was established since 2005, this financial institution has not been able to deliver much to fund for climate change mitigation. But with its existence, it is still hoped that it will open separate windows to support the implementation of decisions taken by SAARC leaders, particularly decisions made on disaster risk reduction, climate change adaptation and mitigation, agriculture and food security (CANSA, November 26, 2014).

Since its establishment, the SAARC Agricultural Centre (SAC) has been providing relevant agricultural research and information networks among the SAARC Member States in order to exchange regionally generated technical information and to strengthen agricultural research, development, and innovations (Muniruzzaman, 2013: 46–73). This is a positive sign since agriculture is one of the most important economic activities of the people in the region, which is also most vulnerable to climate change. Perhaps the member states of SAARC will bargain from this Centre.

The Severe Thunderstorm Observations and Regional Modelling (STORM) Program implemented by the SAARC Meteorological Research Centre (SMRC) is one of the SAARC's most successful endeavours. The STORM Program was initially envisioned for understanding the severe thunderstorms locally known as *Kal Baisakhi* or Nor'westers that affect West Bengal and the northeastern parts of India during the pre-monsoon season (March–May). During this season, a lot of thunderstorms occur over northeast India, Bangladesh, Nepal, and Bhutan. They are called Nor'westers since they usually spread from the northwest to the southeast. The Nor'westers cause loss of human lives and damage to properties worth millions of dollars (De, Dube and Rao, 2005: 173–187). Recognizing the importance of the extreme weather events and the socio-economic effects, India's Department of Science and Technology started the nationally coordinated STORM Program in 2005. Under the STORM a network of weather stations has been set up across four countries to monitor severe thunderstorms like the Nor'westers (*Kal Baisakhi*). *Kal Baisakhi*, or Nor'westers (a mass of thick black clouds or *kal*), a term used in West Bengal, are thundershowers known to arrive from the north or northwest direction, bringing good rain with squally winds during early summers. When other South Asian countries are affected by the Nor'westers, the STORM Program has been expanded to cover all the countries under SAARC. The STORM Program covers all the SAARC countries in three phases. In the first phase, the focus is on Nor'westers that form over the eastern and northeastern parts of India, Bangladesh, Nepal, and Bhutan. In the second phase, the dry convective storms/dust storms and deep convection that occur in the western parts of India, Pakistan, and Afghanistan will be investigated. Similarly, in the third phase, the maritime and continental thunderstorms over southern parts of India, Sri Lanka, and the Maldives will be investigated. Thus, overall the STORM Program of SAARC will cover investigations about formation, modelling, and forecasting, including now forecasting of severe convective weather in the pre-monsoon season over South Asia. Therefore, the STORM is a coordinated effort on understanding severe thunderstorms through observations and regional modelling by eight South Asian countries (Das *et al.*, 2014). SAARC has implemented the South Asia Disaster Knowledge Network (SADKN) (2009–12), which is mainly funded by the WB's Global Facility for Disaster Reduction and Recovery through UNISDR. The SADKN is a major platform for sharing data and information about disaster risk reduction management in the region, though cooperation within the SADKN exists only within a bilateral level and through alternative regional configurations (Krampe, 2018). On seeing the efforts put forth by the members, it can be stated that the prospects of SAARC are welcoming.

Owing to the lack of finance, the governments of SAARC countries are generating funds from their own resources for adaptation and mitigation. Besides the funds that they are generating on their own, several donors such as the civil society organizations and NGOs have contributed to the regional efforts. International donors such as the WB and the ADB have sincerely contributed since resilience is closely linked to their development mission. The ADB formulates Climate Change Implementation Plans for all developing countries in the region. It is also supporting agricultural research centres and exploring CDM opportunities in the subcontinent (Islam, Hove and Parry, 2011). Another creditable initiative is the one taken by the International Centre for Integrated Mountain Development (ICIMOD), a regional intergovernmental knowledge centre serving countries of the Himalayas that includes all the SAARC countries except the Islands. Some of the schemes the ICIMOD is assisting the communities to understand and adapt to changes in the fragile mountain ecosystem are:

i) New initiatives and activities have introduced in the River Basins regional programme. Through such initiatives, ICIMOD is helping people understand the impact of changes in the water resources and to support them in preparing and adapting to such changes.
ii) A flood information system is established at the regional level, and ICIMOD is working both to enhance the regional capacity to monitor flood risks as well as to ensure that these messages reach communities at risk in a timely manner.
iii) It is in the year 2013 where the ICIMOD laid down the groundwork for the Himalayan Adaptation, Water, and Resilience (HI-AWARE) initiative, a consortium that will conduct large-scale research on the impact of climate change on the poor communities in the Indus, Ganges, and Brahmaputra river basins.

(ICIMOD, 2013: 2–3)

The ICIMOD supports transboundary programmes and addresses up-stream and down-stream issues which are highly sensitive subjects in South Asia. Working on transboundary programmes such as the addressing of up-stream and down-stream issues (floods and droughts), which is supported by ICIMOD, provides South Asian countries an opportunity to work together to find new solutions to emerging challenges. The Asian Disaster Preparedness Centre prepares vulnerable people for disasters through skills development and awareness building. Environmental concerns of groups such as the World Wide Fund for Nature (WWF) are supporting adaptation efforts. The WWF's Living Himalayas Initiative in Bhutan, India, and Nepal has been working to ensure the Eastern Himalayas are conserved and managed properly to sustain the livelihood of local people (Khatun and Hossain, 2012: 30–31). The WFP, which is the UN frontline agency mandated to combat global hunger, has been active in providing food supply to the SAARC countries. On seeing the contribution and commendable works done by SAARC and the involvement and help that other international agencies offer to the region, it can be stated that regional cooperation for climate change has a good prospect. In view of the previous discussion, it may be stated that although problems exist the prospects for regional cooperation while addressing the issue of climate change are not entirely bleak.

Recognizing the consequences of climate change in the Himalayas can no longer be ignored, four Himalayan nations – Bhutan, India, Nepal, and Bangladesh – attended the regional Climate Summit for a Living Himalayas in Bhutan's capital, Thimphu, in November 2011. The Climate Summit adopted and endorsed a ten-year road map for adaptation to climate change in the Eastern Himalayan subregion. The statement of the Climate Summit for a Living Himalayas (Gurung, 2011; Tornikoski, 2015; MoAF, 2011) says that it reached the consensus on food security and securing livelihoods, with the deal covering adaptive approaches to improving and sustaining food production. The officials who attended the Summit agreed upon a regional "Framework of Cooperation" which aims at building regional resilience to the negative effects of climate change in the Himalayas together. The objectives of the Summit are:

i) Securing Biodiversity and Ensuring its Sustainable Use;
ii) Securing the Natural Freshwater Systems of the Himalayas;
iii) Ensuring Energy Security and Enhancing Alternative Technologies; and
iv) Ensuring Food Security and Securing Livelihoods.

(Tornikoski, 2015)

The eight countries of South Asia have enforced environmental action plans to mitigate the impact of climate change in the region. Despite its economic challenges, Bhutan committed to going carbon neutral at the 2009 Copenhagen Climate Conference, and has, to date, kept its promise. It is now carbon negative. This is possible due to its vast forest reserves that have the potential to sequester 6.3 million tonnes (Mt) of CO_2 annually, easily eclipsing the country's estimated year 2013 emission total of 2.2 Mt of CO_2 equivalents (Munawar, 2016). This country is committed to keeping 60 per cent of its territory forested. The latest National Forestry Inventory of Bhutan indicates a forest cover of 71 per cent (MoAF, 2017). Nepal aims to reduce reliance on fossil fuels by 50 per cent by 2050 and achieve 80 per cent electrification through renewable energy sources. It plans to maintain 40 per cent of its total area under forest cover. Initiatives like "Zero Carbon Nepal – Vision 2030" have been launched under the National Planning Commission with approval of the Confederation of Nepalese Industries to promote green economy and low carbon development by developing a "Made in Zero Carbon Nepal" label for every Nepalese product that not only strengthens their economy but also establishes Nepal's identity as a carbon-neutral country (Pandey and Karki, 2016). Besides national strategies, both Sri Lanka and Bangladesh aim for a reduction in GHG emissions of 7 and 5 per cent respectively from business-as-usual levels by 2030, or up to 23 per cent and 15 per cent with international support (see Table 6.2). Even Afghanistan and the Maldives are committing to 13.6 per cent and 10 per cent emission cuts by 2030 respectively (Table 6.2). The EU under the "Sustainable Growth and Jobs" component of its Development Assistance Program to Afghanistan provides support to renewable energies. Together with its member states, EU supports multiple initiatives in irrigation and water management, infrastructure, natural resources management, and reforestation (EEAS, 2018). Rather than leaving the islands, the Maldivian government is concentrating on reclaiming as well as building artificial islands to accommodate its population (Dauenhauer, 2017). Besides, there are "citizen-scientists" or volunteer marine biologists who have received training to monitor marine biodiversity (Gunasekera, 2016). The marine biodiversity is not only supporting major fisheries but also acting as buffer shorelines against waves, storms, and floods, helping to prevent loss of life, property damage, and land erosion. Pakistan, which has set up an exclusive Ministry of Climate Change, has provided a comprehensive menu of capacity building required in the crucial areas of energy, transport, and agriculture (Lama, 2018). Its Nationally Determined Contribution (NDCs) submitted to the Paris Agreement promised to reduce up to 20 per cent of its 2030 projected GHG emissions (Jaffery, 2018).

On June 17, 2015, India's Cabinet meeting revised the cumulative targets for grid-connected solar-power projects initially envisioned from 20,000 MW by 2021–2022 to an ambitious 100,000 MW by 2021–2022 (Rattani, 2018: 11). Solar parks and Ultra Mega Solar Power Projects will be set up by 2019–2020 with the central government's financial support of Rs 81,000 million. The total capacity, when operational, will generate 64 billion units of electricity per year that will lead to an abatement of around 55 million tonnes of CO_2 per year over its life cycle (Rattani, 2018: 11). The WB, Kreditanstalt für Wiederaufbau (KFW), ADB, and New Development Bank (NDB) sanctioned US$1,300 million for the State Bank of India (SBI), Punjab National Bank (PNB), Canara Bank, and Indian Renewable Energy Development Agency (IREDA) to fund solar rooftops at less than 10 per cent interest (Rattani, 2018: 12). The *Swachh Bharat Abhiyaan* (Clean India Mission) of India, a grand design which was started on October 2, 2014, to ensure India becomes open defecation free (ODF) by 2019 has a success story. As of March 2018, more than 64.32 million toilets have been built across the country. This is according to data tabled in the *Lok Sabha* (Lower House of the Parliament) in March 2018, and 11 states and two Union territories have been declared ODF (Sharma, 2018).

VII Mutual understanding could be good for a change

While the world community is unable to construct global governance for climate change and to regulate it in an effective way, it is seen the SAARC countries have realized they have no choice but to turn to other forms of cooperation at the regional level. This is due to the reason that in dealing with transborder issues, governments often look at cooperative decision-making as a crucial means to strengthen and to exercise shared authority in the framework of regional cooperation (Farrell, 2005: 4; Hey, 2007: 750; Govin, 2013: 103). Numerous geographers, political scientists, environmentalists, and other social scientists share a common interest in Regional Environmental Governance (REG) (Balsiger and Debarbieux, 2011). It may be due to this reason that the thrust for regional cooperation is rooted in the failure of the global movement (Conca, 2012: 127). All states may not be interested in promoting regionalism in the form of regional cooperation (Fawcett, 2005b: 21–22). Nevertheless, regional problems more often do invite regional solutions. For example, when state power alone is inadequate and existing global institutions face severe burdens or whose agendas are heavily skewed to favour key states, regional cooperation is both desirable and necessary (Fawcett, 2005b: 21–22).

When all the South Asian countries face common threats from climate change, the national governments are compelled to cooperate because individual action alone will not be effective. In this region, climate change is no longer a national or subnational problem. In fact, it may be justifiably called a widespread, collective, or regional concern. As a regional cooperation, SAARC may have faced many difficulties to deal with climate change; at the same time it has shown its effectiveness better than its single member. In some way regional cooperations appear to be quite successful in providing high-level forums (Tussie, 2006: 7; Alagappa, 2000: 265). That is because collective action through the framework of regional cooperation can play an important role in providing incentives for improved environmental protection (Kay and Jacobson, 1983; Hveem, 2006: 134).

Most regional cooperations may not be so blessed and have been much less effective, especially in the implementation of certain agendas. This should not be understood to mean that SAARC will not play a significant role or to have a prospect in climate change adaptation. This regional cooperation may be less effective on other issues, but with reference to climate change there is a sense of collective will and cohesively better options. It possessed good regional policies which take into consideration climate change effects in a fundamental way where it can be pursued by its members. In addition, the progress it has made is an indication of bright prospects.

7 Conclusion

This study spells out climate change as one of the most pressing issues affecting food security, floods, and droughts that necessitated SAARC to adopt collective measures through functionalism to cope with the issue and its drastic consequences in South Asia. The reason is that climate change will bring further damages and miseries to the region in the near future. The main effects of climate change focused on in the study are the following.

Rise in temperature

The rising temperature of the globe causes tremendous weather events resulting in various disasters. For example, warmer surface temperatures result in the accelerated melting of ice and snow or the melting of the Hindu Kush and Himalayan mountains' glaciers. The melting of huge glaciers from these gigantic mountains ultimately lead to increasing risk due to the volumetric flow of GLOFs and the increase in the size of rivers and lakes fed by glaciers. With rising temperatures South Asia consequently faces a new or unfamiliar climate change–related threat that is sea level rise.

Sea level rise

Sea level rise threatens the coastal belts and river banks, important tourist infrastructures, beaches, and fishing industry. Sea level rise is of particular concern to the island countries of the Maldives and Sri Lanka, the former of which faces the danger of almost total submergence due to rising sea levels. Coastal South Asian countries, such as Bangladesh, India, the Maldives, Pakistan, and Sri Lanka, have each emphasized the vulnerability of their coastal zones to the rise of sea level. The effects of sea level rise to these countries include seawater intrusion, coral bleaching, and disappearance of beaches. Millions of people living in the low-lying areas and in a few cities of the region such as Mumbai, Kolkata, Dhaka, and Karachi are expected to be flooded on an annual basis. It is worth noting that the urban areas of these cities which are home to millions of people face the greatest risk of flood-related damage over the next century. The crisis of water scarcity in the Himalayan River basins will be intensified by rising sea levels. Saltwater will intrude into freshwater reservoirs. Moreover, heedless anthropogenic activities such as the unrestrained exploitation of surface and groundwater will draw saltwater into the aquifers along the coastal areas of the region. In a flood plain country such as Bangladesh, as the sea levels rise the saltwater intrusion into groundwater tables increases, thereby reducing the utility potential of the groundwater. Worse is that the process of salinity intrusion into groundwater tables is irreversible and as a result will further restrict the freshwater availability in the river basins.

Floods

One of the hardest-hit areas of the world affected by floods is found to be South Asia. This region has a high number of people concentrated in flood-prone areas. Due to poverty, the region also has lower overall flood protection and preparedness, which means that more severe floods pose a greater risk to the people. In fact, millions of people are affected on average every year in India, Bangladesh, Bhutan, Nepal, Pakistan, Sri Lanka, and Afghanistan. Flash floods and volumetric flow of GLOFs cause severe damages in terms of loss of lives, hastening erosion, and damaging property, agricultural land, and important infrastructure in the valley. In addition, the melting of snow and the escalation of monsoons contribute to flood disasters in Himalayan catchments. That is why there is an increasing number of flood victims every year.

Droughts

Due to the melting of the Hindu-Kush Himalayan glaciers, the supply of water is projected to increase in the coming decades as the perennial covering of snow and ice decreases. In the long run, however, the continued reduction of glaciers will ultimately shrink water supply to the glacial lakes and rivers. In addition, monsoons are becoming more erratic. It is also becoming shortened with heavier rainfall in shorter periods of time. This is likely to intensify the phenomenon of droughts as lesser quantities of water are absorbed into groundwater aquifers. Meanwhile, due to erratic rainfall the supply of water during the dry season is expected to decline. In fact, droughts have affected nearly all countries in South Asia. Although droughts do not often result in structural damage and may not claim too many lives or do not bring large-scale destruction, they have considerable effects on livelihood, mainly for the communities dependent on the agricultural sector. Droughts result in income reduction for the people dependent on farming and livestock, and lead to a decline in nutrition and health status and increased incidences of debt. For instance, droughts have often compelled many farmers in India to commit suicide.

Food security

Due to numerous environmental challenges such as sea level intrusion, flooding, erratic rainfall, temperature rise, and groundwater resources depletion, the water situation in many areas of South Asia could become increasingly precarious. Thus, agriculture, the primary source of employment in the region, is likely to face a severe blow not just in terms of economic gaps in demand and supply of food production, but also with regard to loss of livelihood and nutrient consumption. It is to be noted that the agricultural sector in South Asia is almost entirely dependent on the annual monsoon season and the subsequent replenishment of surface and groundwater resources. But when irrigation becomes more and more difficult as groundwater tables recede and annual river flows fluctuate, the lives of people dependent on agriculture become susceptible due to the question of water scarcity. When water becomes a scarce commodity, people will have to relocate their abodes to grow food and to find accessible water to drink. Over the last two decades most of the SAARC members have experienced relatively rapid GDP growth but this did not always translate into significant improvement in food security, with Nepal and Bangladesh already facing this problem. Thus, hunger is a common problem for the South Asian population. Both floods and droughts which have interrupted agricultural production in recent times will intensify the escalating rise of food prices owing to the decline of food production. When income levels are low for the poor and food expenditures are high, they will be in a precarious position to afford food. Likewise, the rich will also face the problem to access food due to the decline of food supply. A hike in the cost of food such as rice and wheat could push millions more South Asian people into extreme poverty. In this way, food insecurity may further escalate hunger to millions of South Asians who are already malnourished. When the competition to gain access to food intensifies and livelihoods are threatened, human security will also be increasingly threatened. Such problems could be the basis for the mass of hungry and impoverished people to become desperate. In this way, the economy of the region will eventually be disturbed. Hunger could even lead to an increase in migration.

Refugees and security

Climate disruption is a major challenge to regional security and sustainability with severe consequences such as population displacements, undermining of development, and an increase in the competition for natural resources that will ultimately implicate social conflict and instability (Podesta, 2019: 3; Ratna, 2015; Bhattacharyya and Werz, 2012). Any variant in water flow, whether an increase or a decrease, will certainly have an effect on the people living in and around the river basins. For instance, the perpetual inundation of land will ultimately deny farmers the ability to plough their fields. With less water during winters and the dry season, those already stressed by the lack of water resources will feel even more pressure, perhaps forcing them to migrate. Accordingly, the intrusion of seawater will force many in Bangladesh and other coastal areas of South Asia to migrate elsewhere as the farmers will not have enough cultivable lands and the attendant problem gets magnified because of the soaring population. On the other hand, drought arises when places are in the grip of water scarcity or when the rivers are reaching dead storage and groundwater is plummeting. This is potentially leading to large migrations. Due to severe misery, drought will trigger migration. People from Bangladesh and other coastal areas of the region will have to migrate both within and outside their respective countries to escape complications. The Maldives is suffering from the brunt of rising sea levels since it does not have enough land to accommodate its population. In future, the sea level will rise by at least 1 m, which raises the possibility that all the Maldivians become refugees and that the Maldives will disappear from the face of the earth forever. Many people in other coastal regions of South Asia will be affected. In addition, the cultural and linguistic differences of the refugees could complicate the situation as these people will find it difficult to adjust to a new environment. Such a situation will create a non-military threat because migrants will have the potential to create law-and-order problems in the adjoining countries and impose economic and social burdens. Further, internal migration of environmental refugees will be a challenge to these countries because, as outsiders being different in ethnicity, linguistic, and religious background, there is the possibility that bitterness and ill-feelings towards the new migrants will rise.

The risk that climate change will bring from its effects such as drought, floods, temperature rise, or a rise in greater numbers of hot days and food shortages will eventually affect human health substantially. Climate change also raises the possibility of hampering the achievement of many of the SDGs, including those on poverty eradication, child mortality, malaria, other diseases as well as environmental sustainability. Taking into consideration the threats faced by the SAARC countries, a need for regional cooperation to achieve greater importance becomes imperative. More so, because the national security in the region is no longer about belligerent forces and military warfare alone. It is also relates to food and water shortages caused by temperature rise that simultaneously results in GLOFs, floods, and droughts. The challenges posed by climate change illustrate the requirement for the nation-states to act together because governments have realized their inability to adapt to its transboundary nature individually. Mutual cooperation among nation-states is required. Its intricate effects call for innovative thinking and transnational creativity.

The study adopts functionalism as an approach for regional cooperation in dealing with climate change in the geographical contiguous South Asia since it focuses on common interests and needs shared by states in the process of agreeable regional cooperation. Resolutions of the inherent differences between the SAARC nations are quite unlikely in the foreseeable future. However, this study has revealed that they can be persuaded to join in a partnership when their interests dictate. The cooperative management for addressing the issue of climate change based on a functional approach that utilizes the common and pressing interests of the SAARC countries is one such example where members have adopted a thematic and problem-oriented approach to regional challenges by participating in extensive functional cooperation on various issues related to climate change. Climate change in a way, therefore, has the potential to unite the SAARC countries so long as the member states recognize climate change as a threat and cooperate to achieve a policy that is both coordinated and inclusive.

Adaptation measures

Besides declarations, all the SAARC Environmental Ministers are organizing environmental meetings. These series of programmes are their idea. The experts (ministers) dealing with such problems are drawn together from all the SAARC members. Functionalism assumes that any technical problem has a solution that can be agreed upon by the objectives created by the technical experts. It can be assumed that the SAARC Environmental Ministers may be considered as the technical experts in their respective fields. Regional cooperation in South Asia attempting to address the issue of climate change rests on the fundamental belief that such a subject can be best addressed by bypassing national territory, that is by giving the task to perform to a new centre (SAARC) which comprises the eight countries of South Asia.

Various initiatives have been taken up by SAARC to strengthen the scope for regional cooperation in the areas of climate change adaptation. One such example is that despite the fact that India's position on GHG emission is different from the other SAARC members, as a regional body it has taken a common stand that it is not in a position to take up the same responsibility like the developed countries since reduction of GHG emission would entail greater economic consequences for SAARC countries. The fact that the SAARC Environment Ministers have been meeting regularly since 1992 shows the importance attached to enhancing regional cooperation in the area of climate change.

Successive SAARC Summits and Meetings of the SAARC Environment Ministers demonstrate the continued momentum for strengthening and intensifying regional cooperation in the area of climate change. They designate the areas of possible actions that will be taken up by them. It is through their mutual understanding that climate change–related natural disasters have been prioritized and its members called for the early implementation of its policies/initiatives. While drawing on the thematic problem-oriented agenda and to functionally manage the challenges confronting the region, a number of projects have been identified. Various committees and working groups have been instituted to deal with floods and droughts as well as in attending to food shortages.

Right from the early 1980s, the members of SAARC have been addressing the issue of climate change by undertaking several measures. Besides several SAARC Summits, the SAARC Ministerial Meetings on climate change have been held on a regular basis. The SAARC Environment Action Plan (1997); Dhaka Declaration and SAARC Action Plan on Climate Change (2008); the Comprehensive Framework on Disaster Management (2006–2015); the Thimphu Statement on Climate Change (2010); the SAARC Convention on Cooperation on Environment (2010); and the SAARC Agreement on Rapid Response to Natural Disasters (2011) are some of the most significant initiatives so far taken through the efforts of its members to strengthen regional cooperation in the areas of climate change adaptation. The 2010 SAARC Summit in Thimphu is a distinctive moment in the history of South Asian regional cooperation and regional policies towards environmental protection and climate change preventive measures. In addition to the fact that the very theme of the meeting was openly addressing the environmental cause as its main goal – "Towards a Green and Happy SouthAsia" – it was also remarkable because the previous SAARC communiqués and pronunciations regarding good practices on environmental issues finally took the first step in the direction of becoming more binding agreements, especially in light of the SAARC Convention in Cooperation on Environment that was agreed to by all the countries. It also marked the shift from the previous soft approach adopted by SAARC on environmental issues to a more committed one, which expressed the conscious will of extending the SouthAsian agenda to international organizations and global negotiations on the issue such as the UNFCCC. To address the issue of food security, SAARC countries have launched the South Asia Food Security Programme for the improvement of crop production and nutrition in the region by pooling together scientific and natural resources. There is also a SAARC Food Bank, which means that the SAARC members are becoming more selective on the functional issue to cooperate on something which would do none of them harm and do all of them good. SAARC members are, therefore, embracing flexibility and openness through SAARC, particularly in climate change adaptation which has led to progress in regional cooperation. All of these existing policies/initiatives/action plans of SAARC have been undertaken to address the climate change–related disasters. When the horrific earthquake hit Nepal in April 2015, India was quick in offering immediate assistance. In this way, a sense of helping each other in a time of crisis is seen, which is indicative of a positive commitment among the members.

Integrating low carbon

The SAARC in general and its individual countries in particular have introduced initiatives to reduce GHG emissions. The Dhaka Declaration on Climate Change (2008) is about cooperation in capacity building including the development of CDM projects and for enhancing south-south cooperation on technology development and transfer. The SAARC Action Plan on Climate Change (2008) covers cooperation on adaptation; sharing on best practices on addressing mitigation; cooperation over technology sharing and transfer; creating a financing mechanism to support climate change actions; and creating public awareness. The commitments of SAARC Action Plan on Climate Change were reiterated in Thimphu Statement on Climate Change (2010), in which SAARC agreed to establish an IGEG.CC to develop clear policy direction and guidance for regional cooperation; undertake an assessment of climate risks in the region; commission a study to explore the feasibility of establishing a SAARC mechanism to provide capital for projects that promote low-carbon technology and renewable energy; and establish a low-carbon research and development institute at a South Asian university. Although the follow-up has not been as strong, this collective declaration is a reflection of the commitment of SAARC to work towards addressing GHG emissions. In Chapter 5, it is seen that all countries in the region have taken parallel initiatives to address emissions by promoting alternative technologies, including hydropower and small renewable power installations serving rural populations (Sharma, Kishan and Doig, 2014: 22). There is also progress in developing power grid integration among countries in the region. The power trading arrangements which have been realized to date are between Nepal-India, Bhutan-India, and India-Bangladesh. Similarly, India and Sri Lanka will soon have a capacity of 500 megawatts (MW), which could be upgraded to 1,000 MW (ESCAP, 2018: 7).

Obstacles

Many of these policies/initiatives/action plans are at a nascent stage of implementation. At the meetings of the SAARC Agriculture Ministers and the SAARC leaders, the enormity of the challenges related to food insecurity and poverty eradication in South Asia was acknowledged. In practice, however, the SAARC Food Bank that had been established to provide emergency supplies to a nation facing a food crisis resulting from the cyclones, floods, droughts, and earthquakes failed to live up to its expected goal. The April 2015 earthquake in Nepal raised questions on the preparedness of the SAARC to meet such challenges. Although SAARC's Secretariat positions were in Kathmandu, which was moreover headed by a Nepali national during the earthquake, yet as a regional body, SAARC could not offer much humanitarian and technical assistance to Nepal immediately after an earthquake struck the Himalayan nation. Further, while Nepal was the chair of SAARC at the time, it had failed to utilize its position to activate SAARC to respond to Nepal's tragedy. Similarly, in spite of the existence of the SDMC, the Nepal earthquake brought to the fore the difficulties and ineffectiveness of this body to rise to the occasion. So, it raised questions of the preparedness of SAARC to meet such challenges in the future in the region in time of crisis. Therefore, despite the positive indications of the endeavour of the SAARC members to address climate change, their attempts have not risen to the expectation. As climate change remains an issue, the recurrence of climate-related disasters at higher intensity is expected, and for this reason, the region must step up its disaster management practices by accelerating the implementation of policies and action plans. Therefore, a strong disaster management regional framework, along with regular regional disaster relief simulation exercises, needs to be developed for effective action.

Efforts have been made at the national level by its members through the development of NAPAs in their own capacity. Bhutan, the Maldives, Afghanistan, and Nepal released their NAPAs as per the UNFCCC directive. These include Afghanistan's joint NCSA and NAPA; Bangladesh's BCCSAP; Bhutan's NAPA, its Vision 2020, and Forest and Nature Conservation rules (2000); the NAPCC and the NGT of India; the NAPA of the Maldives; the NAPA of Nepal; the NCCAS of Sri Lanka; and the National Climate Change Policy of Pakistan and Pakistan Climate Change Act. To address adaptation concerns as part of their national development plans, the explicit focus on disaster risk in the region is observed. Pakistan and Sri Lanka have integrated disaster risk reduction into educational institutions. The Safe Island programme of the Maldives is an integrated effort on addressing vulnerability by a strategic planning for climate change adaptation in the Maldives. Similarly, coastal zone management efforts in India, Pakistan, and Sri Lanka are other cases in this direction. What can be concluded is that, although individual SAARC countries have adopted policy initiatives in addressing the issue of climate change, they have done so based on their individual liabilities to climate change. Though governments in many countries have regulatory procedures for implementing environmental safeguards to minimize or mitigate problems related to climate change, compliance of these safeguards is often low. As such, compliance with these laws and processes must be monitored to ensure that they are not simply a bureaucratic exercise.

Similarly, the progress in developing power grid integration has been uneven. Encouraging progress concentrated only in the northeastern portion of the region has been made through bilateral interconnections. Ongoing political tensions between India and Pakistan prevent the benefits of power trade between the two regional rivals (ESCAP, 2018: 7).

In most of these South Asian countries, climate change–related policies and strategies that have existed for so long have not been revised or updated by their respective governments, despite serious discussion on SDGs taking place in recent times. While some may have updated their approaches to environmental problems in their periodic national plans and poverty reduction strategies and have also set targets in pursuit of the SDGs, these do not follow a common pattern as each country has its peculiar issues with diverse impacts (Zafarullah and Huque, 2018: 28). Given South Asia's vulnerability to the impacts of climate change, its large share of world population and extreme poverty, the world will not be able to achieve SDGs without this region achieving them. Recognizing the multifaceted impacts of climate change in the region, it is essential for SAARC countries to focus on the enhancement of environmental sustainability because a sustainable future is critical to meet the needs of current and future generations.

A large number of cross-border migration in this region is "undocumented" or "clandestine", ranging from simple border crossings to organized trafficking and smuggling of people (Wickramasekara, 2011; Srivastava and Pandey, 2017: 45; Saidi, 2017; Garrote-Sanchez, 2017: 3; Sood, 2017: 35–36; Lal, 2016; Werz and Hoffman, 2015: 99–108; Regmi and Paudyal, 2009, 1–13; Butler, 2009; Sharma and Thapa, 2013). Such illicit movement of people has resulted in exploitation (including sexual exploitation), the violation of migrants' human rights, and even ruthless expulsion (Majaw, 2016: 35, 48; Majaw, 2014: 135; Kajjo and Jedinia, 2019; Lazarus, 2018; Gowen, 2018). The SAARC is undertaking a consultative process to protect the interests of migrants. Its member states are also initiating a range of policies related to emigration (Kathmandu Declaration, 2014; Srivastava and Pandey, 2017: 9; Rabbani, Shafeeqa and Sharma, 2016: 23). In contrast, there is no comprehensive policy on migration related to environmental disasters. Such a lack of strategy measures has been recognized by different scholars (Gill, 2003a; Srivastava, 2012: 48–57; Srivastava *et al.*, 2014). While there may be no legally binding international regimes that address and protect climate migrants, SAARC cannot wait for the world community to tackle this problem. The reason is that climate-induced migration is a cross-cutting issue, relevant to most if not to all of the SDGs in South Asia. It is significantly related to economic, social, and political aspects in the region. In order to achieve the SDGs and ensure that they are carried out inclusively, climate migration must be acknowledged as a complex and multifaceted phenomenon that impacts all areas of governance. The SAARC cannot simply neglect the role of climate change in aggravating human displacement in the region. Therefore, it is vital for the SAARC in general and its members in particular to start inclusive mechanisms to address climate change–induced cross-border migration in the region and to look into myriad environmental factors that play a role.

Confidence building is an ongoing objective of the Summits and Meetings between SAARC members, upon which different layers of functional cooperation have been added at various stages to address the issue of climate change. Although the numerous attempts at addressing the issue is a positive indication, these endeavours, however, have not resulted in yielding major outcomes since the individual SAARC countries have taken a different approach in preparing and implementing their national strategies.

It seems that there is disengagement between the institutional and legislative systems which have been developed to address disaster risk and those developed to address climate change. It is understood that adaptation and its practice in the region is still in the early stages. Zia-ul Hoque Mukta, the Regional Policy Coordinator of Oxfam Asia, said that while each country in the region has been developing policies and programmes on climate change, the closely linked economic, political, and social ties among them make it impossible for each government to implement any measures alone if sustainability is the common goal (Wijenayake, 2013). India's neighbours also maintain that India doesn't take their concerns on board as it is busy being a big power rather than focusing on the needs of other neighbours (Goswami, 2014). Some of the countries of the region receive more assistance than the others from industrialized countries. Further, international donors who provide funds to them represent only a small portion, which is inadequate to initiate capacity building for adaptation action. The study therefore suggests that actions at the national level remain inadequate and calls for national robustness to be strengthened through cooperation and collaboration at the regional level.

Despite the fact that the numerous attempts made by SAARC at addressing the issue are positive indications, these endeavours, however, have not resulted in yielding major outcomes. This is so because the individual SAARC countries have taken a different approach in preparing and implementing their national strategies. Given the differences between the countries in South Asia in terms of their varying size, their diverse economies, and the different levels of their socio-economic development, it is not surprising that their priorities are also different. Preferred adaptation options in the region also differ based on the views and areas of vulnerabilities of the individual members. It is seen that some countries are concentrating adaptation measures in the coastal areas while others are not. Such divergent views may obstruct adaptive actions. Further, for poorer countries like Afghanistan, Nepal, Bhutan, Sri Lanka, the Maldives, and Bangladesh, the priority is to access the "adaptation fund" (the funds that finance tangible adaptation projects in developing countries), while for Pakistan and India it is primarily to access the Clean Development Mechanism (CDM) fund. The CDM fund covers programs that focus on GHG reductions through several activities, such as afforestation and reforestation (often known as sinks), solar electrification, recovery of energy (biogas) from waste, installation of more energy-efficient boilers, and introduction of cleaner transportation methods. The study therefore suggests that actions at the national level remain inadequate and calls for national robustness to be strengthened through cooperation and collaboration at the regional level.

Moreover, the members of SAARC are at the same time the active members of other climate-negotiating blocs. While the SAARC members are from the same region, there are vast differences among them in their socio-economic conditions. These differences influence their positions in many international negotiations including the UNFCCC. For instance, the eight countries of SAARC negotiate climate change effects based their stand on the country's vulnerabilities and interest: India as a BASIC (Brazil, South Africa, India, and China), the Maldives as AOSIS (Alliance of Small Island States), Sri Lanka and Pakistan as LMDC (Like Minded Developing Countries), Bhutan,

Nepal, and Bangladesh as LDCs (Least Developed Countries). The members of these blocs are also determined to battle for the interests of their respective blocs' members outside SAARC. They have their own opinions which can be different from SAARC's points of view. According to the stand of the BASIC, the mounting risk of climate change is rooted in a lack of action from developed countries, which continue to overlook their UNFCCC commitments. Thus, the BASIC promotes itself as unifying and strengthening the position of developing countries in the UNFCCC and urges the rich countries to act on climate change (Blaxekjær and Nielsen, 2014: 751–766). The LMDC stress for a fair, universal, and rule-based regime (a strengthened Kyoto agreement), in which Annex I countries launch and pursue ambitious mitigation actions as well as support developing countries in their contribution to tackling climate change (Almassy, 2014: 33). In addition, the LMDC emphasizes the need for balancing mitigation and adaptation actions, stressing that the new agreement must appropriately address the issue of financing and technology transfer by developed countries (Almassy, 2014: 33). The LDC stresses more on the responsibility of developed countries, but in addition advocates for all countries to take climate actions, that is balancing between adaptation and mitigation (Almassy, 2014: 34). The LDCs such as Bangladesh and Nepal have on occasion demanded that India do more to address climate change (Goswami, 2014). The AOSIS battles for the welfare of Small Island Developing States (SIDs), which is most affected by the rise of sea level due to global warming. The AOSIS demands that developed countries must take ambitious mitigation targets, but it supports quantifiable contributions of developing countries as well. In the meetings in 2013, the AOSIS stressed the principle of common but differentiated responsibilities and respective capabilities, highlighted means of implementation, and called for further work on linkages between existing institutions (Herold *et al.*, 2013: 19–99). These different groupings are promoting the viewpoints and welfare of their member countries as well as playing a major role in building consensus by supporting each other groups' view. This sort of having loyalty to other climate-negotiating blocs may sometimes generate a tendency to take a bi-partisan approach in dealing with environmental problems, bypassing the regional body.

Unlike ASEAN, SAARC does not have an official anthem which is supposed to be an expression of the unity of its members. SAARC needs an anthem which need not necessarily replace the national anthems of its members, but one that will honour and respect the values that all the members share and as an expression of their unity in diversity in the region. It may only be symbolic in nature, but the existence of an anthem can strengthen the sense of SAARC unity and identity and boost a sense of belonging among the people of the region. SAARC has no standard flag to portray South Asia as a community. Perhaps SAARC needs a flag having a motto which can be a visual representation of its people and to distinguish it from other nations. Flying a flag is also a way to show the people's pride and ownership of

SAARC. The existence of the flag may help to unite the SAARC countries better.

Another problem faced by SAARC is that it has no common funding agencies with regard to climate change. It depends on international donors which is often insufficient. Therefore, for countries such as Bangladesh and the Maldives, in addition to the fact that they lack sufficient funding, their ability to adapt to vulnerability is dependent on the rate and magnitude of climate change as well.

Consequently, SAARC countries will have to find alternate ways to generate funds instead of solely depending on the developed countries. One way would be to establish a Special Fund to Climate Change with clear sources of funding to implement its important policies/initiatives. Similarly, efforts will have to be made for research and development in technologies that will reduce GHG emissions. For this, there is the need to build green infrastructure (Green Infrastructure is a network providing the "ingredients" for solving urban and climatic challenges by building with nature) to restore ecosystems, reduce atmospheric carbon, protect and expand carbon sinks. A carbon sink is anything that absorbs more carbon that it releases, whilst a carbon source is anything that releases more carbon than it absorbs. Forests, soils, oceans, and the atmosphere all store carbon and this carbon moves between them in a continuous cycle. Consequently, forests can act as sources or sinks at different times.

The Environment Fund with a plan to bring capital to areas that need the most attention, such as protecting the forests and wildlife, educating the public and policymakers on critical issues affecting the environment, is also required. The reason is that the policymakers and people in general are the important pillars to support in building up the wisdom of environmental safeguards in the region. In the long run citizen engagement with compliance issues will help to make the overall regulatory system more proactive and responsive to the concern about climate change. Thus, a sentiment of green consciousness must be imparted to their awareness through seminars and training programmes etc.

Further, what is stated in law often does not correlate with the actual adversities that affected people are forced to deal with. Since climate change combat is often in the hands of policymakers, for the policy to be effective, the experience and traditional practices of the indigenous community who are repositories of decades of observational data and experimental knowledge must be included.

To achieve this end, work is required by *Track One* to promote functional cooperation and jointly address the issue. The work involved should not be underestimated. It requires identifying the particular function for cooperation, defining the area for joint management, identifying the specific mission of the joint venture, finding a formula to share costs and resources, and establishing a management body. *Track Two* can provide vital assistance in progressing this work. Rather than establishing new forums, ongoing support is required for the existing measures that are based on board as well as the adoption of climate-smart approaches that rest on increased dialogue with and participation of local communities and smallholders.

The SAARC members must realize that the region's tradition of love, respect, and reverence for nature has existed from time immemorial. Historically, the protection of nature and wild life form an ardent piece of faith reflected in the daily lives of the people and also enshrined in the myths, folklore and religion, art and culture. This had been practised even by those rulers of various kingdoms in pre-colonial India who were conscious of environmental conservation. Similarly, there is the pronouncement of Asoka who views the protection of animals' life and the preservation of plants as one of the duties of the king, and forbade killing and hunting of animals for food and game which constitute the teachings of Buddhism and Jainism. Should the members of SAARC honour its Fifteenth Summit, wherein it was expressed to try to restore harmony with nature by drawing on the ancient South Asian cultural values and traditions of environmental responsibility and sustainability, the SAARC region will be safer.

Another important requirement is for all its members and governments to take measures to build a consensus for regional cooperation and to channel increasing proportions of resources towards the protection of the environment as well as to build adaptive capacity. In the end it can be said that climate change in a way has the potential to unite the SAARC countries so long as the member states recognize climate change as a threat and cooperate to achieve a policy that is coordinated and inclusive. In this the ability of SAARC to evolve its food system and capacity to deal with floods and droughts in light of climate change, will be the key indicators of its future progress. Otherwise SAARC ministerial summits and subsequent declarations will have little significance.

References

6th SAES Theme Paper. (2013). *Managing Climate Change, Water Resources, and Food Security in South Asia*. South Asia Watch on Trade Economics and Environment (SAWTEE), Nepal, for Plenary Session 2 of the 6th South Asia Economic Summit, September 2–4.

Aazim, Mohiuddin. (2018). "Challenges in Implementing Food Security Policy." *Dawn*, April 16.

Abeysinghe, Achala. and Prolo, Caroline. (2016). *Entry into Force of the Paris Agreement: The Legal Process*. Issue Paper. London: International Institute for Environment and Development (IIED).

Abhayasinghe, K.R. (2007). "Climate." In: *The National Atlas of Sri Lanka*. Colombo, Sri Lanka: Survey Department of Sri Lanka.

Abhiyan, Wada Na Todo. (2017). *Sustainable Development Goals: Agenda 2030*. A Civil Society Report. Green Park Ext, New Delhi, July 6.

Abid, Muhammad., Schilling, Janpeter., Scheffran, Jürgen. and Zulfiqar, Farhad. (2016). "Climate Change Vulnerability, Adaptation and Risk Perceptions at Farm Level in Punjab, Pakistan." *The Science of the Total Environment*, Vol. 547.

Adams, Barbara., Bissio, Roberto., Ling, Chee Yoke., Judd, Karen., Martens, Jens. and Obenland, Wolfgang. (2016). *Spotlight on Sustainable Development 2016-Report of the Reflection Group on the 2030 Agenda for Sustainable Development*. Bonn: Friedrich Ebert Stiftung.

Adams, Sophie., Baarsch, Florent., Bondeau, Alberte., Coumou, Dim., Donner. Reik., Frieler, Katja., Hare, Bill., Menon, Arathy., Perette, Mahe., Piontek, Franziska., Rehfeld, Kira., Robinson, Alexander., Rocha, Marcia., Rogelj, Joeri., Runge, Jakob., Schaeffer, Michiel., Schewe, Jacob., Schleussner, Carl-Friedrich., Schwan, Susanne., Serdeczny, Olivia., Svirejeva-Hopkins, Anastasia., Vieweg, Marion. and Warszawski, Lila. (2013). *Turn Down the Heat: Climate Extremes, Regional Impacts, and the Case for Resilience: Full Report (English)*. Washington, DC: World Bank.

Address by His Excellency Mr. Maumoon Abdul Gayoom, President of The Republic of Maldives before the Forty Second Session of The United Nations General Assembly on the Special Debate On Environment and Development. (October 19, 1987).

Adhikary, S.K., Das, S.K., Saha, G.C. and Chaki, T. (2013). "Groundwater Drought Assessment for Barind Irrigation Project in Northwestern Bangladesh." In: Adhikary, S.K., Das, S.K., Saha, G.C. and Chaki, T. (eds.) *Groundwater Drought Assessment for Barind Irrigation Project in Northwestern Bangladesh*. Adelaide, Australia: 20th International Congress on Modelling and Simulation.

Adoption of the Paris Agreement, Annex. (2015). ("Paris Agreement"), Art. 4(9): "Each Party Shall Communicate a Nationally Determined Contribution Every Five Years in Accordance with [the Decision of COP-21 Adopting the Paris Agreement]," December 12.

Afghanistan National Development Strategy (ANDS). (2008). Government of Islamic Republic of Afghanistan.

Afghanistan Statement. (2010). The Asia Conference of the Global Climate Change Alliance (GCCA). Dhaka, Bangladesh.

Aftab, Noor. (2012). "Much awaited 'National Climate Change Policy' Approved." *International: The News*, March 18. Management Division.

Agarwal, Anil., Narain, Sunita. and Sharma, Anju. (eds.). (1999). *Green Politics: State of Global Environmental Negotiations*. New Delhi: Centre for Science and Environment.

Agenda 21 Was Revealed at the United Nations Conference on Environment and Development (UNCED) or Earth Summit. (1992). Rio de Janeiro.

Agnihotri, V.K. (2008). Climate Change: Challenges to Sustainable Development in India. Occasional Paper Series (3). New Delhi: Research Unit (LARRDIS), Rajya Sabha Secretariat, October.

Agrawala, Shardul., Ota, Tomoko., Ahmed, Ahsan Uddin., Smith, Joel. and Aalst, Maarten van. (2003a). *Development and Climate Change in Bangladesh: Focus on Coastal Flooding and the Sundarbans*. Environment Directorate Development Co-Operation Directorate. Working Party on Global and Structural Policies Working Party on Development Co-operation and Environment. Paris: Organization for Economic Co-operation and Development.

Agrawala, Shardul., Raksakulthai, Vivian., van Aalst, Maarten., Larsen, Peter., Smith, Joel. and Reynolds, John. (2003b). *Development and Climate Change in Nepal: Focus on Water Resources and Hydropower*. Environment Directorate Development Co-Operation Directorate, Working Party on Global and Structural Policies Working Party on Development Co-operation and Environment.

Ahmad, Munir., Iqbal, Muhammad. and Farooq, Umar. (2015). Food Security and Its Constraining Factors in South Asia: Challenges and Opportunities. MPRA Paper No. 72868.

Ahmad, Q.K., Chowdhury, A.K.A., Imam, S.H. and Sarker, M. (eds.). (2000). *Perspectives on Flood 1998*. Dhaka: The University Press Limited (UPL).

Ahmad, S., Bari, A. and Muhammad, A. (2003). *Climate Change and Water Resources of Pakistan: Impact, Vulnerabilities, and Coping Mechanisms, Proceedings of Year End Workshop*. APN, START, Fred J Hansen Institute for World Peace, Kathmandu, Nepal: Asianics.

Ahmad, Shahid., Hussain, Zahid., Qureshi, Asaf Sarwar., Majeed, Rashida. and Saleem, Mohammad. (2004). *Drought Mitigation in Pakistan: Current Status and Options for Future Strategies*. Drought Series, Paper 3, Working Paper 85.

Ahmadzai, Saifullah. (2013). "Non-Traditional Security Issues in Afghanistan." In: Nayak, Nihar. (ed.) *Cooperative Security Framework for South Asia*. New Delhi: IDSA Pentagon Press.

Ahmed, A.U. (2004). A Review of the Current Policy Regime in Bangladesh in Relation to Climate Change Adaptation. Reducing Vulnerability to Climate Change Project (RVCC), Khulna: CARE Bangladesh.

Ahmed, Mahfuz. and Suphachalasai, Suphachol. (2014). *Assessing the Costs of Climate Change and Adaptation in South Asia*. Manila: Asian Development Bank.

Ahmed, Nafeez Mosaddeq. (2011). "The International Relations of Crisis and the Crisis of International Relations: From the Securitisation of Scarcity to the Militarisation of Society." *Global Change, Peace & Security*, Vol. 23, No. 3.

Aich, Valentin., Akhundzadah, Noor Ahmad., Knuerr, Alec., Jamshed Khoshbeen, Ahmad., Hattermann, Fred., Paeth, Heiko., Scanlon, Andrew. and Paton, Eva Nora. (2017). "Climate Change in Afghanistan Deduced from Reanalysis and Coordinated Regional Climate Downscaling Experiment (CORDEX)- South Asia Simulations." *Climate*, Vol. 5, No. 38.

Akbari, Mohammad Zahir. (2018). "Afghanistan: Drought and climate Change Effects." *Daily Outlook*, February 12, Afghanistan.

Alagappa, Muthiah. (2000). "Environmental Governance: The Potential of Regional Institutions: Introduction." In: Chasek, Pamela S. (ed.) *The Global Environment in the Twenty-First Century: Prospects for International Cooperation*. Tokyo: United Nations University.

Alagh, Yoginder K. (2001). "Water and Food Security in South Asia." *International Journal of Water Resources Development*, Vol. 17, No. 1.

Alam, Mozaharul. and Regmi, Bimal Raj. (2004). *Adverse Impacts of Climate Change on Development of Nepal: Integrating Adaptation into Policies and Activities*. CLACC Working paper No. 3. Dhaka: International Institute for Environment and Development, and Bangladesh Center for Development Studies.

Alam, Mozaharul. and Tshering, Dago. (2004). *Adverse Impacts of Climate Change on Development of Bhutan: Integrating Adaptation into Policies and Activities, Capacity Strengthening in the Least Developed Countries (LDCs) For Adaptation to Climate Change (CLACC), CLACC*. Working Paper No. 2. Bangladesh Centre for Advanced Studies (BCAS).

Alderson, Kai. and Hurrel, Andrew. (eds.). (2003). *Hedley Bull on International Society*. London: Palgrave Macmillan.

Aldy, Joseph E. (2017). "Real World Headwinds for Trump Climate Change Policy." *Bulletin of the Atomic Scientists*, October 16, doi:10.1080/00963402.2017.1388673.

Ali, Akhtar. (2013). *Indus Basin Floods: Mechanisms, Impacts, and Management*. Environment, Natural Resources, and Agriculture/Pakistan/2013, Philippines: Asian Development Bank.

Ali, Shahjahan., Alam, Khandaker Jahangir., Islam, Shafiul. and Hossain, Morshed. (2015). "An Empirical Analysis of Population Growth on Economic Development: The Case Study of Bangladesh." *International Journal of Economics, Finance and Management Sciences*, Vol. 3, No. 3.

Aljazeera. (2014). "Maldives Capital in Crisis as Water Supply Dries Up," December 8. Available at <http://america.aljazeera.com/articles/2014/12/8/maldives-water-crisis.html>. Accessed on February 14, 2020.

Allan, Jennifer., Bhandary, Rishikesh Ram., Bisiaux, Alice., Chasek, Pamela., Jones, Natalie., Luomi, Mari., Schulz, Anna., Verkuijl, Cleo. and Woods, Bryndis. (2017). *From Bali to Marrakech: A Decade of International Climate Negotiations*. New York: The International Institute for Sustainable Development.

Alley, Richard B. (2000). *The Two-Mile Time Machine: Ice Cores, Abrupt Climate Change, and Our Future*. New York: Princeton University Press.

Allison, I. (2009). *The Copenhagen Diagnosis Updating the World on the Latest Climate Science*. Sydney: UNSW Climate Change Research Centre.

Almassy, Dora. (2014). *Handbook for ASEAN Government Officials on Climate Change and the UN Sustainable Development Goals*. Hanoi: Hanns Seidel Foundation Viet Nam.

Anderegg, W.R.L., Prall, J.W., Harold, J. and Schneider, S.H. (2010). "Expert Credibility in Climate Change." *Proceeding of National Academy of Sciences*, No. 107, U.S.A.

Anglia Ruskin University. (2014). *Climate Change, Resource Scarcity & Conflict: Case Studies of Shared Water Resources in the Indian Subcontinent.* Cambridge: Global Sustainability Institute,

Anglo, E.G., Bolhofer, W.C., Erda, L., Huq, S., Lenhart, S., Mukherjee, S.K., Smith, J. and Wisniewski, J. (1996). *Regional Workshop on Climate Change Vulnerability and Adaptation in Asia and the Pacific: Workshop Summary.* Dordrecht: Kluwer Academic Publishers.

Ansorg, Thomas. and Donnelly, Thomas. (2008). "Climate Change in Bangladesh: Coping and Conflict." *European Security Review*, No. 40, September.

Anthoff, David., Nicholls, Robert J. and Tol, Richard, S. J. (2010). "The Economic Impact of Substantial Sea Level Rise." *Mitigation and Adaptation Strategies for Global Change*, Vol. 15, No. 4.

Antos, David. (2017). "India, Climate Change and Security in South Asia." *BRIEFER*, The Center for Climate and Security, No. 36, May 3.

Anup, K.C. (2017). "Climate Change and Its Impact on Tourism in Nepal." *Journal of Tourism and Hospitality Education*, Vol. 7, https://doi.org/10.3126/jthe. v7i0.17688.

APPRO. (2014). *Climate Change and Food Security in Afghanistan: Evidence from Balkh, Heart and Nangarhar.* Occasional Paper. Afghanistan Public Policy Research Organization.

Aranzadi, Javier. (2006). *Liberalism against Liberalism Theoretical analysis of the works of Ludwig von Mises and Gary Becker.* New York: Routledge.

Area of Cooperation- Environment, Climate Change and Natural Disasters. (2012). Available at <www.saarc-sec.org/userfiles/Webupdate-AreaofCooperation-ENB Nov2014.pdf>. Accessed on February 28, 2015.

Arif, Abdullah Al. and Karim, Md. Ershadul. (2015). "A Research Guide on the South Asian Association for Regional Cooperation (SAARC)." *Globalex*, June. Available at <www.nyulawglobal.org/globalex/SAARC.html>. Accessed on February 14, 2020.

Armington, Stan. (2002). *Bhutan.* Franklin: Lonely Planet Publishers.

Arndt, C., Chinowsky, P., Robinson, S., Strzepek, K., Tarp, R. and Thurlow, J. (2012). "Economic Development under Climate Change." *Review of Development Economics*, Vol. 16.

Arrow, Kenneth. (2007). "Global Climate Change: A Challenge to Policy." *Economists' Voice Art*, Vol. 4, No. 3.

Arruda, Pedro Lara. (2012). "Unconditional Consciousness." *The Framework of the XVI SAARC Summit's Environmental Approach*, February 19.

Aryal, Jeetendra Prakash., Sapkota, Tek B., Khurana, Ritika., Khatri-Chhetri, Arun., Rahut, Dil Bahadur. and Jat, M.L. (2019). "Climate Change and Agriculture in South Asia: Adaptation Options in Smallholder Production Systems." *Environment, Development and Sustainability.* https://doi.org/10.1007/s10668-019-00414-4.

Asaduzzaman, M. (1994). *Economic and Social Impacts of Climate Change: A Case Study of Bangladesh Coastal Zones.* Mimeo. Dhaka: Bangladesh Institute of Development Studies.

Ashfaq, M., Shi, Y., Tung, W.W., Trapp, R.J., Gao, X., Pal, J.S. and Diffenbaugh, N.S. (2009). "Suppression of South Asian Summer Monsoon Precipitation in the 21st Century." *Geophysical Research Letters*, Vol. 36, L01704, 3 January.

Ashraf, Arshad., Naz, Rozina. and Roohi, Rakhshan. (2012). "Glacial Lake Outburst Flood Hazards in Hindukush, Karakoram and Himalayan Ranges of Pakistan: Implications and Risk Analysis." *Geomatics, Natural Hazards and Risk*, Vol. 3, No. 2.

Asian Development Bank (ADB). (2003). *Country Strategy and Programme Update, 2004–2006*. Manila: ADB.

Asian Development Bank (ADB). (2006). *Technical Assistance for the Development Partnership Program for South Asia*. Manila: ADB.

Asian Development Bank (ADB). (2010). *Climate Change in South Asia: Strong Responses for Building a Sustainable Future*. Manila: ADB.

Asian Development Bank (ADB). (2013). *Bhutan Transport 2040 Integrated Strategic Vision*. Manila: ADB.

Asian Development Bank (ADB). (2015). *The Asian Development Bank and the Climate Investment Funds: Country Fact Sheets*. Manila, Philippines: ADB Climate Change and Disaster Risk Management Division.

Asian Development Bank (ADB). (2017). *Country Poverty Analysis (Detailed) Nepal, Country Partnership Strategy: Nepal, 2013–2017*. Bangkok: Asian Development Bank.

Asif, Muhammad. (2013). Climatic Change, Irrigation Water Crisis and Food Security in Pakistan. Examensarbete i HållbarUtveckling 170, Master Thesis. Department of Earth, Uppsala University Sciences.

Atkinson, A. and Davis, J. (2001). "Donella Meadows, Lead Author of 'The Limits to Growth', Has Died." *Ecological Economics*, Vol. 38, No. 2.

Austen, Ian. (2016). "Wildfire Empties Fort McMurray in Alberta's Oil Sands Region." *New York Times*, 3rd May.

Averchenkova, Alina. (2010). *The Outcomes of Copenhagen: The Negotiations & the Accord*. UNDP Environment & Energy Group Climate Policy Series.

Axelrod, Robert. (1984). *Evolution of Cooperation*. New York: Basic Books, Inc Publishers.

Ayeb-Karlsson, Sonja. (2017). "Facing Disasters: Lessons from a Bangladeshi Island." *Down to Earth*, August 28.

Ayers, Jessica M. and Huq, Saleemul. (2008a). *Adaptation Funding and Development Assistance: Some FAQs*, IIED Briefing. London: IIED.

Ayers, Jessica M. and Huq, Saleemul. (2008b). "The Value of Linking Mitigation and Adaptation: A case study of Bangladesh." *Environmental Management*, Vol. 43, No. 5.

Azimi, Ali. and McCauley, David. (2003). *Afghanistan's Environment in Transition*. Philippines, Manila: Asian Development Bank.

Baba, Nazran. (2010). "Sinking the Pearl of the Indian Ocean: Climate Change in Sri Lanka." *Global Majority E-Journal*, Vol. 1, No. 1, June.

Baghel, Ravi., Stepan, Lea. and Hill, Joseph K.W. (2016). *Water, Knowledge and the Environment in Asia: Epistemologies, Practices and Locales (Earthscan Studies in Water Resource Management)*. London: Routledge.

Bahadur, Tejendra., Air, Anju., Uprety, Batu K. and Midha, Nish. (2017). *Nepal's Approach to Climate Change Adaptation with Local Adaptation Plans for Action (LAPAs): A Water Resource Perspective*. Water Resource Perspective.

Baishyal, Reshu. (2017). "Climate Change Impacts in Water Resources: A Case of Nepal." June 22. Available at <http://climatetracker.org/climate-change-impacts-water-resources-case-nepal/>. Accessed on February 14, 2020.

Bajracharya, Samjwal Ratna. and Shrestha, Basanta. (eds.). (2011). *The Status of Glaciers in the Hindu Kush-Himalayan Region*. Kathmandu, Nepal: International Centre for Integrated Mountain Development (ICIMOD).

Balaji, J. (2010). "Lok Sabha Passes Green Tribunal Bill." *The Hindu*, May 1.

The Bali Action Plan: Key Issues in the Climate Negotiations: Summary for Policy Makers. (2008). An Environment & Energy Group Publication. Chad Carpenter, UNDP, September.

Balling Jr., R.C. (1992). *The Heated Debate: Greenhouse Predictions Versus Climate Reality*. San Francisco, CA: Pacific Research Institute for Public Policy.

Balsiger, Jorg. and Debarbieux, Bernard. (eds.). (2011). "Regional Environmental Governance: Interdisciplinary Perspectives, Theoretical Issues, Comparative Designs." *Procedia-Social and Behavioral Sciences*, Vol. 14, Amsterdam: Elsevier.

Bandara, J. S., and Cai, Y. (2014). "The Impact of Climate Change on Food Crop Productivity, Food Prices and Food Security in South Asia." *Economic Analysis and Policy*, Vol. 44, No. 4.

Bandyopadhyay, Chandrani. (2007). *Disaster Preparedness for Natural Hazards: Current Status in India*. Kathmandu: ICIMOD.

Banerjee, Shrestha. (2015). "Vanishing Islands of Bangladesh." *Down to Earth*, August 17.

Banerjee, Shrestha. and Juneja, Sugandh. (2015). "Living on the Edge." *Down to Earth*, August 17.

Bangladesh Ministry of Environment and Forests (BMEF). (2005). *National Adaptation Programme of Action*. Bangladesh.

Bangladesh Ministry of Environment and Forests (BMEF). (2009). *Bangladesh Climate Change Strategy and Action Plan*.

Bansal, Alok. and Datta, Sreeradha. (eds.). (2012). *South Asian Security: 21st Century Discourses*, Routledge Contemporary South Asia Series. London: Routledge.

Banuri, Tariq. (1993). *The Pakistan National Conservation Strategy: A Plan of Action for the 1990s*. Policy Paper Series # 3, A publication of the Sustainable Development Policy Institute (SDPI).

Bartlett, Ryan., Bharati, Luna., Pant, Dhruba., Hosterman, Heather. and McCornick, Peter. (2010). *Climate Change Impacts and Adaptation in Nepal*. IWMI Working Paper 139. Colombo: International Water Management Institute.

Barnett, J. (2003). "Security and Climate Change." *Global Environmental Change*, Vol. 13, No. 1.

Barnett, J. and Adger, N. (2007). "Climate Change, Human Security and Violent Conflict." *Political Geography*, Vol. 26.

Barnett, Jon. and Campbell, John. (2010). *Climate Change and Small Island States Power- Knowledge and the South Pacific*. London: Earthscan.

Barnett, T.P., Adam, J.C. and Lettenmaier, D.P. (2005). "Potential Impacts of Warming Climate on Water Availability in Snow-Dominated Regions." *Nature*, 438.

Barua, Anamika., Narain, Vishal. and Vij, Sumit. (eds.). (2018). *Climate Change Governance and Adaptation: Case Studies from South Asia*. Boca Raton, FL: CRC Press.

Basnayake, B.R.S.B. (2007). "Climate Change." In: *The National Atlas of Sri Lanka*. Colombo: Survey Department of Sri Lanka.

Basnet, R. (2009). "Carbon Ownership in Community Managed Forests." *Journal of Forest and Livelihood*, Vol. 8, No. 1.

Bates, B.C., Kundzewicz, Z.W., Wu, S. and Palutikof, J.P. (eds.). (2008). *Climate Change and Water*. Technical Paper of the Intergovernmental Panel on Climate Change. Geneva: IPCC Secretariat.

Baylis, John. and Smith, Steve. (2005). *The Globalization of World Politics: An Introduction to International Relations*. New Delhi: Oxford University Press.

Baylis, John. and Smith, Steve. (2011). *The Globalization of World Politics: An Introduction to International Relations*. New Delhi: Oxford University Press.

Baynes, Chris. (2019). "South Asia Floods Force Millions to Flee Homes as Death Toll Rises to 300." *Independent*, July 22.

BBC China. (2011a). "Drought Turns into Floods, Serious Flooding in the South." *BBC China*, June 6.

BBC China. (2011b). "Mid and Lower Stretches of Yangtze Face Drought and Flood." *BBC China*, June 6.

BBC China. (2011c). "Drought Turns into Floods, Serious Flooding in the South". *BBC China*, June 14.

Beaumont, P. (2015). "Deadly Everest Avalanche Triggered by Nepal Earthquake." *The Guardian*, April 26.

Beder, Sharon. (2007). *Environmental Principles and Policies: An Interdisciplinary Introduction*. Sydney: UNSW Press.

Bengali, Kaiser. and Jury, Allan.(2010). "Hunger Vulnerability and Food Assistance in Pakistan: The World Food Program Experience." In: Kugelman, Michael. and Hathaway, Robert M. (eds.) *Hunger Pains: Pakistan's Food Insecurity*. Washington, DC: Woodrow Wilson International Center for Scholars.

Beniston, M. (2010). "Climate Change and Its Impacts: Growing Stress Factors for Human Societies." *International Review of the Red Cross*, Vol. 92.

Berthier, E., Arnaud, Y., Rajesh, Y., Sarfaraz, A., Wagnon, P. and Chevallier, P. (2007). "Remote Sensing Estimates of Glacier Mass Balances in the Himachal Pradesh (Western Himalaya, India)." *Remote Sensing of Environment*, Vol. 108, No. 3.

Beukering, Pieter van. and Vellinga, Pier. (1996). "Climate Change: From Science to Global Policies." In Sloep, Peter. and Blowers, Andrew. (eds.) *Environmental Policy in an International Context: Environmental Problems as Conflict of Interest*. London: Arnold, a member of Hodder Headline group.

Bhalla, Nita. (2009). "Disaster-Prone Bangladesh Trials Early Warning Cell Phone Alerts." *Reuters*, June 23.

Bhatt, Ramesh Prasad. and Khanal, Sanjay Nath. (2010). "Environmental Impact Assessment System and Process: A Study on Policy and Legal Instruments in Nepal." *African Journal of Environmental Science and Technology*, Vol. 4, No. 9.

Bhattacharyya, Arpita and Werz, Michael. (2012). *Climate Change, Migration, and Conflict in South Asia Rising Tensions and Policy Options across the Subcontinent*. Heinrich Boll Stiftung, Centre for American Progress.

Bhattarai, Rudra., Pokhrel, Meena., Singh, Trijan. and Rupakheti, Deepakar. (2015). "SAARC's Policies and Programs on Food Security and Nutrition, Food and Seed Banks." *Researches on the Situation of Seeds in Selected SAARC Countries*. AFA Research Report, National Land Rights Forum Nepal, March.

Bhushal, Yubaraj. (2013). "NPC at Work on 3-Year Plan." *NEW SPOTLIGHT*, Vol. 6, No. 15, January 15.

Bhutan Observer. (2012). "Pemagatshel Facing Drought without Rain." Thimphu, June 16.

Bhutan Observer. (2010). "COP16 Hope for Bhutan." December 17.

Bialos, Jeffrey P., Koehl, Stuart L., Catarious, David M. and Spaulding, Suzanne E. (2008). *Ideas for America's Future: Core Elements of a New National Security Strategy.* Washington, DC: Center for Transatlantic Relations, The Johns Hopkins University.

Bidwai, Praful. (2012). *The Politics of Climate Change and the Global Crisis: Mortgaging Our Future.* Hyderabad: Orient Blackswan Private Limited.

Black, R., Arnell, N.W., Adger, W.N., Thomas, D. and Geddes, A. (2013). "Migration, Immobility and Displacement Outcomes Following Extreme Events." *Environmental Science & Policy*, Vol. 27, Nos. S32–S43.

Blaxekjær, Lau Øfjord. and Nielsen, Tobias Dan. (2014). "Mapping the Narrative Positions of New Political Groups under the UNFCCC." *Climate Policy*, Vol. 15, No. 6.

Blood, Peter R. (ed.). (2001). *Afghanistan: A Country Study.* Washington: GPO for the Library of Congress.

Bloom, David E. and Rosenberg, Larry. (2011). *The Future of South Asia: Population Dynamics, Economic Prospects, and Regional Coherence.* PGDA Working Paper No. 68. Program on the Global Demography of Aging.

BMCI. (2016). *Bhutan: Climate Change: Handbook.* Thimphu, Bhutan: Bhutan Media and Communications Institute.

Bodansky, Daniel. (1999). "The Legitimacy of International Governance: A Coming Challenge for International Environmental Law?" *Journal of International Law*, Vol. 93, No. 3.

Bodansky, Daniel. (2001). "Bonn Voyage: Kyoto's Uncertain Revival." *The National Interest*, No. 65, Fall.

Bodansky, Daniel. (2015). "Legally-Binding vs. Non-Legally Binding Instruments." In: Barrett, Scott., Carraro, Carlo. and Melo, Jaime de. (eds.) *Towards a Workable and Effective Climate Regime.* London: Centre for Economic Policy Research Press.

Bodansky, Daniel. (2016). "The Legal Character of the Paris Agreement." *Review of European, Comparative, and International Environmental Law (RECIEL)*, Vol. 25, No. 2.

Böhringer, Christoph. and Rutherford, Thomas F. (2017). *US Withdrawal from the Paris Agreement: Economic Implications of Carbon-Tariff Conflicts, The Harvard Project on Climate Agreements*, Discussion Paper 17–89, August.

Bolch, T., Buchroithner, M.F., Peters, J., Baessler, M. and Bajracharya, S. (2008a). "Identification of Glacier Motion and Potentially Dangerous Glacial Lakes in the Mt. Everest Region/ Nepal Using Spaceborne Imagery." *Natural Hazards and Earth System Sciences*, Vol. 8, No. 6.

Bolch, T., Buchroithner, M.F., Pieczonka, T. and Kunert, A. (2008b). "Planimetric and Volumetric Glacier Changes in the Khumbu Himal, Nepal, since 1962 using Corona, Landsat TM and ASTER Data." *Journal of Glaciology*, Vol. 54, No. 187.

Bolch, T., Kulkarni, A., Kääb, A., Huggel, C., Paul, F., Cogley, J.G., Frey, H., Kargel, J.S., Fujita, K., Scheel, M., Bajracharya, S. and Stoffel, M. (2012). "The State and Fate of Himalayan Glaciers." *Science*, Vol. 336, No. 6079.

Bolin, Bert. (1997). "International Scientific Networks." In: Rolen, Mats., Sjoberg, Helen. and Svedin, Uno. (eds.) *International Governance on Environmental Issues.* Dordrecht: Kluwer Academic Publishers.

Bose, Sahana. (2013). "Sea-Level Rise and Population Displacement in Bangladesh: Impact on India." *Maritime Affairs: Journal of the National Maritime Foundation of India*, Vol. 9, No. 2.

Bousquet, P., Tyler, S.C., Peylin, P., Van Der Werf, G.R., Prigent, C., Hauglustaine, D.A., Dlugokencky, E.J., Miller, J.B., Ciais, P., White, J., Steele, L.P., Schmidt, M., Ramonet, M., Papa, F., Lathière, J., Langenfelds, R.L., Carouge, C. and Brunke, E.G. (2006). "Contribution of Anthropogenic and Natural Sources to Atmospheric Methane Variability." *Nature*, Vol. 443, No. 7110.

Boyle, Rebecca. (2012). "Maldivian Leaders Might Move the Entire Nation to Australia If Sea Keeps Rising." *Science*, January 10.

Brack, Duncan. and Gray, Kevin. (2003). *Multilateral Environmental Agreements and the WTO, Report*. London: The Royal Institute of International Affairs.

Brammer, Hugh. (2009). "Climate Refugees: A Rejoinder." *Economic and Political Weekly*, Vol. 44, No. 29.

Brecher, Michael. (1966). *The New States of Asia: A Political Analysis*. London: Oxford University Press.

Brechin, Steven R. (2003). "Comparative Public Opinion and Knowledge on Global Climatic Change and the Kyoto Protocol: The U.S. Versus the World?" *The International Journal of Sociology and Public Policy*, Vol. 23, No. 10.

Breidenich, Clare., Magraw, Daniel., Rowley, Anne. and Rubin, James W. (1998). "The Kyoto Protocol to the United Nations Framework Convention on Climate Change." *American Journal of International Law*, Vol. 92, No. 2.

Bridge to India. (2017). "Renewable Capacity Addition Catches up with Thermal Power in India." October 20. Available at <www.bridgetoindia.com/renewable-capacity-addition-catches-thermal-power-india/>. Accessed on February 14, 2020.

Brouwer, R., Akter, S., Brander, L. and Haque, E. (2007). "Socioeconomic Vulnerability and Adaptation to Environmental Risk: A Case Study of Climate Change and Flooding in Bangladesh." *Risk Analysis*, Vol. 27.

Brown, Antje. and Kutting, Gabriela. (2008). "The Environment." In: Salmon, Trevor C. and Imber, Mark F. (eds.) *Issues in International Relations*. New York: Routledge.

Brown, Oli. (2008). *Migration and Climate Change*. Geneva, Switzerland: International Organization for Migration (IOM), No. 31.

Bruce, J.P., Lee, H. and Haites, E.F. (eds.). (1996). *Climate Change 1995: Economic and Social Dimensions of Climate Change*. Contribution of Working Group III to the Second Assessment Report of the Intergovernmental Panel on Climate Change. United Kingdom and New York: Cambridge University Press.

Brunner, Ronald D. (2001). "Science and the Climate Change Regime." *Policy Sciences*, Vol. 34, No. 1.

Bull, Hedley. (1997). *The Anarchical Society: A Study of Order in World Politics*. London: Macmillan.

Bull, S.R., Bilello, D.E., Ekmann, J., Sale, M.J. and Schmalzer, D.K. (2007). "Effects of Climate Change on Energy Production and Distribution in the United States." In: Wilbanks, T.J., Bhatt, V., Bilello, D.E., Bull, S.R., Ekmann, J., Horak, W.C., Huang, Y.J., Levine, M.D., Sale, M.J., Schmalzer, D.K. and Scott, M.J. (eds.) *Effects of Climate Change on Energy Production and Use in the United States*. Washington, DC: U.S. Climate Change Science Program.

Bunsha, Dionne. (2005). "Rain Havoc in Gujarat." *Frontline*, Vol. 22, No. 15.

Buob, Seraina. and Stephan, Gunter. (2013). "On the Incentive Compatibility of Funding Adaptation." *Climate Change Economics*, Vol. 4, No. 2.

Burck, Jan., Hagen, Ursula., Marten, Franziska., Höhne, Niklas. and Bals, Christoph. (2019). *Climate Performance Index: Results 2019*. Bonn, Germany: Germanwatch.

Burck, Jan., Marten, Franziska. and Bals, Christoph. (2017). *Results 2017*, Climate Change Performance Index. Bonn, Germany: German Watch.

Burke, J. and Walker, P. (2014). "Nepal Blames Cheap Tourists for Falling Victim to Snowstorm in Himalayas." *TheGuardian*, October 17.

Burkett, V.R. and Davidson, M.A. (eds.). (2012). *Coastal Impacts, Adaptation and Vulnerability: A Technical Input to the 2012 National Climate Assessment*. Cooperative Report to the 2013 National Climate Assessment. Washington: Island Press.

Burroughs, William James. (2007). *Climate Change: A Multidisciplinary Approach*. Cambridge: Cambridge University Press.

Burt, T.P. and Weerasinghe, K.D.N. (2014). "Rainfall Distributions in Sri Lanka in Time and Space: An Analysis Based on Daily Rainfall Data." *Climate*, Vol. 2.

Burton, Ian. (1996). "The Growth of Adaptation Capacity: Practice and Policy." In: *Adapting to Climate Change: An International Perspective*. New York: Springer-Verlag.

Busby, J.W. (2007). *Climate Change and National Security: An Agenda for Action*. New York: Council on Foreign Relations.

Busby, John. (2016). "The Paris Agreement: When is a Treaty not a Treaty?" *Global Policy*, April 26.

Business Bhutan. (2011). Quoted in *Bhutan Weekly News- Roundup*. October 1–7. Available at <www.bhutan-research.org>. Accessed on June 11, 2015.

Business Recorder. (2012). "Pakistan Should Take Strong Position on Global Climate Change Issues: Advisor (Climate Change)." October 19.

Business Standard. (2018). "Kerala Floods: Sikh Community Provides Food to Victims." August 20.

Butler, Phil. (2009). "Melting Glaciers of the Himalayas Threaten All of Us, But Asia First." *Everything PR*, The Public Relations News Portal.

Butler, R.A. (2007). "Damage to Yangtze 'Irreversible' Says China." *San Francisco: Mongabay Rainforests*, April 16.

Butt, Atif. (2015). "Karachi May Sink into the Ocean by 2060, Senate Warns." *Dawn*, March 13.

Butterworth, M.K., Morin, C.W. and Comrie, A.C. (2017). "An Analysis of the Potential Impact of Climate Change on Dengue Transmission in the Southeastern United States." *Environmental Health Perspectives*, Vol. 125.

Byravan, Sujatha. and Rajan, Sudhir Chella. (2009). "The Social Impacts of Climate Change in South Asia." *JMRI*, Vol. 5, No. 3.

Cai, Kevin G. (2010). *The Politics of Economic Regionalism: Explaining Regional Economic Integration in East Asia*. New York: Palgrave Macmillan.

Calder, Gideon. and McKinnon, Catriona. (2012). "Introduction: Climate Change and Liberal Priorities." In: Calder, Gideon. and McKinnon, Catriona. (eds.) *Climate Change and Liberal Priorities*. London: Routledge.

Caldwell, Lynton K. (1974). "From Stockholm to Nairobi to Caracas: Route toward a New International Law?" *IUSTITIA*, Vol. 2, No. 2, Article 5.

Callaghan, A. and Thapa, R. (2015). "An Oral History of Langtang, The Valley Destroyed by Nepal Earthquake." September 28.

Calvin, William H. (2002). *A Brain for All Seasons: Human Evolution and Abrupt Climate Change*. Chicago: University of Chicago Press.

Canuto, Otaviano. (2013). "South Asia and the Geography of Poverty." *HUFFPOST*, March 18.

Carlsnaes, Walter., Risse, Thomas. and Simmons, Beth A. (2002). *Handbook of International Relations*. London: Sage Publications Ltd.

Carr, E.H. (2001). *The Twenty Years' Crisis, 1919–1939: An Introduction to the Study of International Relations*. New York: Palgrave Macmillan.

Carrington, Damian. (2015). "14 of the 15 Hottest Years on Record Have Occurred since 2000, UN Says." *The Guardian*, February 2.

Carson, Rachel. (1965). *Silent Spring*. Hemsworth: Penguin.

Cash, R.A., Halder, S.R., Husain, M., Islam, M.S., Mallick, F.H., May, M.A., Rahman, M. and Rahman, M.A. (2013). "Bangladesh: Innovation for Universal Health Coverage." *The Lancet*, Vol. 382, No. 9910.

Castle, S. (2002). *Environmental Change and Force Migration: Making Sense of the Debate*. New Issues in Refugees Research Working Paper 70. Geneva: United Nations High Commissioner for Refugees.

Catton, W. (1982). *Overshoot: The Ecological Basis of Revolutionary Change*. Chicago: University of Illinois Press.

CDKN. (2014). The IPCC's Fifth Assessment Report: What's in Is for South Asia? UK Department for International Development.

Census. (2011). *Population and Housing Census: Preliminary Results*. Dhaka, Bangladesh.

Central Bank of Sri Lanka. (2010). *Annual Report*. Colombo.

Central Intelligence Agency (CIA). (2015). *Maldives. The CIA World Fact Book*. Available at <www.cia.gov/library/publications/the-world-factbook/geos/mv.html>.

Chadburn, O. (2007). "Storing Rainwater in Rajasthan." *Footsteps*, Vol. 70, March.

Chandra, Shekhar. (2019). "Indian Cities Are Becoming Urban Heat Islands." *CITYLAB*, August 23. Available at <www.citylab.com/environment/2019/08/heat-wave-india-urban-island-effect-climate-global-warming/596371/>. Accessed on February 14, 2020.

Chandran, Suba. (2012). "Disaster Management in South Asia: A Regional Approach." In: Delinic, Tomislav. and Pandey, Nishchal N. (eds.) *Regional Environmental Issues: Water and Disaster Management*. Kathmandu, Nepal: Centre for South Asian Studies.

Chandrapala, L. (2007a). "Temperature." In: *The National Atlas of Sri Lanka*. Colombo, Sri Lanka: Survey Department of Sri Lanka.

Chandrapala, L. (2007b). "Rainfall." In: *The National Atlas of Sri Lanka*. Survey Department of Sri Lanka.

Chang, J.H. (1969). "The Indian Summer Monsoon." *Geographical Review*, Vol. 57, No. 3.

Charter of the South Asian Association for Regional Cooperation. (1985). Kathmandu, SAARC Secretariat.

Chatterjee, Bipul. and Khadka, Manbar. (eds.). (2011). *Climate Change and Food Security in South Asia*. Jaipur: CUTS International.

Chaudhry, Qamar Uz Zaman. (2017). *Climate Change Profile of Pakistan*. Philippines, Manila: Asian Development Bank.

Chaudhary, Q.Z., Mahmood, A., Rasul, G. and Afzaal, M. (2009). "Climate Change Indicators of Pakistan." In: *Technical Report No. PMD-22/2009*. Islamabad: Pakistan Meteorological Department.

Chaudhary, S., Jimee, G.K. and Basyal, G.K. (2015). Trend and Geographical Distribution of Landslides in Nepal Based on Nepal DesInventar Data. Paper Presented at The Fourteenth International Symposium on New Technologies for Urban Safety of Mega Cities in Asia, Kathmandu, Nepal, October 29–31.

Cheema, Pervaiz Iqbal. (1986). "Threat Perceptions in South Asia and Their Impact on Regional Cooperation." In: Sen Gupta, Bhabani. (ed.) *Regional Cooperation and Development in South Asia.* New Delhi: South Asian Publishers.

Chella, Rajan Sudhir. (2008). *Blue Alert: Climate Migrants in South Asia: Estimate and Solution.* Bangalore: Greenpeace India, Society.

Chestnoy, S. and Gershinkova, D. (2017). "USA Withdrawal from Paris Agreement: What Next?" *International Organisations Research Journal,* Vol. 12, No. 4.

Chhetri, Dipika. (2010). *Bhutan Today,* Thimphu, Posted February 1.

Chhibber, Bhrati. (2004). *Regional Security and Regional Cooperation: A Comparative Study of SAARC and ASEAN.* New Delhi: New Century Publications.

Chophel, Tshering. and Dorji, Tashi. (2015). *Mainstreaming: Gender, Environment, Climate-Change, Disaster, and Poverty, into the Development Policies, Plans and Programmes in Bhutan, Experiences, Challenges and Lessons.* Department of Local Governance Ministry of Home and Cultural Affairs. Thimphu: Royal Government of Bhutan.

Choudhury, A.M., Quadir, D.A., Neelormi, S. and Ahmed, A.U. (2003). "Climate Change and Its Impacts on Water Resources of Bangladesh." In: Muhammed, A. (ed.) *Climate Change and Water Resources in South Asia.* Islamabad: Asianics Agro Dev International.

Chugh, Nishtha. (2019). "Why the Melting of the Hindu Kush and Himalayan Glaciers Matters: The Entire Region Is Headed for an Immense Climate Crisis by 2100, Scientists Warn." *The Diplomat,* May 8.

Church, John A. and White, Neil J. (2011). "Sea-Level Rise from the Late 19th to the Early 21st Century." *Surveys in Geophysics,* Vol. 32, Nos. 4–5.

Clark, Duncan. (2009). "Maldives First to Go Carbon Neutral." *Observer,* March 15.

Clark, Duncan. (2011). "How Will Climate Change Affect Rainfall?" *The Guardian,* December 15.

Cleland, E.E., Chuine, I., Menzel, A., Mooney, H.A. and Schwartz, M.D. (2007). "Shifting Plant Phenology in Response to Global Change." *Trends in Ecology and Evolution,* Vol. 22, No. 7.

Clewett, Paul. (2015). Redefining Nepal: Internal Migration in a Post-Conflict, Post-Disaster Society. Migration Policy Institute, June 18.

Climate Action Network South Asia (CANSA). (2014). SAARC Governments Must Act Together and Quickly On Climate Change: CANSA. November 26.

Climate Change Adaptation and Disaster Risk Reduction Institutional and Policy Landscape in Asia and Pacific, Appendix 1. (2010). Descriptions of Regional Institutions and Policies International Strategy for Disaster Risk Reduction (ISDR)-United Nations, Draft, August 5.

Climate Investment Funds (CIF). (2010). *Strategic Program for Climate Resilience in Bangladesh,* PPCR/SC.7/5, October 25.

Cohen, Luc. (2016). "Bolivia Declares State of Emergency Due to Drought, Water Shortage." *Reuters,* November 21.

Coleman, William D. and Underhill, Geoffrey R.D. (eds.). (1998). *Regionalism and Global Economic Integration: Europe, Asia, and the Americas.* London: Routledge.

The Colombo Declaration of the Heads of State or Government of the Member Countries of South Asian Association for Regional Cooperation issued on December 21, 1991.

Colombo Declaration on a Common Environment Programme. (1998). cited in SAARC. (1999). *SAARC Ministerial Meetings April 1986–August 1999*, SAARC Secretariat, Kathmandu.

Comiso, J.C. and Hall, D.K. (2014). "Climate Trends in the Arctic as Observed from Space." *WIREs Climate Change*, Vol. 5.

Common SAARC Position. (2010). United Nations Climate Change Conference (COP 16/CMP 6). Cancun, Mexico, November 29–December 10, 2010, to be presented at COP 16 by the Chair of SAARC (Bhutan) as decided at the Sixteenth SAARC Summit, Thimphu, April 28–29, 2010.

Compact2025. (2016). *Bangladesh: Ending Hunger and Undernutrition: Challenges and Opportunities*. Washington, DC: Compact2025.

Conca, Ken. (2012). "The Rise of the Region in Global Environmental Politics." *Global Environmental Politics*, Vol. 12, No. 3, August.

Conference of the Parties (CoP). (2015: 26). Twenty-First Session, Paris, November 30–December 11.

The Constitution of the Islamic Republic of Afghanistan. (2004). Translated by Sayed Shafi Rahel for the Secretariat of the Constitutional Commission, January 3.

Consulting, Samuel Hall. (2014). Displacement Dynamics: IDP Movement Tracking, Needs and Vulnerability Analysis, Herat and Helmand, Commissioned by the International Organization for Migration (IOM), Kabul.

COP 16/CMP 6. (2010). Common SAARC Position on Climate Change, Cancun, Mexico, to be presented at COP 16 by the Chair of SAARC (Bhutan) as decided at the Sixteenth SAARC Summit (Thimphu, April 28–29, 2010). November 29–December 10.

Costa, Anna da. (2018). "India Steps Up Climate Change Efforts." Worldwatch Institute, Washington, DC.

Couloumbis, Theodore A. and Wolfe, James H. (1981). *Introduction to International Relations: Power and Justice*. New Delhi: Prentice-Hall of India.

Council on Foreign Relations. (2013). "The Global Climate Change Regime." June 19. Available at <www.cfr.org/report/global-climate-change-regime>. Accessed on February 14, 2020.

Couzens, Ed. and Honkonen, Tuula. (2011). *International Environmental Lawmaking and Diplomacy Review 2010*. University of Eastern Finland–UNEP Course Series 10.

Cox, John D. (2007). *Climate Crash: Abrupt Climate Change and What It Means for Our Future*. Washington, DC: Joseph Henry Press.

Cronin, C. (ed.). (1995). *Uncommon Ground: Rethinking the Human Place in Nature*. New York: W. W. Norton & Co.

Cruz, Rex Victor., Harasawa, Hideo., Lal, Murari. and Wu, Shaohong. (2007). "Asia." In: Parry, Martin., Canziani, Osvaldo., Palutikof, Jean., Linden, Paul van der. and Hanson, Clair., (eds.) *Climate Change 2007: Impacts, Adaptation and Vulnerability*. Contribution of Working Group II to the Fourth Assessment Report of the IPCC, New York, US: Cambridge University Press.

Daily Mirror. (2017). "Climate Literacy in Sri Lanka's Development Pathway." April 28.

Daily News. (2004). "MoU between SAARC and SACEP Signed." Sri Lanka. July 20.

Daily News. (2009). "Sri Lanka to Present Common SAARC Positions at COP 15." October 24.

Daily Outlook. (2012). "India Ratifies SAARC Agreement on Natural Disasters, Afghanistan." August 23.

The Daily Star. (2010a). "Combating Climate Change: Agenda for Bangladesh." May 8.

The Daily Star. (2010b). "SAARC Vows Climate Unity." April 29.

The Daily Star. (2011). "17th SAARC Summit Adopts 'Addu Declaration, Bangladesh: Dhaka." November 12.

Daily Times. (2007). "Climate Change Affecting Pakistan's Environment: Faisal Saleh." September 26. Available at <www.dailytimes.com.pk>. Accessed on February 14, 2020.

Damodaran, A. (2010). *Encircling the Seamless: India, Climate Change, and the Global Commons.* New York: Oxford University Press.

Dann, Christine R. (1999). From Earth's Last Islands: The Global Origins of Green Politics. PhD Thesis of Lincoln University. Pennsylvania.

DARA. (2010). *Climate Vulnerability Monitor 2010 the State of the Climate Crisis.* Madrid, Spain: DARA/Climate Vulnerable Forum.

Darragh, I. (1998). *A Guide to Kyoto: Climate Change and What It Means to Canadians.* Winnipeg, Manitoba: The International Institute for Sustainable Development (IISD).

Darwall, Rupert. (2015a). "Paris: The Treaty That Dare Not Speak Its Name." *National Review,* December 14.

Darwall, Rupert. (2015b). "Paris Climate-Conference Deal: The West Will Commit to Paying Billions to Developing Nations." *National Review,* December 8.

Das, A. (2009). Integrated Farming: A Climate Change Adaptation Strategy for Small and Marginal Farmers in Low Lands of Sundarbans. 3rd International Conference on Community Based Adaptation to Climate Change. Organized by Bangladesh Centre for Advanced Studies, IIED & The Ring, Dhaka, February 18–24.

Das, Kasturi. and Bandyopadhyay, Kaushik Ranjan. (2015). "Climate Change Adaptation in the Framework of Regional Cooperation in South Asia." *Carbon & Climate Law Review,* Vol. 9, No. 1.

Das, Krishna N. (2015). "India Minister Blames Climate Change for Deadly Heatwave, Weak Monsoon." *Reuters,* June 2.

Das, Someshwar., Mohanty, U.C., Tyagi, Ajit., Sikka, D.R., Joseph, P.V., Rathore, L.S., Habib, A., Baidya, S., Sonam, K. and Sarkar, A. (2014). "The SAARC STORM A Coordinated Field Experiment on Severe Thunderstorm Observations and Regional Modeling over the South Asian Region." *Bulletin of the American Meteorological Society,* April, doi:10.1175/BAMS-D-12-00237.1.

Dasgupta, Kum Kum. (2008). "Less on Our Plate." *The Hindustan Times,* December 30.

Dasgupta, Manas. (2005). "Vadodara Flood Situation Grim." *The Hindu,* July 2.

Dasgupta, S., Laplante, B., Meisner, C., Wheeler, D. and Yan, J. (2007). Sea-Level Rise and Storm Surges: A Comparative Analysis of Impacts in Developing Countries. *World Bank Policy Research.* Working Paper 4136.

Dasgupta, S., Laplante, B., Murray, S. and Wheeler, D. (2009). Climate Change and the Future Impacts of Storm-Surge Disasters in Developing Countries. *Center for Global Development.* Working Paper 182.

Dasgupta, S., Laplante, B., Murray, S. and Wheeler, D. (2010). Climate Change and the Future Impacts of Storm-Surge Disasters in Developing Countries. *World Bank Policy Research.* Working Paper 4901.

Dasgupta, Susmita. and Meisner, Craig. (2009). *Climate Change and Sea Level Rise: A Review of the Scientific Evidence.* Environment Department Papers No. 118. Climate change series. Washington, DC: World Bank.

Dash, Sarat. (2016). "Foreword." In: Rabbani, Golam., Shafeeqa, Fathimath. and Sharma, Sanjay. (eds.) *Assessing the Climate Change Environmental Degradation and Migration Nexus in South Asia*. Bangladesh, Dhaka: International Organization for Migration (IOM).

Dauenhauer, NenadJarić. (2017). "On Front Line of Climate Change as Maldives Fights Rising Seas." *New Scientist*, March 20.

Dawn. (2006). "SAARC to Set up Food Bank." December 15.

Dawn. (2007a). "Afghanistan Inducted as the 8th Member: 14th SAARC Summit Begins." April 4. Available at <www.dawn.com/news/240651>. Accessed on February 14, 2020.

Dawn. (2007b). "BD Wants SAARC Food Bank Made Functional." December 26.

Dawn. (2010). "The Never-Ending Flood." Lahore. 22–23 July.

De, U.S., Dube, R.K. and Rao, G.S. Prakasa. (2005). "Extreme Weather Events over India in the Last 100 Years." *Journal of Indian Geophysical Union*, Vol. 9.

de Almeida, Beatriz Azevedo. and Mostafavi, Ali. (2016). "Resilience of Infrastructure Systems to Sea-Level Rise in Coastal Areas: Impacts, Adaptation Measures, and Implementation Challenges." *Sustainability*, Vol. 8, 1115.

Declarations of SAARC Summits, 1985–1998. (2001). Kathmandu: South Asian Association for Regional Cooperation.

The Declaration of the Seventh SAARC Summit of the Heads of State or Government of Member Countries of the South Asian Association for Regional Cooperation issued on April 11, 1993.

The Declaration of the Tenth SAARC Summit of the Heads of State or Government of the Member Countries of the South Asian Association for Regional Cooperation issued on July 31, 1998, Colombo.

The Declaration of the Thirteenth SAARC Summit. (2005). Dhaka, November 13.

The Declaration of the Twelfth Summit. (2004). Pakistan: Islamabad, January 4–6.

DeConto, R.M. and Pollard, D. (2016). "Contribution of Antarctica to Past and Future Sea-Level Rise." *Nature*, Vol. 531, No. 7596.

Dehlavi, Ali., Gorst, Ashley., Groom, Ben. and Zaman, Farrukh. (2015). *Climate Change Adaptation in the Indus Ecoregion: A Microeconometric Study of the Determinants, Impacts, and Cost Effectiveness of Adaptation Strategies*. Islamabad: World Wide Fund for Nature (WWF) Pakistan.

De Leo, Giulio, Rizzi, L., Caizzi, A. and Gatto, Marino. (2001). "The Economic Benefits of the Kyoto Protocol." *Nature*, Vol. 413, October 4.

Delhi Statement on Cooperation in Environment. (2009). SAARC Ministerial Statement on Cooperation in Environment (Final). *Draft for Discussion Purposes Only Embargoed until 12.30hrs*, October 20.

DeMint, Jim. (2016). "Obama Sidesteps Senate Approval of Paris Agreement." *The Heritage Foundation*, April 22.

Denissen, Anne-Katrien. (2012). "Climate Change & Its Impacts on Bangladesh." *NCDO*, 3 April. Available at <www.ncdo.nl/artikel/climate-change-its-impacts-bangladesh>. Accessed on September 2, 2019.

Denman, K.L., Brasseur, G., Chidthaisong, A., Ciais, P., Cox, P.M., Dickinson, R.E., Hauglustaine, D., Heinze, C., Holland, E., Jacob, D., Lohmann, U., Ramachandran, S., da Silva Dias, P.L., Wofsy, S.C. and Zhang, X. (2007). "Couplings between Changes in the Climate System and Biogeochemistry." In: *Climate Change 2007: The Physical Science Basis. Contribution of Working Group I to the Fourth Assessment Report of the Intergovernmental Panel on Climate Change*. Cambridge, UK and New York: Cambridge University Press.

Dent, C.M. (2009). *East Asian Regionalism.* London and New York: Routledge.

Department of Geology and Mine (DoGM). (2009). "Government of India." Available at <www.mti.gov>. Accessed on February 3, 2013.

Depledge, Joanna. (2000). United Nations Framework Convention on Climate Change (UNFCCC) Technical Paper: Tracing the Origins of the Kyoto Protocol: An Article-by-Article Textual History. November.

Depledge, Joanna. (2006). *The Organisation of Global Negotiations-Constructing Climate Change Regime.* London: Earthscan.

Desai, Bharat H. (2010). *Multilateral Environmental Agreements: Legal Status of the Secretariats.* Cambridge and Delhi: Cambridge University Press.

Dessler, Andrew E. and Parson, Edward A. (2006). *The Science and Politics of Global Climate Change: A Guide to the Debate.* Cambridge: Cambridge University Press.

Dev, S. Mahendra. and Sharma, Alakh N. (2010). *Food Security in India: Performance, Challenges and Policies.* Oxfam India Working Papers Series, OIWPS-VII, September. New Delhi: OXFAM INDIA.

Dewan, Tanvir H. (2015). "Societal Impacts and Vulnerability to Floods in Bangladesh and Nepal." *Weather and Climate Extremes,* Vol. 7.

Dhaka Declaration on Climate Change. (2008). *Adopted by a SAARC Ministerial Meeting on Climate Change,* Dhaka, July 3.

Dhaka Tribune. (2015). *Disaster Monitoring System for SAARC Soon,* April 27.

Dhakal, K., Silwal, S. and Khanal, G. (2010). *Assessment of Climate Change Impacts on Water Resources and Vulnerability in Hills of Nepal: A Case Study of DhareKhola Watershed of Dhading District.* National Adaptation Program of Action to Climate Change.

Di Cesare, M., Bhatti, Z., Soofi, S.B., Fortunato, L., Ezzati, M. and Bhutta, Z.A. (2015). "Geographical and Socioeconomic Inequalities in Women and Children's Nutritional Status in Pakistan in 2011: An Analysis of Data from a Nationally Representative Survey." *Lancet Global Health,* Vol. 3.

DiMento, Joseph F. C. and Doughman, Pamela M. (2014). *Climate Change: What It Means for Us, Our Children, and Our Grandchildren (American and Comparative Environmental Policy).* 2nd Edition. New York: The MIT Press.

Ding, Ting. and Ke, Zongjian. (2015). "Characteristics and Changes of Regional Wet and Dry Heat Wave Events in China during 1960–2013." *Theoretical and Applied Climatology,* Vol. 122, Nos. 3–4.

Ding, T., Qian, W. and Yan, Z. (2010). "Changes in Hot Days and Heat Waves in China during 1961–2007." *International Journal of Climatology,* Vol. 30, No. 10.

Disaster Management Act. (2012). Government of the People's Republic of Bangladesh.

Disaster Management Law. (2012). Kabul: The Islamic Republic of Afghanistan.

Dixit, Ajaya. (2010). "Climate Change in Nepal: Impacts and Adaptive Strategies." Institute for Social and Environmental Transition, Kathmandu, Nepal.

Dolsak, Nives. (2001). "Mitigating Global Climate Change: Why Are Some Countries More Committed Than Others?" *Policy Studies Journal,* Vol. 29, No. 3.

Donnelly, Jack. (2000). *Realism and International Relations.* Cambridge, UK: Cambridge University Press.

Dorji, Chhimmi. (2013). "Climate Change as a Security Issue: A Case Study of Bhutan." In: Nayak, Nihar. (ed.) *Cooperative Security Framework for South Asia.* New Delhi: IDSA, Pentagon Press.

Douglas, Ian. (2009). "Climate Change, Flooding and Food Security in South Asia." *Food Security,* Vol. 1, No. 2.

Doyle, Michael W. (1986). "Liberalism and World Politics." *The American Political Science Review*, Vol. 80, No. 4.

Doyle, Timothy. and McEachern, Doug. (2001). *Environment and Politics*. London: Routledge.

Duan, H.X., Wang, S.P. and Feng, J.Y. (2013). "The National Drought Situation and Its Impacts and Causes in the Summer 2013." *Journal of Arid Meteorology*, Vol. 31, No. 3.

Duba, Sangay., Gurung, Tayan Raj. and Ghimiray, Mahesh. (2006). *Assessment of SARD-M Policies in the Hindu Kush-Himalayas: The case of Land Use Policies in Bhutan*. Project for Sustainable Agriculture and Rural Development in Mountain Regions (SARD-M), RNRRC Bajo, December.

Dunne, Tim. and Schmidt, Brian C. (2005). "Realism." In: Baylis, John. and Smith, Steve. (eds.) *The Globalization of World Politics: An Introduction of International Relations*. New Delhi: Oxford University Press.

Durbar, Singha. (2002). *Sustainable Development Agenda for Nepal*. Kathmandu: His Majesty's Government of Nepal Singha Durbar.

Dutt, A.K. and Geib, M.M. (1998). "Nepal." In: *Atlas of South Asia*. Oxford: Oxford University Press.

Dutta, Saptarshi. (2019). "Four Years of Swachh Bharat Abhiyan: With over 9 Crore Toilets, India Inches Towards Becoming ODF." NDTV, August 8. Available at <https://swachhindia.ndtv.com/4-years-of-swachh-bharat-abhiyan-india-inching-towards-open-defecation-free-9-crore-toilets-25117/>. Accessed on April 4, 2020.

Dyer, Gwynne. (2015). "In Maldives, Politics, Greed Trump Climate Change." *The Japan Times*, December 2.

Easterling, D.R., Kunkel, K.E., Arnold, J.R., Knutson, T., LeGrande, A.N., Leung, L.R., Vose, R.S., Waliser, D.E. and Wehner, M.F. (2017). "Precipitation Change in the United States." In: Wuebbles, D.J., Fahey, D.W., Hibbard, K.A., Dokken, D.J., Stewart, B.C. and Maycock, T.K. (eds.) *Climate Science Special Report: Fourth National Climate Assessment*. Volume 1. Washington, DC: U.S. Global Change Research Program. http://dx.doi.org/10.7930/J0H993CC.

Ebenstine, William. (1960). *Great Political Thinkers: Plato to the Present*. New Delhi: Oxford & IBH Publishing Co. Pvt Ltd.

Ebi, K.L., Balbus, J.M., Luber, G., Bole, A., Crimmins, A., Glass, G., Saha, S., Shimamoto, M.M., Trtanj, J. and White-Newsome, J.L. (2018). "Human Health." In: Reidmiller, D.R., C.W. Avery, D.R. Easterling, K.E. Kunkel, K.L.M. Lewis, T.K. May-cock, and B.C. Stewart (eds.) *Impacts, Risks, and Adaptation in the United States: Fourth National Climate Assessment*. Volume 2. Washington, DC: U.S. Global Change Research Program.

Eckersley, Robyn. (2004). *The Green State: Rethinking Democracy and Sovereignty*. Cambridge: MIT Press.

Eckersley, Robyn. (2010). "Green Theory." In: Dunne, Tim., Kurki, Milja. and Smith, Steve. (eds.) *International Relations Theories: Discipline and Diversity*. Oxford: Oxford University Press.

Eckstein, David., Hutfils, Marie-Lena. and Winges, Maik. (2018). *Global Climate Risk Index 2019*. Bonn: Germanwatch.

Economic and Political Weekly (EPW). (2009). "Climate Refugees." June 6.

The Economic Times. (2010). "Neighbouring Countries Can Create Problem for India over Water." July 17. Available at <https://economictimes.indiatimes.com/news/politics-and-nation/neighbouring-countries-can-create-problem-for-india-over-water/articleshow/6181142.cms>. Accessed on February 14, 2020.

The Economic Times. (2014a). "18th SAARC Summit Begins in Kathmandu." November 26. Available at <http://articles.economictimes.indiatimes.com>. Accessed on January 26, 2015.

The Economic Times. (2014b). "Over 35,000 Marooned by Bad Weather and Flash Floods in Sri Lanka." December 21. Available at <https://economictimes. indiatimes.com/news/international/world-news/over-35000-marooned-by-bad-weather-and-flash-floods-in-sri-lanka/articleshow/45592749.cms>. Accessed on February 14, 2020.

Edame, G.E., Ekpenyong, A.B., Fonta, W.M. and Duru, E.J.C. (2011). "Climate Change, Food Security and Agricultural Productivity in Africa: Issues and Policy Directions." *International Journal of Humanities and Social Science*, Vol. 1, No. 21.

Edenhofer, Ottmar., Pichs-Madruga, Ramón., Sokona, Youba., Minx, Jan C., Farahani, Ellie., Kadner, Susanne., Seyboth, Kristin., Adler, Anna., Baum, Ina., Brunner, Steffen., Eickemeier, Patrick., Kriemann, Benjamin., Savolainen, Jussi., Schlömer, Steffen., von Stechow, Christoph. and Zwickel, Timm. (eds.) (2014). *IPCC, 2014: Climate Change 2014: Mitigation of Climate Change*. Contribution of Working Group III to the Fifth Assessment Report of the Intergovernmental Panel on Climate Change. New York: Cambridge University Press.

Edwards, Michael. (2004). *Future Positive: International Co-Operation in the 21st Century*. London: Earthscan.

EEAS. (2018). *Third EU Serena Dialogue: Climate Change in Afghanistan: Common Challenge, Collective Response*. Brussel: European Union External Action, EEAS Building, 9A Rond Point Schuman.

Ehrlich, Paul. (1968). *The Population Bomb*. New York: Ballantine.

The Eight Declaration of the Heads of State or Government of the Member Countries of South Asian Association for Regional Cooperation, New Delhi, May 4, 1995.

Eighteenth SAARC Summit. (2014). Kathmandu, Nepal. November, 26–27.

The Eight Meeting of Ministers of Environment. (2009). India: New Delhi, October 20–21.

Ekins, P. (1993). "'Limits to Growth' and 'Sustainable Development': Grappling with Ecological Realities." *Ecological Economics*, Vol. 8, No. 3.

Eklund, Lars. (2009). "SASNET visit to Maldives 6e8 February 2009, Report by Lars Eklund." Available at <www.sasnet.lu.se/maldives09.html>. Accessed on January 13, 2014.

Eldho, Sreeja., Sreeja, K.G. and Madhusoodhanan, G. (2016). "Climate Change Impact Assessments on the Water Resources of India under Extensive Human Interventions." *Ambio*, Vol. 45, No. 6.

The Eleventh SAARC Summit held at Kathmandu. (January 4–6, 2002).

Elrick-Barr, C., Glavonic, B.C. and Kay, K. (2015). "A Tale of Two Atoll Nations: A Comparison of Risk, Resilience and Adaptive Response of Kiribati and the Maldives." In: Glavonic, B., Kelly, M., Kay, R. and Travers. (eds.) *Climate Change and the Coast: Building Resilient Communities*. Boca Raton, FL: CRC Press.

Elsner, James B., Kossin, James P. and Jagger, Thomas H. (2008). "The Increasing Intensity of the Strongest Tropical Cyclones." *Nature*, Vol. 455, No. 7209.

Emanuel, Kerry. (2007). "Environmental Factors Affecting Tropical Cyclone Power Dissipation." *Journal of Climate*, Vol. 20, No. 22.

EM-DAT. (2011). *Natural Disasters in Indonesia*, Table created on: 1 October, 2011, Data version: v12.07. Brussels: Center for Research on the Epidemiology of Disasters (CRED).

Emergency Plan of Action (EPoA). (2014). *Maldives and South Asia: Water Crisis.* International Federation of Red Cross and Red Crescent Societies-IFRC.

Ensor, J. and Berger, R. (2009). *Understanding Climate Change: Lessons from Community-Based Approaches.* Rugby, UK: Practical Action Publishing.

Epatko, Larisa. (2017). "South Asia Floods Have Killed More Than 1,400 as Communities Brace for More." *PBSO News Hour*, September 1.

Ericson, J.P., Vorosmarty, C.J., Dingman, S.L., Ward, L.G. and Meybeck, M. (2005). "Effective Sea-Level Rise and Deltas: Causes of Change and Human Dimension Implications." *Global Planetary Change*, Vol. 50.

Eriyagama, Nishadi., Smakhtin, Vladimir., Chandrapala, L. and Fernando, K. (2010). *Impacts of Climate Change on Water Resources and Agriculture in Sri Lanka: A Review and Preliminary Vulnerability Mapping.* IWMI Research Report 135. Colombo, Sri Lanka: International Water Management Institute.

ESCAP. (2018). *Integrating South Asia's Power Grid for a Sustainable and Low Carbon Future.* Bangkok: Economic and Social Commission for Asia and the Pacific (ESCAP).

Esty, Daniel C. and Mendelsohn, Robert. (1998). "Moving from National to International Environmental Policy." *Policy Science*, Vol. 31, 225.

Esty, Daniel C. and Moffa, Anthony L.I. (2012). "Why Climate Change Collective Action Has Failed and What Needs to be Done Within and Without the Trade Regime." *Journal of International Economic Law*, Vol. 15, No. 3.

European Commission. (2003). *Country Strategy Paper: Bhutan and the European Community Co-Operation Strategy 2002–2006.* European Commission External Relations Directorate General, Brussels.

European Commission. (2017). "EU Climate Action." Available at <https://ec.europa.eu/clima/citizens/eu_en>. Accessed on February 14, 2020.

Evans, Graham. and Newnham, Jeffrey. (1998). *The Penguin Dictionary of International Relations.* London: Penguin Books.

Extra-Ordinary Meeting of the SAARC Agriculture Ministers. (2008). India: New Delhi, cited in *SAARC Ministerial Meetings- 2000–2012.* (2012) SAARC Secretariat: Kathmandu, November 5.

Fairclough, A.J. (1991). "Global Environment and Natural Resource Problems: Their Economic, Political and Security Implications." *Washington Quarterly*, Vol. 14, No. 1, Winter.

Faizi, S. (2018). "The Kerala Deluge: Global Warming's Latest Act." *Down to Earth*, September 8.

Falk, Richard. (2016). "Voluntary International Law and the Paris Agreement." January 17. Available at <https://zcomm.org/znetarticle/voluntary-international-law-and-the-paris-agreement/>. Accessed on February 14, 2020.

FAO (Food and Agricultural Organisation of United Nations). (2002). *Report of FAO-CRIDA Expert Group Consultation on Farming System and Best Practices for Drought-prone Areas of Asia and the Pacific Region.* Hyderabad, India: Published by Central Research Institute for Dryland Agriculture.

FAO. (2005). *The World's Mangroves 1980–2005. A Thematic Study Prepared in the Framework of the Global Forest Resources Assessment 2005.* Rome: FAO Forestry Paper 153, Food and Agriculture Organization of the United Nations.

FAO. (2011). *Maldives and FAO-Achievements and Success Stories.* FAO Representation in Maldives, Rome, Italy.

FAO. (2016a). *FAOSTAT.* Available at <http://faostat3.fao.org/download/Q/QC/E>. Accessed on February 2016.

FAO. (2016b). *Land Cover Atlas of the Islamic Republic of Afghanistan*. Kabul, Afghanistan: FAO.

FAO. (2017a). *Evaluation of FAO's Contribution to the Islamic Republic of Pakistan 2012–2017*. Country Programme Evaluation Series. Rome: Food and Agriculture Organization of the United Nations Office of Evaluation.

FAO. (2017b). *Special Report: FAO/WFP Crop and Food Security Assessment Mission to Sri Lanka*. Rome: Food and Agriculture Organization of the United Nations World Food Programme.

FAOSTAT. (2006). *Food and Agriculture Indicators, 2006*, Data are reported for 2004. Available at <www.fao.org>. Accessed on July 10, 2014.

Farooqi, Anjum Bari., Khan, Azmat Hayat. and Mir, Hazrat. (2005). "Climate Change Perspective in Pakistan." *Pakistan Journal of Meteorology*, Vol. 2, No. 3, March.

Farrell, Mary. (2005). "The Global Politics of Regionalism: An Introduction." In: Farrell, Mary., Hettne, Björn. and Langenhove, Luk Van. (eds.) *Global Politics of Regionalism: Theory and Practice*. London: Pluto Press.

Faruque, Abdullah A. L. (2010). "Protection of Environment through Judicial Activism in Bangladesh." *South Asian Journal*, July–September.

Fawcett, Louise. (2005a). "The Regional Dimension in International Relations Theory." In: Farrell, Mary., Hettne, Björn. and Langenhove, Luk Van. (eds.) *Global Politics of Regionalism: Theory and Practice*. London: Pluto Press.

Fawcett, Louise. (2005b). "Regionalism from an Historical Perspective." In: Farrell, Mary., Hettne, Björn. and Langenhove, Luk Van. (eds.) *Global Politics of Regionalism: Theory and Practice*. London: Pluto Press.

Feil, Moira., Klein, Diana. and Westerkamp, Meike. (2009). Regional Cooperation on Environment, Economy and Natural Resource Management: How Can It Contribute to Peace Building? Synthesis Report. Adelphi Research.

Femia, Francesco. and Werrell, Caitlin. (2012). "Pakistan's New Climate Policy." *The Center for Climate & Security*, October 3.

Field, C.B., Barros, V., Stocker, T.F., Dahe, Q., Dokken, D.J., Ebi, K.L., Mastrandrea, M.D., Mach, K.J., Plattner, G.-K., Allen, S.K. and Tignor, M. (eds.). (2012). *Managing the Risks of Extreme Events and Disasters to Advance Climate Change Adaptation*. A Special Report of Working Groups I and II of the Intergovernmental Panel on Climate Change. Cambridge, London and New York: Cambridge University Press.

Fifteenth SAARC Summit. (2008). Sri Lanka: Colombo, August 2–3.

The Fifth Meeting of Ministers of Environment. (2002). Thimphu, August 10–11, cited in SAARC. (2012). *SAARC Ministerial Meetings 2000–2012*, Secretariat of the South Asian Association for Regional Cooperation: Kathmandu, 111–116.

The Fifth Summit Malé Declaration of the Heads of State or Government of the Member Countries of South Asian Association for Regional Cooperation issued on November 23, 1990.

FitzRoy, Felix R. and Papyrakis, Elissaios. (2010). *An Introduction to Climate Change Economics and Policy*. London: Earthscan.

Flagg, J.A. (2015). "Aiming for Zero: What Makes Nations Adopt Carbon Neutral Pledges?" *Environmental Sociology*, Vol. 1, No. 1.

Fleming, E., Payne, J., Sweet, W., Craghan, M., Haines, J., Hart, J.F., Stiller, H. and Sutton-Grier, A. (2018). "Coastal Effects." In: Reidmiller, D.R., Avery, C.W., Easterling, D.R., Kunkel, K.E., Lewis, K.L.M., Maycock, T.K. and Stewart, B.C. (eds.)

Impacts, Risks, and Adaptation in the United States: Fourth National Climate Assessment. Volume 2. Washington, DC: U.S. Global Change Research Program.

F.No.8–14/2004-FP. (2012). Government of India Ministry of Environment & Forests Forest Policy Division, Lodhi Road, New Delhi, December 27.

Food and Agriculture Organization (FAO). (2013). *Irrigation in Central Asia in Figures: AQUASTAT Survey-2012* ("FAO AQUASTAT Survey-2012"). Available at <www.fao.org/docrep/018/i3289e/i3289e.pdf.>. Accessed on February 14, 2020.

Food and Agriculture Organization of the United Nations (FAO). (2015). *Bhutan and FAO.* Available at <http://bit.ly/2p6vbKB>. Accessed on May 20, 2016.

Food Planning and Monitoring Unit (FPMU). (2015). National Food Policy Plan of Action and Country Investment Plan, Monitoring Report, Ministry of Food and Disaster Management. Dhaka: Government of the People's Republic of Bangladesh.

Forsyth, Tim. (2000). *Vulnerability to Climate Change: Theoretical Concerns and a Case Study from Thailand: Draft Belfer Center for Science and International Affairs (BCSIA).* Discussion Paper. Cambridge, MA: Environment and Natural Resources Program, Kennedy School of Government, Harvard University.

Fourteenth SAARC Summit. (2007). Kathmandu, SAARC Secretariat.

The Fourteenth Summit Meeting of the South Asian Association for Regional Cooperation (SAARC) held in New Delhi, India on April 3–4, 2007.

Fox News. (2004). "Maldives Devastated by Tsunami." December 29.

Franceschet, A. (2002). "Moral Principles and Political Institutions: Perspectives on Ethics and International Affairs." *Millennium Journal of International Studies*, Vol. 31, No. 2.

Frankel, B. (1996). "Restating the Realist Case: An Introduction." *Security Studies*, Vol. 5, No. 3.

Frumhoff, P.C., McCarthy, J.J., Melillo, J.M., Moser, S.C. and Wuebbles, D.J. (2007). *Confronting Climate Change in the U.S. Northeast: Science, Impacts, and Solutions.* Synthesis Report of the Northeast Climate Impacts Assessment. Cambridge, MA: Union of Concerned Scientists.

Frykman, H. and Seiron, Per-Olof. (2009). The Effects of Climate Induced Sea Level Rise on the Coastal Areas in the Hambantota District, Sri Lanka: A Geographical Study of Hambantota and an Identification of Vulnerable Ecosystems and Land Use along the Coast. Lund: Physical Geography and Ecosystems Analysis Lund University.

Gaan, Narottam., Acharya, Nivedita. and Mohapatra, Sonali. (2013). "Climate Change: A Threat to Human Security in Nepal." *Regional Studies: Quarterly Journal of the Institute of Regional Studies*, Vol. 31, No. 2, Spring.

GACGC. (2006). *The Future Oceans Warming Up, Rising High, Turning Sour.* Berlin: German Advisory Council on Climate Change.

Ganesan, Narayanan. (2010). "Regionalism in International Relations Theory." In: Ganesan, Narayanan. and Dürkop, Colin. (eds.) East Asia Regionalism. Papers of the Workshop on "Differing Perspectives on East Asian Regionalism" organized by the Konrad-Adenauer-Stiftung and the Hiroshima Peace Institute in Hiroshima. Tokyo: Konrad Adenauer Foundation.

Gardner, R.N. (1990). "The Comeback of International Liberalism." *The Washington Quarterly*, No. 13, Vol. 3.

Garrote-Sanchez, Daniel. (2017). International Labor Mobility of Nationals: Experience and Evidence for Afghanistan at Macro Level, BGP 2 A Background Paper to the World Bank Project on "Afghanistan: Managed International Labor Mobility as Contribution to Economic Development and Growth". Washington, DC.

Gautier, Catherine. and Fellous, Jean-Louis. (2008). "Introduction." In: Gautier, Catherine. and Fellous, Jean-Louis. (eds.) *Facing Climate Change Together*. Cambridge: Cambridge University Press.

Gayle, Damien. (2015). "Number of Nepal Earthquake Victims from Abroad Still Unknown." *The Guardian*, May 5.

Gelmo, Dawa. (2015). "Weather Fluctuations Wreak Havoc on Bhutan's Crops." *The Third Pole Net*, November 27.

Georgakakos, A., Fleming, P., Dettinger, M., Peters-Lidard, C., Richmond, T.C., Reckhow, K., White, K. and Yates, D. (2014). "Ch. 3: Water Resources." In: Melillo, J.M., Richmond, T.C. and Yohe, G.W. (eds.) *Climate Change Impacts in the United States: The Third National Climate Assessment*. Washington, DC, USA: U.S. Global Change Research Program.

German Advisory Council on Global Change (WBGU). (2008). *Climate Change as a Security Risk*. London: Earthscan.

German Watch Report. (2017). "Global Climate Risk Index 2017." Available at <https://germanwatch.org/en/download/16411.pdf>. Accessed on August 10, 2019.

Gettleman, Jeffrey. (2017). "The New York Times: More Than 1,000 Died in South Asia Floods This Summer." *The New York Times*, September 1.

Ghani, A. and Muhammad, A. (2017). "Climate Change Implications for Food Security; Pakistan Perspective." *Agricultural Research and Technology*, Vol. 6, No. 3.

Ghatak, Mriganka., Kamal, Ahmed. and Mishra, O.P. (eds.). (2012). *Background Paper Flood Risk Management in South Asia*. Proceedings of the SAARC Workshop on Flood Risk Management in South Asia.

Ghiasy, Richard., Zhou, Jiayi. and Hallgren, Henrik. (2015). *Afghanistan's Private Sector: Status and Ways Forward*. Stockholm: Stockholm International Peace Research Institute (SIPRI).

Ghimire, Bhumika. (2009). "Nepal: Taking on the Challenge of Climate Change." *Global Voices*, September 1.

Ghimire, Ramesh., Ferreira, Susana. and Dorfman, Jeffrey H. (2015). "Flood-Induced Displacement and Civil Conflict." *World Development*, Vol. 66(C).

Ghina, Fathimath. (2003). "Sustainable Development in Small Island Developing States: The Case of the Maldives." *Environment, Development and Sustainability*, Vol. 5, No. 1.

Ghosh, Jayati. (2014). "Why Asia Is Probably Poorer Than We Think." *The Guardian*, September 9.

Giddens, Anthony. (1994). *Beyond Left and Right*. Cambridge: Polity Press.

Giddens, Anthony. (2011). *The Politics of Climate Change*. Cambridge: Polity Press.

Gill, G.J. (2003a). *Annex 7: Food Security in Nepal, 2003*. Available at <http://www.odi.org.uk/publications/working paper/wp231/wp231-refrences.pdf>; <http://id_cntre.apic.org/apic/influenza/avianflue/news/aug112005tibet.html>

Gill, G.J. (2003b). *Seasonal Labour Migration in Rural Nepal: A Preliminary Overview*. Working Paper 218. London: Overseas Development Institute.

Gill, Paramjit Kaur. (2005). "Regional Cooperation in South Asia: The Search for Strategy." In: Singh, Gopal. and Chauhan, Ramesh. (eds.) *South Asia Today*. New Delhi: Anamika Publishers & Distributors Pvt Ltd.

Gilpin, Robert G. (1975). U.S. *Power and the Multinational Corporation*. New York: Basic Books.

Ginnetti, J. (2015). *Disaster-Related Displacement Risk: Measuring the Risk and Addressing Its Drivers.* Geneva, Switzerland: International Displacement Monitoring Centre, Norwegian Refugee Council.

Glantz, M.H. (ed.). (2002). *Water, Climate, and Development Issues in the Amudarya Basin.* Report of Informal Planning Meeting held 18–19 June 2002 in Philadelphia, Pennsylvania. Boulder, CO: Environmental and Societal Impacts Group, NCAR.

Gleick, Peter H. (1989). "Climate Change and International Politics: Problems Facing Developing Countries." *Ambio*, Vol. 18, No. 6.

Global Environment Facility (GEF). (2009). "Integration of Climate Change Risks into the Maldives Safer Island Development Programme: Information about Fund." Available at www.apan-gan.net/adaptationpractices/integration-climate-change-risks-maldives-safer-island-development-programme.

Global Facility for Disaster Reduction and Recovery (GFDRR). (2011). *Climate Risk and Adaptation Country Profile.* Washington, DC: World Bank.

Glum, Julia. (2015). "Pakistan Heat Wave 2015: Death Toll Exceeds 1,200 as Karachi Struggles with Continued Extreme Weather during Ramadan." *International Business Times*, June 27.

Goel, O.P. (2004). *India and SAARC Engagements.* Volume 1. New Delhi: Isha Books.

GoI. (2012a). *Twelfth Five Year Plan (2012–2017): Faster, More Inclusive and Sustainable Growth.* Volume 1. New Delhi: Planning Commission, Government of India.

GoI. (2012b). *Twelfth Five Year Plan (2012–2017): Economic Sectors.* Volume 2. New Delhi: Planning Commission, Government of India.

Goldsmith, E. (1972). *A Blueprint for Survival.* Harmondsworth: Penguin.

Goldstein, J.S. and Pevehouse, Jon C. (2008). *International Relations.* New Delhi: Pearson.

Gonselves, Eric. (1995). "South Asian Cooperation: An Agenda and A Vision for the Future." In: Mehotra, L.L., Chopra, H.S. and Kueck, Gert, W. (eds.) *SAARC 2000 and Beyond.* New Delhi: Omega Scientific Publishers.

Goodwin, Jean. (2009). *Working Draft: The Authority of the IPCC First Assessment Report and the Manufacture of Consensus.* Chicago: National Communication Association.

Gopal, Krishan. (1990). Geo-Political Relations and Regional Cooperation in South Asia with Special Reference to India's Role. PhD Thesis of Meerut University, Meerut.

Gopalakrishnan, Tarun. and Gopal, Padmini. (2018). "BASIC Ministerial on Climate Change: India Calls for Developed Countries to Step up at Katowice." *Down to Earth*, November 22.

Gore, Al. (2007). *An Inconvenient Truth: The Crisis of Global Warming.* New York: Viking Juvenile.

Goswami, Subhojit. (2017). "Climate Change Impact on Agriculture Leads to 1.5 per cent Loss in India's GDP." *Down to Earth*, May, 18.

Goswami, Urmi. (2014). "Lima Climate Conference: India Plans an Outreach with SAARC Countries." *The Economic Times*, December 6.

Goulder, Lawrence H. and Nadreau, Brian M. (2001). "International Approaches to Reducing Greenhouse Gas Emissions." January 15. Available at <https://web.stanford.edu/~goulder/Papers/Published%20Papers/Intl%20Approaches%20for%20GHG%20Redux%20-%20Goulder-Nadreau.pdf>. Accessed on February 14, 2020.

Government of Bangladesh (GoB). (1997). *National Food and Nutrition Policy*. Dhaka: Government of Bangladesh.

Government of Bangladesh (GoB). (1999). *National Water Policy 1999*. Dhaka: Government of People's Republic of Bangladesh.

Government of Bangladesh (GoB). (2011). *Operational Plan for National Nutrition Services*. Dhaka: Government of Bangladesh.

Government of Bangladesh (GoB). (2012a). *Introduction to Ministry of Environment and Forest*. Ministry of Environment and Forest.

Government of Bangladesh (GoB). (2012b). *Perspective Plan of Bangladesh 2010–2021: Making Vision 2021 a Reality*. Dhaka, Bangladesh: Planning Commission, Government of Bangladesh.

Government of Bhutan (GoB). (2009). *Biodiversity Persistence and Climate Change in Bhutan, Biodiversity Action Plan*. Thimphu: Ministry of Agriculture, Royal Government of Bhutan.

Government of India (GoI). (2008). *National Action Plan on Climate Change* (NAPCC), Prime Minister's Council on Climate Change. New Delhi, India, June 30.

Government of India (GOI). (2009). *India's National Action Plan on Climate Change*. New Delhi.

Government of India (GoI). (2011). *Faster, Sustainable and More Inclusive Growth: An Approach to the Twelfth Five Year Plan (2012–17)*. New Delhi, October.

Government of India (GoI). (2012). *Environment, Forests, Wildlife & Climate Change*. Report of the Steering Committee for the Twelfth Five Year Plan (2012–2017). New Delhi: Planning Commission.

Government of India (GoI). (2015). "India's Intended Nationally Determined Contribution: Working towards Climate Justice." Available at <http://ceew.in/pdf/ceew-india-indcs-re-and-the-pathway-to-p.pdf>. Accessed on July 23, 2018.

Government of the Islamic Republic of Afghanistan (GIRA). (2005a). *Millennium Development Goals: Islamic Republic of Afghanistan, a Country Report 2005: Vision 2020*.

Government of the Islamic Republic of Afghanistan (GIRA). (2005b). Decree No. 4, 1384.

Government of the Islamic Republic of Afghanistan (GIRA). (2008). *Water Resource Management 1387–1391 (2007/08–2012/13): Afghanistan National Development Strategy*, Vol. 2.

Government of the Islamic Republic of Afghanistan (GIRA). (2009). *The National Capacity Needs Self-Assessment for Global Environmental Management (NCSA) and the National Adaptation Programme of Action for Climate Change (NAPA) Final Report*. Nairobi: United Nations Environment Program (UNEP).

Government of the Islamic Republic of Afghanistan (GIRA). (2011). *Afghanistan Strategic National Action Plan (SNAP) for Disaster Risk Reduction: Towards Peace and Stable Development*. Afghanistan, Kabul: Afghanistan National Disaster Management Authority (ANDMA).

Government of the Islamic Republic of Afghanistan (GIRA). (2015). *Intended Nationally Determined Contribution*. Submission to the United Nations Framework Convention on Climate Change. Kabul: Islamic Republic of Afghanistan.

Government of Nepal (GoN). (2011a). *National Framework on Local Adaptation Plans for Action*. Singha Durbar: Ministry of Environment.

Government of Nepal (GoN). (2011b). *Climate Change Policy, 2011* (Unofficial translation approved by the Government of Nepal, January 17).

Government of Nepal (GoN). (2012a). *Multi-Sector Nutrition Plan, for Accelerating the Reduction of Maternal and Child Under-Nutrition in Nepal 2013–2017(2023)*. Kathmandu, Nepal: National Planning Commission, September.

Government of Nepal (GoN). (2012b). *Nutrition and Food Security Secretariat*. National Planning Commission. Concept Note. Kathmandu, Nepal.

Government of Pakistan (GoP). (1993). *The Pakistan National Conservation Strategy*. Plan of Action, 1993–1998.

Government of Pakistan (GoP). (2005). *National Environmental Policy*. Ministry of Environment.

Government of Pakistan (GoP). (2010). *Task Force on Climate Change: Final Report*. Islamabad: Planning Commission of Pakistan.

Government of Pakistan (GoP). (2010a). *Pakistan Economic Survey 2009–2010*. Islamabad: Ministry of Finance, Economic Advisor Wing.

Government of Pakistan (GoP). (2010b). *Final Report of the Task Force on Climate Change*. Islamabad: Planning Commission.

Government of Pakistan (GoP). (2012). *National Climate Change Policy*. Islamabad, Pakistan: Ministry of Climate Change, September.

Government of Pakistan (GOP). (2013). *Pakistan Economic Survey, 2012–13*. Islamabad, Pakistan: Finance Division, Economic Advisor's Wing.

Government of Pakistan (GOP). (2016). *Pakistan Economic Survey*. Islamabad: Ministry of Finance.

Government of Pakistan (GoP). (2017a). *Pakistan: Climate Change Financing Framework: A Road Map to Systemically Mainstream Climate Change into Public Economic and Financial Management*. United Nations Development Programme, Islamabad, Pakistan.

Government of Pakistan (GoP). (2017b). *National Food Security Policy*. Draft. Islamabad, Pakistan: Ministry of National Food Security and Research, June.

Government of the People's Republic of Bangladesh (GoB). (1997). *The Fifth Five Year Plan: 1997–2002*. Dhaka: Planning Commission, Ministry of Planning (MOP).

Govin, Paul. (2013). "International Environmental Institutions." In: Alam, Shawkat., Bhuiyan, Johid Hossain., Chowdhury, Tareq M.R. and Techera, Erika J. (eds.) *Routledge Handbook of International Environmental Laws*. New York: Routledge.

Gowda, P., Steiner, J.L., Olson, C., Boggess, M., Farrigan, T. and Grusak, M.A. (2018). "Agriculture and Rural Communities." In: Reidmiller, D.R., C.W. Avery, D.R. Easterling, K.E. Kunkel, K.L.M. Lewis, T.K. Maycock, and B.C. Stewart (eds.) *Impacts, Risks, and Adaptation in the United States: Fourth National Climate Assessment*. Volume 2. Washington, DC: U.S. Global Change Research Program.

Gower, Annie. (2018). "India's Crackdown on Illegal Immigration Could Leave 4 Million People Stateless." *The Washington Post*, July 30.

Gramling, Carolin. and Hamers, Laurel. (2018). "Here's How Much Climate Change Could Cost the U.S." *Science News*, November 28.

Grieco, Joseph M. (1988). "Anarchy and the Limits of Cooperation: A Realist Critique of the Newest Liberal Institutionalism." *International Organization*, Vol. 42, No. 3, Summer.

Grieser, J. (2013). "Shanghai Sets New All-Time Record (Again) as Heat Wave Bakes Eastern China." *The Washington Post*, August 8.

Griffiths, Martin. (2005). *Encyclopedia of International Relations and Global Politics*. London: Routledge.

Griffiths, Martin. and O'Callaghan, Terry. (2002). *International Relations: The Key Concepts*. London: Routledge.

Groom, A.J.R. (1991). *Approaches to Conflict and Cooperation in International Relations: Lessons from Theory for Practice*. The Ford Foundation Lectures in International Relations Studies, M.S. University Baroda.

Groom, A.J.R. and Taylor, Paul. (eds.). (1975). *Functionalism: Theory and Practice in International Relations*. London: University of London Press, cited in Laferrière, Eric. and Stoett, Peter J. (1999). *International Relations Theory and Ecological Thought-Towards a Synthesis*. London: Routledge.

Groom, B. (2012). *Climate Change Adaptation: The Bangladesh Experience*. Pakistan: World Wide Fund for Nature.

Groves, Steven. (2016). "The Paris Agreement Is a Treaty and Should Be Submitted to the Senate." *The Heritage Foundation*, March 15.

Grubb, Michael., Vrolijk, Christiaan. and Brack, Duncan. (eds.). (1999). *The Kyoto Protocol: A Guide and Assessment*. New York: Royal Institute of National Affairs.

The Guardian. (2011). "What Is the Kyoto Protocol and Has It Made Any Difference?" March 11.

The Guardian. (2014). "Flooding in Afghanistan Kills 80 People and Leaves Thousands Homeless." June 7.

Guérin, Emmanuel. and Wemaere, Matthieu. (2009). *The Copenhagen Accord: What Happened? Is It a Good Deal? Who Wins and Who Loses? What Is Next?*08/09 IDDRI 1. December. Available at <www.iddri.org>. Accessed on February 14, 2020.

Guleria, S. and Gupta, A.K. (2018). *Heat Wave in India Documentation of State of Telangana and Odisha*. New Delhi: National Institute of Disaster Management.

Gunasekera, Passanna. (2016). "Climate Change Is Starting to Create Waves among Maldives' Citizens." *Frontline*, March/April.

Gunter, Marc. (2016). "Rich Countries Have Pledged Billions in Climate Aid: Why Has Progress Been So Slow?" *Vox*, May 8.

Gupta, Joyeeta. (2010a). "A History of International Climate Change Policy." *Wiley Interdisciplinary Reviews: Climate Change*, Vol. 1, No. 5, September/October.

Gupta, Joyeeta. (2010b). "Mainstreaming Climate Change: A Theoretical Exploration." In: Gupta, Joyeeta. and Van Der Grijp, Nicolien. (eds.) *Mainstreaming Climate Change in Development Cooperation: Theory, Practice and Implications for the European Union*. Cambridge: Cambridge University Press.

Gupta, Swati. (2018). "Rescue Teams Wade through Filthy Waters to Help Kerala Flood Victims." *CNN*, August 21.

Gurung, Deo Raj. (2011). *ICIMOD Side Events: Bhutan Climate Summit 2011*. November 14–19.

Gurung, Ghana S. and Rai, S.C. (2005). *The Account of Climate Change Program of WWF Nepal. Climate Change Impacts on Biodiversity and Mountain Protected Areas*. Kathmandu and Nepal: World Wildlife Fund (WWF).

Haas, Ernst B. (1964). *Beyond the Nation-State: Functionalism and International Organization*. Stanford: Stanford University Press.

Haas, P., Keohane, R. and Levy, M. (eds.). (1994). *Institutions for the Earth: Sources of Effective International Environmental Protection*. Cambridge, MA: MIT Press.

Habib, Benjamin. (2011). Climate Change and International Relations Theory: Northeast Asia as a Case Study. Paper Presented at the World International Studies

Committee Third Global International Studies Conference, University of Porto, Portugal, August 17–20.

Habib, Haroon. (2008). "SAARC Action Plan on Climate Change." *The Hindu*, July 5.

Habib, Zaigham and Nawaz, Muhammad. (2010). "Floods in Pakistan: A Brief Overview." *South Asian Journal*, July–September.

Hagg, W., Hoelzle, M., Wagner, S. and Mayr, E. (2013). "Glacier and Runoff Changes in the Rukhk Catchment, Upper Amu-Darya Basin until 2050." *Global and Planetary Change*, Vol. 110.

Hague Declaration on the Environment. (1989a). *International Legal Materials*, Vol. 28, No. 5. Climate Change Declaration-Approved By Delegates Attending the 8thWorld Clean Air Congress at The Hague. (September 1989). *The International Union of Air Pollution Prevention Associations*.

Hague Declaration on the Environment. (1989b). *International Legal Materials*. American Society of International Law, Vol. 28, No. 5, September.

Hai-Bin, Zhang., Han-Cheng, Dai., Hua-Xia, Lai. and Wen-Tao, Wang. (2017). "U.S. Withdrawal from the Paris Agreement: Reasons, Impacts, and China's Response." *Advances in Climate Change Research*, Vol. 8, No. 4.

Haidar, Suhasini. (2010). "The Great Subcontinental Green Game." *The Hindu*, November 1.

Hales, S., Edwards, S. and Kovats, R.S. (2003). "Impacts on Health of Climate Extremes." In: McMichael, A.J., Campbell-Lendrum, D.H., Corvalan, C.F., Ebi, K.L. Githeko, A., Scheraga, J.D. and Woodward, A. (eds.) *Climate Change and Human Health: Risks and Responses*. Geneva: World Health Organization.

Hallegatte, Stephane., Bangalore, Mook., Bonzanigo, Laura., Fay, Marianne., Kane, Tamaro., Narloch, Ulf., Rozenberg, Julie., Treguer, David. and Vogt-Schilb, Adrien. (2016). "Shock Waves: Managing the Impacts of Climate Change on Poverty." *Climate Change and Development*. Washington, DC: World Bank.

Hameed, Jawish. (2007). "The Earthquake That Stunned Maldives: Tsunami 2004." January 6. Available at <www.jawish.org/blog/archives/189-The-earthquake-that-stunned-Maldives-Tsunami-2004.html>. Accessed on February 14, 2020.

Hamlin, Larry. (2014). "UN IPCC AR5 Report Infected with Fatal Technical and Procedural Flaws." *WUWT*, April 16.

Han, Victoria. (2017). "Trump's Promise: Withdrawing from the Paris Climate Agreement." *Environmental Claims Journal*, Vol. 29, No. 4.

Hansen, J., Ruedy, R., Sato, M. and Lo, K. (2010). "Global Surface Temperature Change." *Reviews of Geophysics*, Vol. 48, No. 4.

Hansen, J. and Sato, M. (2012). "Paleoclimate Implications for Human-Made Climate Change." In: *Climate Change Inferences from Paleoclimate and Regional Aspects*. Wien: Springer.

Haq, Mahbub ul. (2015). Food Security in Pakistan: Briefing Paper. Policy Brief No. 2. Lahore University of Management Sciences, Lahore, Pakistan, November.

Haque, ANM Nurul. (2008). "SAARC Food Bank Imperative for the Region." *Daily Star*, August 9.

Haque, Ubydul., Hashizume, Masahiro., Kolivras, Korine N., Overgaard, Hans J., Das, Bivash. and Yamamoto, Taro. (2012). "Reduced Death Rates from Cyclones in Bangladesh: What More Needs to be Done?" *Bull World Health Organ*, Vol. 90. doi:10.2471/BLT.11.088302.

Hardin, G. (1968). "The Tragedy of the Commons." *Science*, Vol. 162, December 13.

Hare, William. (2003). *Assessment of Knowledge on Impacts of Climate Change: Contribution to the Specification of Article 2 of the UNFCCC.* Berlin: Wissen.

Hare, W.L., Cramer, W., Schaeffer, M., Battaglini, A. and Jaeger, C.C. (2011). "Climate Hotspots: Key Vulnerable Regions, Climate Change and Limits to Warming." *Regional Environmental Change*, Vol. 11, No. 1.

Harmeling, S. and Bals, Christoph. (2008). *Making the Adaptation Fund Work for the Most Vulnerable People.* Adaptation Fund/Discussion Paper, 2008, Germanwatch/ Brotfür die Welt, Bonn/Stuttgart, December.

Harrabin, Roger. (2007). "How Climate Change Hits India's Poor." *BBC News*, February 1. Available at <http://news.bbc.co.uk>. Accessed on April 8, 2013.

Harris, Gardiner. (2013). "Index of Happiness? Bhutan's New Leader Prefers More Concrete Goal." *New York Times*, October 4.

Harris, Gardiner. (2014). "Borrowed Time on Disappearing Land Facing Rising Seas." *The New York times*, March 28.

Harris, Jonathan M., Roach, Brian. and Codur, Anne-Marie. (2017). *The Economics of Global Climate Change.* Medford: Global Development and Environment Institute, Tufts University.

Harrison, Frances. (2012). *Still Counting the Dead: Survivors of Sri Lanka's Hidden War.* Toronto: Anansi Press.

Harvey, Fiona. (2012). "Developed World Failing on Climate Funds Pledge, Says Bangladeshi Minister." *The Guardian*, January 2.

Hasan, Syed Rizwana. (2005). "Application and Reform Needs of the Environmental Laws in Bangladesh." *Bangladesh Journal of Law*, Vol. 9, Nos. 1–2.

Hasemann, Anna., Roberts, Erin., Huq, Saleemul. and Singh, Harjeet Singh. (2014). *Climate Induced Loss and Damage in South Asia: Draft for Consultation.* Dhaka: Climate Action Network South Asia.

Hasina, Sheik. (2019). "Bangladesh Is Booming- and Here's Why Says the Prime Minister." *World Economic Forum*, October 4.

Hasnain, S.I. (2000). *Status of the Glacier Research in the HKH Region.* Kathmandu: International Centre for Integrated Mountain Development.

Hasnain, S.J. (2002). "Himalayan Glaciers Meltdown: Impacts on South Asian Rivers." In: van Lanen, H.A.J. and Demuth, S. (eds.) *FRIEND 2002-Regional Hydrology: Bridging the Gap Between Research and Practice.* Volume 274. Wallingford: IAHS Publications.

Hausfather, Zeke. (2019). "State of the Climate: 2019 Set to be Second or Third Warmest Year, Carbon Brief." July 25. Available at <www.carbonbrief.org/state-of-the-climate-2019-set-to-be-second-or-third-warmest-year>. Accessed on February 14, 2020.

Haveeru Daily. (2009). "Climate Change Advisory Council inaugurated." Malé. April 2.

Haveeru Daily. (2010). "Maldives President Launches New Global Report on Climate Crisis." Malé. December 4.

Hayat, Ehsanullah. and Elçi, Ġebnem. (2017). Adopting a Strategic Framework for Transboundary Water Resources Management in Afghanistan. IWA 2nd Regional Symposium on Water, Wastewater and Environment. 22–24 March, Turkey: Çesme-Izmir.

Hazarika, Sanjoy. (1993). *Bangladesh and Assam: Land Pressures, Migration and Ethnic Conflict.* Occasional Paper Series of the Project on Environmental Change and Acute Conflict, No. 3. Cambridge, MA: American Academy of Arts and Sciences.

Hazem, Pamir. (2017). "Getting Ahead of Greenhouse-Gas Emissions in Afghanistan: The Case for Shifting from a Command-and-Control to a Cap-and-Trade Regime." *Vermont Journal of Environmental Law*, Vol. 18, No. 3, Spring.

He, Lichao. (2010). "China's Climate-Change Policy from Kyoto to Copenhagen: Domestic Needs and International Aspirations." *Asian Perspective*, Vol. 34, No. 3.

Headey, D., Hoddinott, J., Ali, D., Tesfaye, R. and Dereje, M. (2015). "The Other Asian Enigma: Explaining the Rapid Reduction of Undernutrition in Bangladesh." *World Development*, Vol. 66, February.

Hein, Philippe. (2004). "Small Island Developing States: Origin of the Category and Definition Issues." *Is a Special Treatment of Small Island Developing States Possible?* United Nations New York and Geneva: United Nations Conference on Trade and Development.

Held, D. and Hervey, A. (2011). "Democracy, Climate Change and Global Governance." In: Held, D., Hervey, A. and Theros, M. (eds.) *The Governance of Climate Change*. Cambridge: Polity.

Helvates Nepal. (2012). *Nepal's CC Policies and Plans: Local Communities' Perspective*. Nepal: Helvates Swiss Intercooperation.

Herold, Anke., Cames, Martin., Siemons, Anne., Emele, Lukas. and Cook, Vanessa. (2013). *Directorate General for Internal Policies Policy Department A: Economic and Scientific Policy*. Brussels: European Union, The Development of Climate Negotiations in View of Warsaw (COP 19).

Hersher, Rebecca. (2016). "India Ratifies Paris Climate Change Agreement." *The Two Way*, October 2.

Hettiarachchi, Chaminda. (2012). "Managing Disasters in South Asia: Sri Lankan Experience." In: Delinic, Tomislav. and Pandey, Nishchal N. (eds.) *Regional Environmental Issues: Water and Disaster Management*. Centre for South Asian Studies-Konrad Adenauer Stiftung (KAS). Kathmandu: Modern Printing Press.

Hey, E. (2007). "International Institutions." In: Bodansky, Daniel., Brunnee, J. and Hey, E. (eds.) *The Oxford Handbook of International Environmental Laws*. Oxford: Oxford University Press.

Hickel, Jason. (2017). "The Paris Climate Deal Won't Save Us: Our Future Depends on De-Growth." *The Guardian*, July 3.

Hickmann, T. (2017). GlobaleKlimapolitiknach Paris, Deutsche Gesellschaft für die VereintenNationene.V., cited in Lippelt, Jana. and Mayer, Lea. (2017). After the Paris Agreement- What's Next? Worldwide Implementation, Spotlight, CESifo Forum 4, Vol. 18, December.

High Representative for *Common Foreign and Security Policy* (CFSP) and the European Commission. (2008). Climate Change and International Security: Paper from the High Representative and the European Commission to the European Council. April 4, 2009, Brussels.

Hijioka, Y., Lin, E., Pereira, J.J., Corlett, R.T., Cui, X., Insarov, G.E., Lasco, R.D., Lindgren, E. and Surjan, A. (2014). "Asia." In: Barros, V.R., Field, C.B., Dokken, D.J., Mastrandrea, M.D., Mach, K.J., Bilir, T.E., Chatterjee, M., Ebi, K.L., Estrada, Y.O., Genova, R.C., Girma, B., Kissel, E.S., Levy, A.N., MacCracken, S., Mastrandrea, P.R. and White, L.L. (eds.) *Climate Change 2014: Impacts, Adaptation, and Vulnerability*. Part B: Regional Aspects. Contribution of Working Group II to the Fifth Assessment Report of the Intergovernmental Panel on Climate Change. Cambridge and New York: Cambridge University Press.

Hillstrom, Kevin. (2010). *U.S. Environmental Policy and Politics: A Documentary History*. Washington, DC: CQ Press.

The Hindu. (2008). "For a Food-Secure South Asia." August 7.

The Hindu. (2007). "Council on Climate Change Constituted." June 6.

The Hindu. (2010). "SAARC Leaders to Project Green and Happy South Asia." April 28.

Hindustan Times. (2015a). "Delhi Will Restrict Cars from Jan 1 to Cut Pollution, May Face Challenge." December 29.

Hindustan Times. (2015b). "Heat Now Second-Largest Natural Killer of Indians, Toll Crosses 2,000." May 31.

Hindustan Times. (2015c). "No New Diesel Vehicles of over 2000cc in Delhi Till March 31: SC." December 16.

Hirji, Rafik., Nicol, Alan. and Davis, Richard. (2017). *South Asia 8S Climate Change Risks in Water Management: Climate Risks and Solutions: Adaptation Frameworks for Water Resources Planning, Development and Management in South Asia.* Report No: AUS14873. Washington, DC: The World Bank.

Hoerling, M., Eischeid, J., Perlwitz, J., Quan, X., Zhang, T. and Pegion, P. (2012). "On the Increased Frequency of Medi-Terranean Drought." *Journal of Climate*, Vol. 25.

Hoffmann, Stanley. (1965). *The State of War: Essays in the Theory and Practice of International Politics.* New York: Praeger.

Hoffmann, Stanley. (1995). "The Crises of Liberal Internationalism." *Foreign Policy*, Vol. 98.

Holsti, Ole R. (1970). *Crisis, Escalation, War.* Montreal: McGill University Press.

Holzinger, Katharina. (2003). *The Problems of Collective Action: A New Approach.* Bonn: Institute of Political Science, University of Hamburg.

Homer-Dixon, T. and Percival, Valerie. (1996). *Environmental Scarcity and Violent Conflict: Briefing Book.* Toronto: University of Toronto.

Hönisch, Bärbel., Ridgwell, Andy., Schmidt, Daniela N., Thomas, Ellen., Gibbs, Samantha J., Sluijs, Appy., Zeebe, Richard., Kump, Lee., Martindale, Rowan C. Greene, Sarah E., Kiessling, Wolfgang., Ries, Justin., Zachos, James C., Royer, Dana L., Barker, Stephen., Marchitto Jr., Thomas M., Moyer, Ryan., Pelejero, Carles., Ziveri, Patrizia., Foster, Gavin L. and Williams, Branwen. (2012). "The Geological Record of Ocean Acidification." *Science*, Vol. 335, No. 6072.

Hossain, Segufta. (2013). "Environment Security Nexus: Bangladesh Perspectives." *Bangladesh Institute of International and Strategic Studies*, Vol. 34, No. 2.

Houghton, J.T., Ding, Y., Griggs, D.J., Noguer, N., Linden, P. J. van der., Xiaosu, D., Maskell, K. and Johnson, C.A. (2001). *Climate Change 2001: The Scientific Basis.* Contribution of Working Group I to the Third Assessment Report of the Intergovernmental Panel on Climate Change. Cambridge: Cambridge University Press.

Houghton, J.T., Filho, L.G. Meira., Callander, B.A., Harris, N., Kattenberg, A. and Maskell, K. (eds.). (1996). *Climate Change 1995: The Science of Climate Change.* Contribution of Working Group I to the Second Assessment Report of the Intergovernmental Panel on Climate Change. Cambridge, United Kingdom and New York, NY, USA: Cambridge University Press.

Houghton, R.A. (2010). "How Well Do We Know the Flux of CO_2 from Land-Use Change?" *Tellus B*, Vol. 62, No. 5.

Huang, J. and Rozelle, S. (2009). *Agriculture, Food Security, and Poverty in China, Past Performance, Future Prospects, and Implications for Agricultural R & D Policy.* Washington: International Food Policy Research Institute.

Human Development Report. (2006). *Beyond Scarcity: Power, Poverty and the Global Water Crisis.* New York: United Nations Development Programme (UNDP).

Humphreys, Stephen. (2008). *Climate Change and Human Rights: A Rough Guide.* Versoix, Switzerland: International Council on Human Rights Policy.

Hunter, John. (2010). "Estimating Sea-Level Extremes under Conditions of Uncertain Sea-Level Rise." *Climatic Change*, Vol. 99, Nos. 3–4.

Hunter, Joke Waller. (2005). *UNFCCC (2005) Climate Change, Small Island Developing States.* Bonn: Issued by the Climate Change Secretariat (UNFCCC).

Huq, Riaz ul. (2018). "New Food Security Policy Aims High." *The Express Tribune*, April 6.

Huq, Saleemul. (2016). "Climate Finance in Bangladesh." *The Daily Star*, April 16.

Huq, Saleemul. and Ayers, Jessica. (2008). *Climate Change Impacts and Responses in Bangladesh.* London: International Institute for Environment and Development.

Huq, Saleemul. and Khan, Mizan R. (2017). *Planning for Adaptation in Bangladesh: Past, Present and Future, Policy Brief.* London/Dhaka: International Institute for Environment and Development (IIED).

Hurrell, Andrew. (1995). "Regionalism in Theoretical Perspective." In: Fawcett, Louise. and Hurrell, Andrew. (eds.) *Regionalism in World Politics: Regional Organisation and International Order.* Oxford and New York: Oxford University Press.

Hussain, Delwar. (2009). "Fencing Off Bangladesh." *The Guardian*, September 5.

Hveem, Helge. (2006). "Global Governance and the Comparative Political Advantage of Regional Cooperation." In: Tussie, Diana. (ed.) *The Environment and International Trade Negotiations Developing Country Stakes.* London: Macmillan Press Ltd.

ICARDA. (2002). *Needs Assessment on Soil and Water in Afghanistan: Future Harvest Consortium to Rebuild Agriculture in Afghanistan.* International Center for Agricultural Research in the Dry Areas, Aleppo, Syria, September.

ICIMOD. (2011). *Glacial Lakes and Glacial Lake Outburst Floods in Nepal.* Kathmandu: International Centre for Integrated Mountain Development.

ICIMOD. (2012). *Climate Change Challenges in the Mountains: Implication to Adaptation Needs of the Hindu Kush Himalayas*, Prepared by International Centre for Integrated Mountain Development (ICIMOD) for Asia Pacific Adaptation Network (APAN), March.

ICIMOD. (2013). *Annual Report 2013: Integration, Innovation, Impact.* Kathmandu: International Centre for Integrated Mountain Development.

ICIMOD. (2016). *Bhutan: Climate + Change: Handbook.* Kathmandu: International Centre for Integrated Mountain Development.

IFAD. (2017). *Smallholder Agribusiness Partnerships (SAP).* Programme Final Project Design Report, Main Report and Appendices. Sri Lanka.

IFPRI. (2010). *Food and Nutritional Security in Nepal: A Stocktaking Exercise.* Kathmandu: Food Policy Research Institute for USAID.

Ikram, Qiyamud Din. (2018). "Climate Change Is Happening in Afghanistan." *Environmental Specialist*, September 6.

ILO. (2018). *International Labour Migration Statistics in South Asia: Establishing a Subregional Database and Improving Data Collection for Evidence-Based Policy-Making.* New Delhi: ILO Decent Work Team for South Asia.

Imam, H. (2008). "The Challenge Is Even Bigger." *Daily Star*, May 28.

Imber, Mark F. (1984). "Re-Reading Mitrany: A Pragmatic Assessment of Sovereignty." *Review of International Studies*, Vol. 10, No. 2, cited in McLaren, Robert I. (1985). "Mitranian Functionalism: Possible or Impossible?" *Review of International Studies*, Vol. 11, No. 2.

Imbulana, K.A.U.S., Wijesekara, N.T.S. and Neupane, B.R. (2006). *Sri Lanka Water Development Report 2010*. Colombo: UNESCO and Ministry of Irrigation and Water Resources Management.

IMF. (2011). *World Economic Outlook 2011: Slowing Growth and Rising Risk*. Washington, DC: International Monetary Fund.

Imtiaz, Saba. and Rehman, Zia ur. (2015). "Death Toll From Heat Wave in Karachi, Pakistan, Hits 1,000." *The New York Times*, June 25.

Inaz, Mohamed., Naeem, Hussein., Jameel, Ahmed. and Zuhair, Mohamed. (2004). *State of the Environment Maldives*. Malé, Maldives: Ministry of Environment & Construction, Government of Maldives.

The Indian Express. (2014). "SAARC Panel Decides to Dissolve 3 Regional Centres." November 22.

India Today. (2019a). "600 People Killed, Over 25 Million Affected by Flooding in India, Bangladesh, Nepal & Myanmar: UN." New Delhi, July 27.

India Today. (2019b). "Monsoon Fury: 169 Dead in Flood, Rain-Related Incidents in South and West India." New Delhi, August 12.

Integrated Development Society Nepal (IDSN). (2014). *Economic Impact Assessment of Climate Change in Key Sectors in Nepal. Practical Action Consulting, & Global Climate Adaptation Partnership*. Nepal: Integrated Development Society Nepal.

Internal Displacement Monitoring Centre (IDMC). (2012). Pakistan: Displacement Caused by Conflict and Natural Disasters, Achievements and Challenges, 1.

Internal Displacement Monitoring Centre (IDMC). (2014). *Global Estimates: People Displaced by Disasters*. Geneva: IDMC, September.

Internal Displacement Monitoring Centre (IDMC). (2015). *Bangladesh IDP Figures Analysis*. Geneva: IDMC.

International Center for Agricultural Research in the Dry Areas (ICARDA). (2002). *Needs Assessment on Soil and Water in Afghanistan*. Future Harvest Consortium to rebuild agriculture in Afghanistan. International Center for Agricultural Research in the Dry Areas.

International Food Policy Research Institute (IFPRI). (2010). "Global Hunger Index: Facts and Findings: Asia." cited in Chatterjee, Bipul. and Khadka, Manbar. (eds.) (2011). *Climate Change and Food Security in South Asia*. Jaipur: CUTS International.

IPCC. (1990). *Working Group III, Strategies for Adaptation to Sea Level Rise: Report of the Coastal Zone Management Subgroup*. Intergovernmental Panel on Climate Change Working Group. The Netherlands: Rijkswaterstatt.

IPCC. (2007). "Summary for Policymakers." In: *Climate Change 2007: The Physical Science Basis*. Contribution of Working Group I to the Fourth Assessment Report of the Intergovernmental Panel on Climate Change. Cambridge, UK and New York: Cambridge University Press.

IPCC. (2013). "Summary for Policymakers." In: Stocker, T.F., Qin, D., Plattner, G.K., Tignor, M., Allen, S.K., Boschung, J., Nauels, A., Xia, Y., Bex, V. and Midgley, P.M. (eds.) *Climate Change 2013: The Physical Science Basis*. Contribution of Working Group I to the Fifth Assessment Report of the Intergovernmental Panel on Climate Change. Cambridge, UK and New York: Cambridge University Press.

IPCC. (2014). *Working Group II Contribution to the Intergovernmental Panel on Climate Change*. Fifth Assessment Report, 2013.

Ipiv, Özlem. and Reinhardt, Dieter. (2010). "Sustainable Green Investments and Policies in Bangladesh: Two Problems-One Solution?" In: Senz, Anja. and Reinhardt,

Dieter. (eds.) *Green Governance: One Solution for Two Problems? Climate Change and Economic Shocks: Risk Perceptions and Coping Strategies in China, India and Bangladesh.* Duisburg Working Papers on East Asian Studies, Published in cooperation with the Institute for Development and Peace, Germany, No. 86.

Iqbal, Muhammad., Ahmad, Munir., Khan, Muhammad Azeem., Samad, Ghulam. and Gill, Muhammad Aslam. (2014). *Review of Environmental Policy and Institutions.* Climate Change Working Papers No. 4. Islamabad: Pakistan Institute of Development Economics.

Iqbal, Muhammad. and Amjad, Rashid. (2012). "Food Security in South Asia: Strategies and Programmes for Regional Collaboration." In: Rahman, Sultan Hafeez., Khatri, Sridhar. and Brunner, Hans-Peter. (eds.) *Regional Integration and Economic Development in South Asia.* Cheltenham and Manila: Edward Elgar Publishing Limited, Asian Development Bank.

Irfan, Muhammad. (2018). "National Food Security to Make Agriculture Sector Resilient: Bosan." *Daily Pakistan*, May 29.

IRIN Humanitarian News and Analysis. (2011). "Sri Lanka: Records Rain Increase Urgency of Climate Change Adaptation." January 12.

Islam, A.K.M. Nazrul., Sultan, Salma. and Afroz. (2010). *Climate Change and South Asia: What Makes the Region Most Vulnerable?* MPRA Paper No. 21875.

Islam, Faisal., Hove, Hilary. and Parry, Jo-Ellen. (2011). *Review of Current and Planned Adaptation Action: South Asia, Afghanistan, Bangladesh, Bhutan, India, Maldives, Nepal, Pakistan and Sri Lanka.* International Institute for Sustainable Development.

Islam, Moinul. (1991). Ecological Catastrophes and Refugees in Bangladesh. Paper Presented at the Conference on Worldwide Refugee Movements. New York: New School for Social Research.

Islam, M.R., Rahman, Z.A., Ghazali, A.S., Arnakim, L.Y.B. and Faruque, C.J. (2016). "Contextual Strategies for Hunger Reduction: A Case of South Asian Countries." *Transylvanian Review*, Vol. 24, No. 5, Special Issue.

Islam, Nurul. and von Braun, Joachim. (2008). *Reducing Poverty and Hunger: Agricultural and Rural Development for Reducing Poverty and Hunger in Asia: Past Performance and Priorities for the Future.* International Food Policy Institute Research, Focus 15, Brief 1 of 15. Manila: ADB.

Islamabad Declaration of the Heads of State or Government of the Member Countries of South Asian Association for Regional Cooperation issued on December 31, 1988, Fourth SAARC Summit.

Issue Brief. (2017). "The Road from Paris: India's Progress toward Its Climate Pledge." November.

Ittyerah, Anil Chandy. (2013). *Food Security in India Food Security in India Issues and Suggestions for Effectiveness.* Theme Paper for the 57th Members' Annual Conference, Indian Institute of Public Administration New Delhi.

Ives, Jack D. Shrestha, Rajendra B. and Mool, Pradeep K. (2010). *Formation of Glacial Lakes in the Hindu Kush-Himalayas and GLOF Risk Assessment.* Kathmandu: ICIMOD.

Jabed, M.J.H. (2016). "Agricultural Transformation." *Kathmandu Post*, March 6.

Jackson, Robert. and Sorensen, Georg. (2003). *Introduction to International Relations: Theories and Approaches.* New York: Oxford University Press.

Jaffery, Rubiya. (2018). "Pakistan's Climate Change Plight." *The Diplomat*, March 21.

Jaitly, Ashok. (2009). "South Asian Perspectives on Climate Change and Water Policy." In: Michel, David. and Pandya, Amit. (eds.) *Troubled Waters: Climate Change, Hydropolitics, and Transboundary Resources*. Washington: Stimson Pragmatic Steps for Global Security.

Jamieson, D. (2014). *Reason in a Dark Time: Why the Struggle against Climate Change Failed-and What it Means for Our Future*. Oxford: Oxford University Press.

Janjua, Haroon. and McVeigh, Karen. (2019). "Chilling Reality: Afghanistan Suffers Worst Floods in Seven Years." *The Guardian*, March 6.

Jaramillo, F. and Destouni, G. (2015). "Local Flow Regulation and Irrigation Raise Global Human Water Consumption and Footprint." *Science*, Vol. 350.

Jarsjö, J., Asokan, S.M., Prieto, C., Bring, A. and Destouni, G. (2012). "Hydrological Responses to Climate Change Conditioned by Historic Alterations of Land-Use and Water-Use." *Hydrology Earth System Sciences*, Vol. 16.

Javed, Sajid Amin. (2016). "Climate Change: Reflections on Issues, Challenges and the Way Forward." *Development Advocate Pakistan*, Vol. 2, No. 4.

Javeline, Debra. (2014). "The Most Important Topic Political Scientists Are Not Studying: Adaptation to Climate Change." *Perspectives on Politics*, Vol. 12, No. 2.

Jayatilake, H.M., Chandrapala, L., Basnayake, B.R.S.B. and Dharmaratne, G.H.P. (2005). "Water Resources and Climate Change." In: Wijesekera, N.T.S., Imbulana, K.A.U.S. and Neupane, B. (eds.) *Proceedings of Workshop on Sri Lanka National Water Development Report*. Paris: World Water Assessment Programme (WWAP).

Jeganaathan, J. (2011). *Energy and Environmental Security: A Regional Dialogue*. IPCS Conference Report, 35. New Delhi: Institute of Peace and Conflict Studies, June.

Jelenkovic, Nadja. (2012). "The 2011 South China Floods: Drought, Three Gorges Dam and Migration." In: Gemenne, François., Brücker, Pauline. and Ionesco, Dina (eds.) *The State of Environmental Migration 2011*. Paris: Institute for Sustainable Development and International Relations (IDDRI).

Jemia, MirianTelma. (2016). "Strongest Drought in 25 Years Hits Bolivia." *Mongabay*, October 3.

Jervis, Robert. (1976). *Perception and Misperception in World Politics*. Princeton, NJ: Princeton University Press.

Jianchu, Xu., Eriksson, Mats., Ferdinand, Jacob. and Merz, Juerg. (2006). *Managing Flash Floods and Sustainable Development in the Himalayas Report of an International Workshop held in Lhasa, PRC, October 23–28, 2005*. Kathmandu: ICIMOD.

Jimenez Cisneros, B.E., Oki, T., Arnel, N.W., Benito, G., Cogley, J.G., Doll, P., Jiang, T. and Mwakalila, S. S. (2014). "Freshwater Resources." In: Field, C.B., Barros, V.R., Dokken, D.J., Mach, K.J., Mastrandrea, M.D., Bilir, T.E., Chatterjee, M., Ebi, K.L., Estrada, Y.O., Genova, R.C., Gimma, B., Kissel, E.S., Levy, A.N., MacCracken, S., Mastrandrea, P.R. and White, L.L. (eds.) *Climate Change 2014: Impacts, Adaptation and Vulnerability*. Part A: Global and Sectoral Aspects, Contribution of Working Group II to the Fifth Assessment Report of the Intergovernmental Panel on Climate Change. Cambridge: Cambridge University Press.

Jodoin, Sébastien. and Lofts, Katherine. (eds.). (2013). *Economic, Social, and Cultural Rights and Climate Change: A Legal Reference Guide*. New Haven, CT: CISDL, GEM & ASAP.

Johnson, Brian. (1973). "The Settlement of Stockholm." *Ecologist*, Vol. 3.

Jones, J.M. (2010). "Conservatives' Doubts about Global Warming Grow." *Gallup Poll*. Available at <www.gallup.com/poll/126563/ conservatives-doubts-global-warming-grow.aspx>. Accessed on February 14, 2020.

Julca, A. and Paddison, O. (2010). "Vulnerabilities and Migration in Small Island Developing States in the Context of Climate Change." *Natural Hazards*, Vol. 55, No. 3.

Kafle, SheshKanta. (2017). "Disaster Risk Management Systems in South Asia: Natural Hazards, Vulnerability, Disaster Risk and Legislative and Institutional Frameworks." *Journal of Geography & Natural Disasters*, Vol. 7, No. 3.

Kahn, Greg. (2003). "The Fate of the Kyoto Protocol under the Bush Administration." *Berkeley Journal of International Law*, Vol. 21, No. 3.

Kajjo, Sirwan. and Jedinia, Mehdi. (2019). "Iran Criticized for Threat to Deport Afghan Refugees." *VOA News*, May 15.

Kaltenborn, B.P., Nellemann, C. and Vistnes, I.I. (eds.). (2010). *High Mountain Glaciers and Climate Change: Challenges to Human Livelihoods and Adaptation*. Arendal: UNEP, GRID.

Karetnikov, D., Lakhey, S., Horin, C., Bell, B., Ruth, M., Ross, K. and Irani, D. (2008). *Economic Impacts of Climate Change on North Dakota*. College Park, MD: Center for Integrative Environmental Research, University of Maryland.

Karim, Mahin. (2013). *The Future of South Asian Security: Prospects for a Nontraditional Regional Security Architecture*. Seattle and Washington, DC: The National Bureau of Asian Research NBR Project Report.

Karim, Md. Manzurul. (2013). "Socio-Economic and Political Development of South Asian Countries: In Bangladesh Perspective." *International Journal of Humanities and Social Science Invention*, Vol. 2, No. 2.

Karim, M.F. and Mimura, N. (2008). "Impacts of Climate Change and Sea-Level Rise on Cyclonic Storm Surge Floods in Bangladesh." *Global Environmental Change*, Vol. 18, No. 3.

Karim, Mohd Aminul. (2014). "South Asian Regional Integration: Challenges and Prospects." *Japanese Journal of Political Science*, Vol. 15, No. 2.

Karim, Z., Ibrahim, A., Iqbal, A. and Ahmed, M. (1990). *Drought in Bangladesh Agriculture and Irrigation Schedules for Major Crops*. No. 34, Dhaka: Bangladesh Agricultural Research Council.

Karl, T.R., Melillo, J.M. and Peterson, T.C. (eds.). (2009). *Global Climate Change Impacts in the United States, U.S. Global Change Research Program (USGCRP)*. Washington, DC: Cambridge University Press.

Karns, Margaret P. and Mingst, Karen A. (2005). *International Organizations: The Politics and Processes of Global Governance*. New Delhi: Viva Book Pvt Ltd.

Karthikeyan, T.C. (2010). "Environmental Challenges for Maldives." *South Asian Survey*, Vol. 17, No. 2.

Katakam, Anupama., Bavadam, Lyla. and Bunsha, Dionne. (2005). "High Water and Hell." *Frontline*, Vol. 22, No. 17.

Kathmandu Declaration of the Heads of State or Government of the Member Countries of SAARC issued on November 4, 1987.

Kathmandu Declaration, SAARC 18th Summit. (2014).

The Kathmandu Post. (2010). "Apa Sherpa Named SAARC Envoy for Climate Change." April 27.

The Kathmandu Post. (2014). "SAARC to Trim Regional Centres." November 22.

Kaur, Rashmeen. (2017). "U.S. Participation in Global Climate Change Resolutions: Analysis of the Kyoto Protocol." *McNair Research Journal SJSU*, Vol. 13, Spring.

Kay, David A. and Jacobson, Harold K. (eds.). (1983). *Environmental Protection: The International Dimension.* Totowa, NJ: Allanheld, Osmun.

Keil, Kathrin. (2014). "The Arctic: A New Region of Conflict?" *Cooperation and Conflict*, Vol. 49, No. 2, 162.

Kellet, Jan. (2014). *Disaster Risk Reduction Makes Development Sustainable: A Call for Action.* New York: UNDP.

Kelly, Annie. (2012). "Gross National Happiness in Bhutan: The Big Idea from a Tiny State That Could Change the World." *The Guardian*, December 1.

Kelly, P.M. and Adger, W.N. (1999). "Social Vulnerability to Climate Change and the Architecture of Entitlements." *Mitigation and Adaptation Strategies for Global Change.* Printed in the Netherlands: Kluwer Academic Publishers, Vol. 4.

Kelly, P.M. and Adger, W.N. (2000). "Theory and Practice in Assessing Vulnerability to Climate Change and Facilitating Adaptation." *Climatic Change.* Printed in the Netherlands: Kluwer Academic Publisher, Vol. 47.

Kelman, Ilan., Orlowska, Justyna., Upadhyay, Himani., Stojanov, Robert., Webersik, Christian., Simonelli, Andrea C., Procházka, David. and Němec, Daniel. (2019). "Does Climate Change Influence People's Migration Decisions in Maldives?" *Climate Change*, Vol. 153, Nos. 1–2.

Kench, P.S., McLean, R.F., Brander, R.W., Nicholl, S.L., Smithers, S.G., Ford, M.R., Parnell, K.P. and Aslam, M. (2006). "Geological Effects of Tsunami on Mid-Ocean Atoll Islands: The Maldives before and after the Sumatran Tsunami." *Geology*, Vol. 34.

Kenneally, Christine. (2004). "Surviving the Tsunami: What Sri Lanka's Animals Knew That Humans Didn't." *Slate Magazine*, December 30.

Keohane, Robert O. (1980). "The Theory of Hegemonic Stability and Changes in International Economic Regimes." In: Holsti, O.R., Siverson, R.M. and George, A.L. (eds.) *Change in the International System.* Boulder, CO: Westview Press.

Keohane, Robert O. (1984). *After Hegemony: Cooperation and Discord in the World Political Economy.* Princeton: Princeton University Press.

Keohane, Robert O. (1994). "Cooperation and International Regimes." In: Little, Richard and Smith, Michael. (eds.) *Perspectives on World Politics.* New York: Routledge.

Khadka, Ram B., Dalal-Clayton, Barry, Mathema, Ajay and Shrestha, Pujan. (2012). *Safeguarding the Future, Securing Shangri-La- Integrating Environment and Development in Nepal: Achievements, Challenges and Next Steps.* London, UK: IIED.

Khaleej Times. (2014). "30 Dead, 1 Million Hit by Sri Lanka Flood." December 31. Available at <www.khaleejtimes.com/article/20141230/ARTICLE/312309944/1028>. Accessed on February 14, 2020.

Khalid, Saadia. (2009). "Pakistan Facing Severe Climate Change Effects: Afridi." *The International News*, Islamabad, June 6.

Khan, A.E., Xun, W.W., Ahsan. H., Vineis, P. (2011). "Climate Change, Sea-Level Rise, & Health Impacts in Bangladesh." *Environment: Science and Policy for Sustainable Development*, Vol. 53, No. 3.

Khan, Danial. (2015). "Pakistan's Water Shortage Creates Dangerous Agriculture Conditions." *America*, March 17.

Khan, F. (2009). "Water, Governance, and Corruption in Pakistan." In: Kugelman, M. and Hathaway, R.M. (eds.) *Running on Empty, Pakistan's Water Crisis.* Washington, DC: Woodrow Wilson International Center for Scholars.

Khan, Farrukh Iqbal. and Munawar, Sadia. (2011). Institutional Arrangements for Climate Change in Pakistan Project Report Series # 19. Sustainable Development Policy Institute (SDPI), July.

Khan, Habibur Rahman. (1997). "Challenges of Economic Liberalisation and Environment in Small States of South Asia: Bhutan and the Maldives." *Bangladesh Institute of International and Strategic Studies*, Vol. 18, No. 4.

Khan, Malik Amin Aslam. (2011). *National Economic and Environment Development Study (NEEDS): Pakistan*. Sustainable Development Policy Institute, Pakistan.

Khan, Mizan R. (2016). "Climate Change, Adaptation and International Relations Theory." In: Sosa-Nunez, Gustavo. and Atkins, Ed. (eds.) *Environment, Climate Change and International Relations*. Bristol: E-International Relations.

Khan, Mohd. Atiqueuzzaman., Shamsuddoha, Md., Al Helal, Abdullah. and Hassan, Asif. (2013). "Climate Change Mitigation Approaches in Bangladesh." *Journal of Sustainable Development*, Vol. 6, No. 7.

Khan, Rina Saeed. (2013). "National Climate Change Policy for a Vulnerable Country." *Dawn*, March 2.

Khan, Rina Saeed. (2017). "Pakistan Passes Climate Change Act, Reviving Hopes–and Skepticism." *Reuters*, March 24.

Khan, Rina Saeed. and Arora, Kabir. (2014). "The Climate Crisis: An Appeal to the Governments of Pakistan and India." *Dawn*, September 22.

Khan, S. (2010). "Promised Fund for Climate Change Adaptation, Mitigation." *The Financial Express*, Dhaka, March 7.

Khan, Tariq Masood Ali. and Rabbani, M.M. (2000). *Sea Level Monitoring and Study of Sea Level Variations along Pakistan Coast: A Component of Integrated Coastal Zone Management*. Karachi: National Institute of Oceanography.

Khanom, Sufia. (2012). "The 17th SAARC Summit: Issues, Outcome and Implications for Bangladesh." *Bangladesh Institute of International and Strategic Studies*, Vol. 33, No. 1.

Khatri, Nam Raj. (2013). "Climate-Change Refugees in Nepal: The Need for Climate-Smart Capacity Building." In: Filho, Walter Leal (ed.) *Climate Change and Disaster Risk Management: Climate Change Management*. Heidelberg and New York: Springer.

Khatun, Fahmida Akter. (2002). *Population and Environment in Bangladesh: Designing a Policy Accounting for Linkages*. Dhaka: Centre for Policy Dialogue.

Khatun, Fahmida. and Hossain, Samina. (2012). *Adapting to Climate Change: Issues for South Asia*. Kathmandu: South Asia Watch on Trade, Economic and Environment (SAWTEE).

Kildow, J.T., Colgan, C.S., Johnston, P., Scorse, J.D. and Farnum, M.G. (2016). *State of the U.S. Ocean and Coastal Economies: 2016 Update*. Monterey, CA: National Ocean Economics Program.

Kindleberger, Charles P. (1973). *The World in Depression, 1929–1939*. Los Angeles: University of California Press.

King, Alexander. and Schneider, Bertrand. (1972). *The First Global Revolution: A Report by the Council of the Club of Rome*. New York: Orient Longman.

King, Ed. (2015). "Kyoto Protocol: 10 Years of the World's First Climate Change Treaty." *Climate Home News*, February 16. Available at <www.climatechangenews.com/2015/02/16/kyoto-protocol-10-years-of-the-worlds-first-climate-change-treaty/>. Accessed on February 14, 2020.

King, Matthew. and Sturtewagen, Benjamin. (2010). *Making the Most of Afghanistan's River Basins: Opportunities for Regional Cooperation*. New York: The East-West Institute.

Kiran, Afifa. and Ain, Qurat ul. (2017). *Climate Change: Implications for Pakistan and Way Forward*. ISSRA PAPERS.

Klein, R.J.T. and Tol, R.S.J. (1997). *Adaptation to Climate Change: Options and Technologies: An Overview Paper*. Technical Paper FCCC/TP/1997/3. Bonn: UNFCCC Secretariat.

Kluger, Jeffrey. (2001). "A Climate of Despair." *Time*, April 9.

Kothari, Uma. (2014). "Political Discourses of Climate Change and Migration: Resettlement Policies in the Maldives." *The Geographical Journal*, Vol. 180, No. 2.

Kovats, R.S., Valentini, R., Bouwer, L.M., Georgopoulou, E., Jacob, D., Martin, E., Rounsevell, M. and Soussana, J.F. (2014). "Europe." In: *Climate Change 2014: Impacts, Adaptation and Vulnerability*. Part B: Regional Aspects, Contribution of Working Group II to the Fifth Assessment Report of the Intergovernmental Panel on Climate Change.

Krampe, Florian. (2018). "Is SAARC Prepared to Combat Climate Change and Its Security Risks?" *The Third Pole Net*, September 6.

Krasner, Stephen D. (1999). *Sovereignty: Organized Hypocrisy*. Princeton: Princeton University Press.

Krasner, Stephen D. (2009). *Power, the State, and Sovereignty: Essays on International Relations*. London: Routledge.

Kreft, S., Eckstein, D., Dorsch, L. and Fischer, L. (2016). *Global Climate Risk Index 2016: Who Suffers Most from Extreme Weather Events? Weather-Related Loss Events in 2014 and 1995 to 2014*. Bonn: Germanwatch.

Kreft, S. and Eckstein, D. (2014). *Global Climate Risk Index 2014: Who Suffers Most from Extreme Weather Events? Weather-Related Loss Events in 2012 and 1993 to 2012*. Bonn: Germanwatch.

Kreft, S., Eckstein, D., Junghans, L., Kerestan, C. and Hagen, U. (2015). *Global Climate Risk Index 2015: Who Suffers Most from Extreme Weather Events? Weather-Related Loss Events in 2013 and 1994 to 2013*. Bonn: Germanwatch e.V.

Kripalani, R.H., Oh, J.H., Sabade, S.S. and Chaudhari, H.S. (2007). "South Asian Summer Monsoon Precipitation Variability: Coupled Climate Simulations and Projections Under IPCC AR4." *Theoretical and Applied Climatology*, Vol. 90, Nos. 3–4, November.

Krishnakumar, R. (2018). "Battling a Deluge." *Frontline*, September 14.

Kruijk, Hans de. (2010). *Poverty Dynamics: The Case of the Maldives*. Rotterdam: Optima Grafische, Communicatie.

Kulshrestha, Umesh. (2017). *Air Pollution and Climate Change in South Asia: Issues, Impact and Initiatives*. London, UK: Athena Academic.

Kumar, Ashwin J. (2018). "Kerala Floods: Armed Forces Rescue over 23,000 People Till Now." *The Times of India*, August 19.

Kumar, K. Rupa., Sahai, A.K., Krishna Kumar, K., Patwardhan, S.K., Mishra, P.K., Revadekar, J.V., Kamala, K. and Kant, G.B. (2006). "High-Resolution Climate Change Scenarios for India for the 21st Century." *Current Science*, February, Vol. 90, No. 3.

Kumar, K.S. and Parikh, J. Kavi. (2001a). "Socio-Economic Impacts of Climate Change on Indian Agriculture." *International Review for Environmental Strategies*, Vol. 2, No. 2.

Kumar, K.S. and Parikh, J. Kavi. (2001b). "Indian Agriculture and Climate Sensitivity." *Global Environmental Change*, Vol. 11.

Kumar, R. (2008). Climate Change and India: Impacts, Policy Responses and a Framework for EU-India Cooperation. European Parliament Report no. IP/A/CLIM/NT/2007–10, PE 400.991.

Kumar, Vineet. (2015). "People's Cure to Climate Woes." *Down to Earth*, June 11.

Kumara, Praduman., Mittal, Surabhi. and Hossain, Mahabub. (2008). "Agricultural Growth Accounting and Total Factor Productivity in South Asia: A Review and Policy Implications." *Agricultural Economics Research Review*, Vol. 21.

Kundi, Brayshna. (2017). "Pakistan's Water Crisis: Why a National Water Policy is Needed." *The Asia Foundation*, November 1.

Kundur, Suresh Kumar. (2012). "Development of Tourism in Maldives." *International Journal of Scientific and Research Publications*, Vol. 2, No. 4.

Kurukulasuriya, P., Zhao, Y., Mao, J. and Guillemot, J. (2016). *Piloting Climate Change Adaptation to Protect Human Health in China*. New York: Adaptation Learning Mechanism.

Kutesa, H.E. Sam. (2015). *Transforming Our World: The 2030 Agenda for Sustainable Development*. New York: All Permanent Representatives and Permanent Observers to tile United Nations, August 11.

The Kyoto Protocol. (1998). *Kyoto Protocol to the United Nations Framework Convention on Climate Change*. New York: United Nations.

Kyoto Protocol to the United Nations Framework Convention on Climate Change. (1997). Kyoto: United Nations, Treaty Series, Vol. 2303, December 11.

Laczko, F. and Collett, E. (2005). *Assessing the Tsuanmi's Effects on Migration*. Geneva: International Organization for Migration.

Ladha, J.K., Fischer, K.S., Hossain, M., Hobbs, P.R. and Hardy, B. (eds.). (2000). *Improving the Productivity and Sustainability of Rice-Wheat Systems of the Indo-Gangetic Plains: A Synthesis of NARS-IRRI Partnership Research*. Discussion Paper, No. 40. International Rice Research Institute (IRRI).

Laferrière, Eric. and Stoett, Peter J. (1999). *International Relations Theory and Ecological Thought-Towards a Synthesis*. London: Routledge.

Lagos, Ricardo. (2009). *Facilitating an International Agreement on Climate Change: Adaptation to Climate Change*. A Proposal of the Global Leadership for Climate Action, June.

Lagos, Ricardo. and Wirth, Timothy E. (2009). *Facilitating an International Agreement on Climate Change: Adaptation to Climate Change: A Proposal of the Global Leadership for Climate Action*, June, this paper has benefited from the work and report of the International Commission on Climate Change and Development.

Lakshmi, Rama. (2009). "India Rejects Calls for Emission Cuts." *The Washington Post*, April 13.

Lal, M. (2003). "Global Climate Change: India's Monsoon and Its Variability." *Journal of Environmental Studies and Policy*, Vol. 6.

Lal, Murari., Nozawa, T., Emori, S., Harasawa, H., Takahashi, K., Kimoto, M. Abe-Ouchi, A., Nakajima, T., Takemura, T. and Numaguti, A. (2001). "Future Climate Change: Implications for Indian Summer Monsoon and Its Variability." *Current Science*, Vol. 81, No. 9.

Lal, Neeta. (2016). "Read: A Precarious Fate for Climate Migrants in India." *Inter Press Service News Agency*. Available at <www.ipsnews.net/2016/05/a-precarious-fate-for-climate-migrants-in-india/>. Accessed on February 14, 2020.

Lal, R., Sivakumar, M.V.K., Faiz, M.A., Mustafizur Rahman, A.H.M. and Islam, K.R. (eds.). (2011). *Climate Change and Food Security in South Asia*. Netherlands: Springer.

Lall, U., Johnson, T., Colohan, P., Aghakouchak, A., Brown, C., McCabe, G., Pulwarty, R. and Sankarasubramanian, A. (2018). "Water." In: Reidmiller, D.R., C.W. Avery, D.R. Easterling, K.E. Kunkel, K.L.M. Lewis, T.K. Maycock, and B.C. Stewart (eds.) *Impacts, Risks, and Adaptation in the United States: Fourth National Climate Assessment*. Volume 2. Washington, DC: U.S. Global Change Research Program.

Lama, Mahendra P. (2015). "Nepal Crisis Shows It's Time SAARC Food Bank Gets Going." *Hindustan Times*, April 30.

Lama, Mahendra P. (2018). "Post Thimphu Summit." *The Kathmandu Post*, January 10.

Lambourne, Helen. (2005). "Tsunami: Anatomy of a Disaster." *BBC News*, March 27.

Lang, Olivia. (2009). "Nepalese Government Holds Cabinet Meeting on Mount Everest." *The Telegraph*, December 4.

Larsen, J.N., Anisimov, O.A., Constable, A., Hollowed, A.B., Maynard, B.N., Prestrud, P., Prowse, T.D. and Stone, J.M.R. (2014). "Polar Regions." In: *Climate Change 2014: Impacts, Adaptation and Vulnerability*. Part B: Regional Aspects, Contribution of Working Group II to the Fifth Assessment Report of the Intergovernmental Panel on Climate Change. Cambridge: Cambridge University Press.

Laszlo, Ervin., Kurtzman, Joel. and Bhattacharya, A.K. (1981). *Regional Cooperation among Developing Countries: The New Imperative of Development in the 1980's*. London: Pergamon Press, cited by Chhibber, Bhrati. (2004). *Regional Security and Regional Cooperation: A Comparative Study of SAARC and ASEAN*. New Delhi: New Century Publications.

Lattanzio, Richard K. (2017). Paris Agreement: U.S. Climate Finance Commitments, Specialist in Environmental Policy, Congressional Research Service, June 19.

Laub, Zachary. (2014). "The Group of Eight (G8) Industrialized Nations." March 3. Available at <www.cfr.org/backgrounder/group-eight-g8-industrialized-nations>. Accessed on February 14, 2020.

Lawrence-Samuel, Amar., Jackson, Rachel Rose. and Thanki, Nathan. (2017). "SDG 13: The Pivot Point: Realizing Sustainable Development Goals by Ending Corporate Capture of Climate Policy." In, Adams, Barbara., Bissio, Roberto., Ling, Chee Yoke., Donald, Kate., Martens, Jens., Prato, Stefano. and Vermuyten, Sandra. (eds.) *Spotlight on Sustainable Development 2017*. Bonn: Friedrich Ebert Stiftung.

Lawson, Fred H. (2008). "Comparing Regionalist Projects in the Middle East and Elsewhere: One Step Back, Two Steps Forward." In: Harders, Cilja. and Legrenzi, Matteo. (eds.) *Beyond Regionalism? Regional Cooperation, Regionalism and Regionalization in the Middle East*. Great Britain: Ashgate Publishing Company.

Lazarus, Neville. (2018). "Millions of Afghan Refugees Facing Deportation Threat from Pakistan." *Sky News*, January 31.

Leach, Anna. (2017). "Mudslides and Floods Cause Devastation in Sri Lanka- in Pictures." *The Guardian*, June 4.

Lee, Bernice. (2009). "Managing the Interlocking Climate and Resource Challenges." *International Affairs*, Vol. 85, No. 6.

Lee, Henry. (2016). "Investing in Climate Adaptation." In: Stavins, Robert N. and Stowe, Robert C. (eds.) *The Paris Agreement and Beyond: International Climate Change Policy Post-2020*. Cambridge: Harvard Project on Climate Agreements.

Leggett, Jane E. (2015). *Greenhouse Gas Pledges by Parties to the United Nations Framework Convention on Climate Change.* Congressional Research Service Report Prepared for members and Committees of Congress, 19th October.

Legrenzi, Matteo. and Harders, Cilja. (2008). "Beyond Regionalism? Regional Cooperation, Regionalism and Regionalization in the Middle East." In: Harders, Cilja. and Legrenzi, Matteo. (eds.) *Beyond Regionalism? Regional Cooperation, Regionalism and Regionalization in the Middle East.* Great Britain: Ashgate Publishing Company.

Leighton, M., Shen, X. and Warner, K. (2011). *Climate Change and Migration: Rethinking Polices for Adaptation and Disaster Risk Reduction.* Publication Series of UNU-EHS. Bonn: United Nations University.

Lenton, T.M., Held, H., Kriegler, E., Hall, J.W., Lucht, W., Rahmstorf, S. and Schellnhuber, H.J. (2008). "Tipping Elements in the Earth's Climate System." *Proceedings of the National Academy of Sciences of the United States of America*, Vol. 105, No. 6.

Le Treut, H., Somerville, R., Cubasch, U., Ding, Y., Mauritzen, C., Mokssit, A., Peterson, T. and Prather, M. (2007). *Historical Overview of Climate Change: Climate Change 2007: The Physical Science Basis: Contribution of Working Group I to the Fourth Assessment Report of the Intergovernmental Panel on Climate Change.* Cambridge, UK and New York: Cambridge University Press.

Liberatore, Angela. (1997). "The Integration of Sustainable Development Objectives into EU Policymaking: Barriers and Prospects." In: Baker, Susan., Kousis, Maria., Richardson, Dick. and Young, Stephen. (eds.) *The Politics of Sustainable Development-Theory, Policy and Practice within the European Union.* London: Routledge.

Liftin, Karen. (ed.). (1998). *The Greening of Sovereignty in World Politics.* Cambridge: MIT Press.

Lindsey, Rebecca. (2018). "Climate Change: Global Sea Level." *Climate*, August 1.

Lingam, Theiva. (2017). "Justice for Asia's Climate Refugees." *The Diplomat*, November 9.

Lippelt, Jana. and Mayer, Lea. (2017). "After the Paris Agreement: What's Next?" *Worldwide Implementation, Spotlight, CESifo Forum 4*, Vol. 18, December.

Lister, Stephen., Girling, Fran., Bhatia, Rita., de Mel, Ruwan. and Musoke, Samm. (2017). Country Portfolio Evaluation- Sri Lanka: An Evaluation of WFP's Portfolio (2011–2015) Evaluation Report. Report number OEV/2016/009.

Locatelli, B. (2010). *Local, Global: Integrating Mitigation and Adaptation.* Perspective Forests/Climate Change No. 3. Cirad: Paris, France.

Long, David. (1963). "International Functionalism and the Politics of Forgetting." *Migrants & Refugees*, Vol. 48, No. 2, Spring, cited in Sterling-Folker, Jennifer. (2002). *Theories of International Cooperation and the Primacy of Anarchy: Explaining U.S. International Policy-Making after Bretton Woods.* New York: University of New York Press.

Lovett, Rick. (2011). "Melting Glaciers Mean Double Trouble for Water Supplies." *National Geographic News*, December 21.

Lowenthal, David. (2003). *George Perkins Marsh: Prophet of Conservation.* Washington: Washington University Press.

Lubna, H. (2012). "Nearly 60 Islands Hit with Water Crisis." *Maldives Independent*, April 8.

Ludi, Eva. (2009). *Climate Change, Water and Food Security.* Background Note. London: Overseas Development Institute.

Luterbacher, Urs. and Sprinz, Detlef F. (2001). *International Relations and Global Climate Change*. London: The MIT Press.

Lynas, Mark. (2015). "The Maldives Cannot Represent Climate Leadership with an Autocrat at the Helm." *The Guardian*, June 3.

Lyu, Hai-Min., Xu, Ye-Shuang., Cheng, Wen-Chieh. and Arulrajah, Arul. (2018). "Flooding Hazards across Southern China and Prospective Sustainability Measures." *Sustainability*, Vol. 10, No. 1682.

Ma, W., Zeng, W., Zhou, M., Wang, L., Rutherford, S., Lin, H., Zhang, Y., Xiao, J., Zhang, Y., Wang, X., Gu, X. and Chu, C. (2015). "The Short-Term Effect of Heat Waves on Mortality and Its Modifiers in China: An Analysis from 66 Communities." *Environ International*, Vol. 75.

Macdonnell, Norman. (1995). Unconditional Threats to Security: Environmental Issues and Sustainable Development. Centre for International Relations. Ontario: Queen's University Kingston.

Mackenzie, Richard. (1993). "Afghanistan's Uneasy Peace." *National Geographic*, Vol. 184, No. 4.

Magnani, Rich., Oot, Lesley., Sethuraman, Kavita., Kabir, Golam. and Rahman, Setara. (2015). *USAID Office of Food for Peace Food Security Country Framework for Bangladesh FY 2015–2019*. Washington, DC: FHI 360/FANTA.

Magness, Dawn R., Lovecraft, Amy Lauren and Morton, John M. (2011). "Factors Influencing Individual Management Preferences for Facilitating Adaptation to Climate Change Within the National Wildlife Refuge System." *Wildlife Society Bulletin*, Vol. 36, No. 3.

Magrin, G.O., Marengo, J.A., Boulanger, J.P., Buckeridge, M.S., Castellanos, E., Poveda, G., Scarano, F.R. and Vicuña, S. (2014). "Central and South America." In: *Climate Change 2014: Impacts, Adaptation and Vulnerability*. Part B: Regional Aspects, Contribution of Working Group II to the Fifth Assessment Report of the Intergovernmental Panel on Climate Change. Cambridge: Cambridge University Press.

Maharjan, S.K., and Maharjan, K.L. (2017). "Review of Climate Policies and Roles of Institutions in the Policy Formulation and Implementation of Adaptation Plans and Strategies in Nepal." *Journal of International Development and Cooperation*, Vol. 23, Nos. 1–2.

Majaw, Baniateilang. (2012). "Climate Change and South Asian Association for Regional Cooperation: A Regional Response." *International Journal of Social Sciences and Humanity Studies*, Vol. 4, No. 2.

Majaw, Baniateilang. (2014). "Meghalaya: Small But Not So Beautiful: A Point Of View." *World Affairs: The Journal of International Issues*, Vol. 18, No. 4.

Majaw, Baniateilang. (2016). "Ending Meghalaya's 'Deadly Occupation': India's National Green Tribunal's Ban on Rat-Hole Mining." *Verfassung und Recht in Übersee: Law and Politics in Africa, Asia and Latin America*, Vol. 16, No. 1.

Majeed, A. and Abdulla, A. (2004). "Economic and Environmental Vulnerabilities of the Maldives and Graduation from LDC Status." In: Briguglio, L. and Kisanga, E. (eds.) *Economic Vulnerability and Resilience of Small States*. Valetta/London: Commonwealth Secretariat and the University of Malta.

Maldives Ministry of Home Affairs, Housing and Environment (MMHAHE). (2001). *First National Communication of the Republic of Maldives to the United Nations Framework Convention on Climate Change*. Malé: Republic of Maldives.

Malé Declaration on Environment in the Aftermath of the Indian Ocean Tsunami Disaster. (2005). Maldives: Malé, June 25. Cited in SAARC. (2012). *SAARC*

Ministerial Meetings 2000–2012, Secretariat of the South Asian Association for Regional Cooperation: Kathmandu, 121–124.

Mall, R.K., Attri, S.D. and Kumar, Santosh. (2011). "Extreme Weather Events and Climate Change Policy in India." *Journal of South Asia Disaster Studies*, Vol. 4. No. 2.

Mall, R.K., Gupta, Akhilesh., Singh, Ranjeet., Singh, R.S. and Rathore, L.S. (2006). "Water Resources and Climate Change: An Indian Perspective." *Current Science*, Vol. 90, No. 12.

Mall, R.K. and Kumar, Santosh. (2014). *Integration of Disaster Risk Reduction and Climate Change Adaptation in SAARC Region*. New Delhi: The United Nations Office for Disaster Risk Reduction.

Manel, Kanthi., Punpuing, Sureeporn. and Perera, Sunethra. (2017). "Social and Economic Differentials among Households of Urban Migrants: Evidence from Kalutara District, Sri Lanka." *Sri Lanka Journal of Social Sciences*, Vol. 40, No. 2.

Mani, Muthukumara., Bandyopadhyay, Sushenjit., Chonabayashi, Shun., Markandya, Anil. and Mosier, Thomas. (2018). *South Asia's Hotspots: The Impact of Temperature and Precipitation Changes on Living Standards*. Washington, DC: International Bank for Reconstruction and Development/The World Bank.

Manne, Robert. (2013). "Climate Change: Some Reasons for Our Failures." *The Guardian*, July 22.

Mansoor, Hasan. (2015). "Heatwave Death Toll in Sindh Tops 1,000." *Dawn*, Lahore, June 25.

Marquina, Antonio. (2010). "From a Socioeconomic Approach to Migration to the Inclusion of Environmentally Induced Migration in the Mediterranean." In: Marquina, Antonio. (ed.) *Global Warming and Climate Change: Prospects and Policies in Asia and Europe*. New York: Palgrave Macmillan.

Martin, Joseph D. (2017). *Sri Lanka Disaster Management Reference Handbook*. Hornet Ave: Center for Excellence in Disaster Management & Humanitarian Assistance.

Martin, Lisa L. (1995). "The Promise of Institutionalist Theory." *International Security*, Vol. 20, No. 1, Summer.

Matthews, Alexandra. (2009). "Bangladesh's Climate Change Emergency." *Center for Strategic and International Studies (CSIS)*. Washington, DC, Vol. 136, December 1.

Matthews, D. (2017). "Donald Trump Has Tweeted Climate Change Skepticism 115 Times. Here's All of It." *Vox*, June 1.

Mattli, Walter. (1999). *The Logic of Regional Integration: European and Beyond*. Cambridge: Cambridge University Press.

Mawilmada, Nayana N. (2010). "Can Sri Lanka Combat Climate Change?" *Sunday Observer*, December 5.

Mazumdar, Hasan. (2012). *A Situation Analysis of Climate Change Adaptation Initiatives in Bangladesh*. Dhaka, Bangladesh: The Asia Foundation.

McAdam, Jane. and Saul, Ben. (2010). "Displacement with Dignity: International Law and Policy Responses to Climate Change Migration and Security in Bangladesh." In: *53 German Yearbook of International Law*, 5–6.

McCarthy, James J., Canziani, Osvaldo F., Leary, Neil A., Dokken, David J. and White, Kasey S. (eds.). (2001). *Climate Change 2001: Impacts, Adaptation, and Vulnerability*. Contribution of Working Group II to the Third Assessment Report of the Intergovernmental Panel on Climate Change. Cambridge: Cambridge University Press.

McDonald, R.I. and Girvetz, E.H. (2013). "Two Challenges for U.S. Irrigation Due to Climate Change: Increasing Irrigated Area in Wet States and Increasing Irrigation Rates in Dry States." *PLoS One*, Vol. 8, No. 6.

McDonnell, Tim. (2019). "Climate Change Creates a New Migration Crisis for Bangladesh." *National Geographic*, January 24.

McGovern, Joe. (2006). *The Kyoto Protocol*. Pittsburgh: Dorrance Publishing.

McGranahan, G., Balk, D. and Anderson, B. (2007). "The Rising Tide: Assessing the Risk of Climate Change and Human Settlements in Low Elevation Coastal Zones." *Environment and Urbanisation*, Vol. 19, No. 1.

McGrath, Matt. (2018). "Climate Change: EU Aims to be 'Climate Neutral' by 2050." *BBC News*, November 28. Available at <www.bbc.com/news/science-environment-46360212>. Accessed on September 19, 2019.

McLaren, Robert I. (1985). "Mitranian Functionalism: Possible or Impossible." *Review of International Studies*, Vol. 11, No. 2.

McLeman, R. and Smit, B. (2004). *Climate Change, Migration and Security*. Commentary No. 86. Ottawa: Canadian Security Intelligence Service.

McLeman, R. and Smit, B. (2006). "Migration as an Adaptation to Climate Change." *Climate Change*.

McLin, Jon. (1972a). *Stockholm: The Politics of Only One Earth*. [American University] Fieldstaff, Reports: West Europe Series 7, Number 4.

McLin, Jon. (1972b). *The United Nations System and the Stockholm Conference*. [American University] Fieldstaff Reports. West Europe Series 7, Number 3.

McMillan, M., Shepherd, A., Sundal, A., Briggs, K., Muir, A., Ridout, A., Hogg, A. and Wingham, D. (2014). "Increased Ice Losses from Antarctica Detected by CryoSat-2." *Geophysical Research Letters*, Vol. 41.

McSweeney, C., New, M. and Lizcano, G. (2008). *UNDP Climate Change Country Profiles: Afghanistan*. United Nations Development Program.

Meadows, Donnela H., Meadows, H., Randers, J. and Behrens, W. (1972). *The Limits to Growth: A Report to the Club of Rome's Project on the Predicament of Mankind*. New York: Universe Books.

Mearsheimer, John J. (1995). "A Realist Reply." *International Security*, Vol. 20, No. 1.

Medina-Ramon, M. and Schwartz, J. (2007). "Temperature, Temperature Extremes, and Mortality: A Study of Acclimitisation and Effect Modification in 50 U.S. Cities." *Occupational and Environmental Medicine*, Vol. 64, No. 12.

Meehl, Gerald A. and Stocker, Thomas F. (2007). "Global Climate Projections." In: Solomon et al. (eds.) *Climate Change 2007: The Physical Science Basis*. Contribution of Working Group I to the Fourth Assessment Report of the Intergovernmental Panel on Climate Change. Cambridge: Cambridge University Press.

Meenawat, H. and Sovacool, B.K. (2011). "Improving Adaptive Capacity and Resilience in Bhutan." *Mitigation and Adaptation Strategies for Global Change*, Vol. 16, No. 5.

Meeting of the SAARC Agriculture Ministers. (2002). Pakistan: Kathmandu, May 15, cited in *SAARC Ministerial Meetings- 2000–2012*. (2012). Kathmandu: SAARC Secretariat.

Meeting of SAARC Agriculture/Food Ministers. (2006). Pakistan: Islamabad, December 14, cited in *SAARC Ministerial Meetings- 2000–2012*. (2012). Kathmandu: SAARC Secretariat.

Mehra, M. (2009). "India Starts to Take on Climate Change." In: *State of the World 2009*. Washington, DC: Worldwatch Institute.

Meier, M.F., Dyurgerov, M.B., Rick, U.K., O'Neel, S., Pfeffer, W.T., Anderson, R.S., Anderson, S.P. and Glazovsky, A.F. (2007). "Glaciers Dominate Eustatic Sea-Level Rise in the 21st Century." *Science*, Vol. 317.

Memon, Naseer. (2012). *Disasters in South Asia: A Regional Perspective*. Karachi: Pakistan Institute of Labour Education and Research.

Memorandum. (2017). *Executive Order 13783, Promoting Energy Independence and Economic Growth*. Washington, DC: United States Environmental Protection Agency, 20460, April 19.

Mendelsohn, R., Morrison, W., Schlesinger, M. and Andronova, N. (2000). "Country Specific Market Impacts of Climate Change." *Climatic Change*, Vol. 45, Nos. 3–4.

Mendelsohn, R., Munasinghe, M. and Niggol, S. Seo. (2005). "Climate Change and Agriculture in Sri Lanka: A Ricardian Valuation." *Journal of Environment and Development Economics*, Vol. 10.

Mengel, Matthias., Levermann, Anders., Frieler, Katja., Robinson, Alexander., Marzeion, Ben. and Winkelmann, Ricarda. (2016). "Future Sea Level Rise Constrained by Observations and Long-term Commitment." *Proceedings of the National Academy of Sciences of the United States of America*, Vol. 113, No. 10.

MHA. (2011). *Disaster Management in India*. Prepared under GoI-UNDP, Disaster Risk Reduction Programme Ministry of Home Affairs, New Delhi, Government of India.

Mia. (2019). "Climate Change and Coral Islands in the Maldives." *The Maldives Expert*, March 2.

Middle East Monitor. (2016). "175,000 Lose Their Jobs Due to Drought in Morocco." August 8.

Mihran, Rozbih. (2011). Rural Community Vulnerability to Food Security Impacts of Climate Change in Afghanistan Evidence from Balkh, Herat, and Nangarhar Provinces. A thesis presented to the University of Waterloo in fulfilment of the thesis requirement for the degree of Master of Environmental Studies in Environment and Resource Studies Waterloo, Ontario, Canada.

Millennium Ecosystem Assessment (MEA). (2005). *Millennium Ecosystem Assessment 2005, Ecosystems and Human Well-Being: Synthesis Report*. Washington, DC: Island Press.

Miller, C.A. and Edward, P.N. (2001). *Changing the Atmosphere: Expert Knowledge and Environmental Governance*. Cambridge: MIT Press.

Milman, Oliver. (2015). "Pacific Nations Beg for Help for Islanders Forced to Flee, Climate Central." *The Guardian*, October 17.

Mimura, N., Nurse, L., McLean, R.F., Agard, J., Briguglio, L., Lefale, P., Payet, R. and Sem, G. (2007). "Small Islands." In: Parry, M.L., Canziani, O.F., Palutikof, J.P., van der Linden, P.J., and Hanson, C.E. (eds.) *Climate Change 2007: Impacts, Adaptation and Vulnerability*. Contribution of Working Group II to the Fourth Assessment Report of the Intergovernmental Panel on Climate Change. Cambridge, UK: Cambridge University Press.

Ministry of Agriculture (MOA). (2011). *National Agriculture Policy (Final Draft): Shared for Parliamentary Review*. Dhaka: Ministry of Agriculture.

Ministry of Agriculture and Cooperatives (MoAC). (2011). Climate Change Adaptation and Disaster Risk Management in Agriculture: Priority Framework for Action 2011–2020.

Ministry of Agriculture and Forests (MoAF). (2011). *Bhutan Climate Summit for a Living Himalayas: National Paper on Water Security*. Thimphu: Royal Government of Bhutan.

Ministry of Agriculture and Forests (MoAF). (2016). *SAPA, RNR Climate Change Adaptation Program.* Thimphu: Royal Government of Bhutan.

Ministry of Agriculture Irrigation and Livestock (MAIL). (2012). FAO Land Cover Atlas of Islamic Republic of Afghanistan.

Ministry of Disaster Management (MoDM). (2005). Towards a Safer Sri Lanka: A Road Map for Disaster Risk Management. Sri Lanka: Ministry of Disaster Management.

Ministry of Disaster Management (MoDM). (2013). Government of Sri Lanka, Draft National Policy on Disaster Management. Sri Lanka: Ministry of Disaster Management.

Ministry of Drinking Water and Sanitation (MoDWS). (2014). *Guidelines for Swachh Bharat Mission (GRAMIN).* New Delhi: Government of India.

Ministry of Education (MoE). (2010). *Annual Report.* Thimphu: Royal Government of Bhutan.

Ministry of Environment (MOE). (2003). *Initial National Communication on Climate Change.* Islamabad, Pakistan: Government of Islamic Republic of Pakistan.

Ministry of Environment (MOE). (2010a). *National Adaptation Programme of Action (NAPA) to Climate Change.* Kathmandu: Government of Nepal, September 30.

Ministry of Environment (MoE). (2010b). *National Climate Change Adaptation Strategy for Sri Lanka: 2011–2016.* Democratic Socialist Republic of Sri Lanka.

Ministry of Environment (MoE). (2011). *National Framework on Local Adaptation Plans for Action.* Singha Durbar, Nepal: Government of Nepal.

Ministry of Environment & Construction (MoEC). (2004). For discussion see *State of the Environment Maldives.* Maldives: Malé.

Ministry of Environment and Energy (MoEEW). (2012). *State of the Environment Report 2011.* Maldives.

Ministry of Environment and Energy (MoEE). (2015). *Maldives Climate Change Policy Framework.* Ministry of Environment and Energy. Malé: Government of Maldives.

Ministry of Environment and Energy (MoEE). (2016). *Second National Communication of Maldives to the United Nations Framework Convention on Climate Change.* Malé: Republic of Maldives.

Ministry of Environment and Forests (MoEF). (2004). *India's Initial National Communication to the United Nations Framework Convention on Climate Change.* New Delhi: Government of India.

Ministry of Environment and Forests (MoEF). (2005). *National Adaptation Programme of Action (NAPA).* Dhaka, Bangladesh: Government of the People's Republic of Bangladesh.

Ministry of Environment and Forests (MoEF). (2009). *Bangladesh Climate Change Strategy and Action Plan (BCCSAP).* Government of the People's Republic of Bangladesh, September.

Ministry of Environment and Forest (MoEF). (2010a). *India First Country in the World to Have Specialized Environment Courts.* New Delhi, March 16.

Ministry of Environment and Forests (MoEF). (2010b). Public Consultations on National Mission for a Green India, Report, Prepared by the Centre for Environment Education (CEE) for Ministry of Environment & Forests, Government of India, August.

Ministry of Environment and Forest (MoEF). (2010c). *National Mission for a Green India,* Draft submitted to Prime Minister's Council on Climate Change. New Delhi: Government of India.

Ministry of Environment and Forests (MoEF). (2010d). *India: Taking on Climate Change Post-Copenhagen Domestic Actions.* Government of India (GoI), June 30.

Ministry of Environment and Forests (MoEF). (2012a). *Bangladesh Climate Change Resilience Fund (BCCRF).* Annual Report. Government of People's Republic of Bangladesh. Prepared by The World Bank.

Ministry of Environment and Forests (MoEF). (2012b). *Rio + 20: National Report on Sustainable Development.* Peoples' Republic of Bangladesh, Dhaka: Bangladesh Secretariat.

Ministry of Environment and Forests (MoEF). (2012c). *Strategic Plan: 2012–13 to 2016–17.* (Aligned with 12thFive year Plan), Paryavaran Bhavan CGO Complex, New Delhi.

Ministry of Environment and Forest (MoEF). (2013a). *Bangladesh Climate Change and Gender Action Plan (CCGAP: Bangladesh).* Bangladesh: Dhaka.

Ministry of Environment and Forests (MoEF). (2013b). *Brief Statement on Activities and Achievements.* New Delhi: Government of India.

Ministry of Environment and Forests (MoEF). (2013c). *National Environmental Policy of Bangladesh 2013,* Government of Bangladesh. Available at <www.moef. gov.bd/html/policy/pdf/Environment_policy.pdf>. Accessed on February 14, 2020.

Ministry of Environment and Renewable Energy (MoERE). (2011). *The National Climate Change Policy of Sri Lanka.* Democratic Socialist Republic of Sri Lanka.

Ministry of Environment, Energy and Water (MoEEW). (2006a). *Climate Risk Profile for the Maldives.* Malé: Republic of Maldives.

Ministry of Environment, Energy and Water (MoEEW). (2006b). *National Adaptation programme of Action: Draft for Comments.* Malé: Republic of Maldives.

Ministry of Environment, Energy and Water (MoEEW). (2006c). *National Adaptation Programme of Action (NAPA).* Malé: Republic of Maldives.

Ministry of Environment, Energy and Water (MoEEW). (2006d). *Live & Learn Environmental Education: A Rapid Assessment of Perceptions into Environmental Management in the Maldives: Volume 1- Environmental Education and Community Mobilisation,* With Assistance from Asian Development Bank, Malé, May.

Ministry of Environment, Energy and Water (MoEEW). (2007). *National Adaptation Programme of Action (NAPA): Maldives.* Malé: Republic of Maldives.

Ministry of Environment, Forests and Climate Change (MoEFCC). (2014). *India's Progress in Combating Climate Change.* Briefing Paper for UNFCCC COP 20 Lima, Peru, Government of India, December.

Ministry of Environment, Science and Technology (MoEST). (2008). *National Capacity Self-Assessment for Global Environment Management.* Nepal National Capacity Self-Assessment Report and Action Plan, Government of Nepal, June.

Ministry of Finance (MoF). (2014–15). Population, Labour Force and Employment. *Pakistan Economic Survey 2014–15.* Government of Pakistan.

Ministry of Forestry & Environment (MoFE). (2000). *Initial National Communication under the United Nations Framework Convention on Climate Change.* Final Draft. Colombo, Sri Lanka: Government of Sri Lanka.

Ministry of Health (MoH). (2014). *Health Master Plan 2016–2025.* Malé: Government of Maldives.

Ministry of Health (MoH). (2016). *National Mental Health Strategic Plan-2016–2021.* Malé: Republic of Maldives, Government of Maldives.

Ministry of Health and Population (MoHP). (2004). *National Nutrition Policy Strategy.* Department of Health Services, Child Health Division, Nutrition Section, Ministry of Health and Population, Government of Nepal, Kathmandu.

Ministry of Health and Population (MoHP). (2006). *National School Health and Nutrition Strategy*. Department of Health Services, Child Health Division, Nutrition Section, Ministry of Health and Population, Government of Nepal, Kathmandu.

Ministry of Health and Population (MoHP). (2010). *Annual Report (2009/2010)*. Department of Health Services, Child Health Division, Nutrition Section, Ministry of Health and Population, Government of Nepal, Kathmandu.

Ministry of Healthcare and Nutrition (MoHN). (2010). *National Nutrition Policy of Sri Lanka*. Colombo, Sri Lanka.

Ministry of Home Affairs (MoHA). (2015). *Nepal Disaster Report 2015, MoHA*. Kathmandu: Government of Nepal and DPNet-Nepal.

Ministry of Home Affairs and Environment (MoHAE). (1994). *National Capacity Needs Self Assessment (NCSA) for Global Environmental Management*. Malé: Republic of Maldives.

Ministry of Home Affairs, Housing and Environment (MHAHE). (2001a). *First National Communication of the Republic of Maldives to the United Nations Framework Convention on Climate Change*. Malé, Maldives.

Ministry of Home Affairs, Housing and Environment (MHAHE). (2001b). *State of the Environment Report*. Malé: Republic of Maldives.

Ministry of Housing, Transport and Environment (MoHTE). (2010a). *National Assessment Report II*. Department of Climate Change and Energy, Government of Maldives.

Ministry of Housing, Transport and Environment (MoHTE). (2010b). *Invitation for Interested Consultant: Develop Guidelines for Climate Risks Resilient Land Use Planning*. Malé: Republic of Maldives, Ref: (IUL) 138-KS1/1/2010/1084.

Ministry of Housing and Urban Affairs (MoHUA). (2018). *Atal Mission for Rejuvenation and Urban Transformation (AMRUT)*. Town and Country Planning Organisation. New Delhi: Government of India.

Ministry of Irrigation, Water Resources and Environment (MoIWRE). (2004). *Strategic Policy Framework for the Water Sector* ("2004 Strategic Policy Framework"). Available at <www.cawater-info.net/afghanistan/pdf/aswf.pdf>.

Ministry of Mahaweli Development and Environment (MoMDE). (2015). National Adaptation Plan for Climate Change Impacts in Sri Lanka- 2016 to 2025, Climate Change Secretariat.

Ministry of Mahaweli Development and Environment (MoMDE). (2016a). Sri Lanka's Intended Nationally Determined Contribution to UNFCCC, Democratic Socialist Republic of Sri Lanka.

Ministry of Mahaweli Development and Environment (MoMDE). (2016b). National Adaptation Plan for Climate Change Impacts in Sri Lanka- 2016–2025, Climate Change Secretariat, Ethul Kotte, Sri Lanka.

Ministry of New and Renewable Energy (MoNRE). (2011). Strategic Plan for New and Renewable Energy Sector for the Period 2011–17. New Delhi: Government of India.

Ministry of New and Renewable Energy (MoNRE). (2017). Government of India, Physical Progress (Achievements): Tentative State-Wise Break-Up of Renewable Power Target To Be Achieved by the Year 2022, September 29.

Ministry of Planning and National Development (MoPND). (2010). *Seventh National Development Plan (2006–2010): Creating Opportunities*. Government of Maldives.

Ministry of Planning, Development, and Reform (MoPDR). (2015a). *Pakistan Vision 2025*. Pakistan Government Official Document.

Ministry of Planning, Development, and Reforms (MoPDR). (2015b). *Annual Plan 2014–2015*. Islamabad: Government of Pakistan.

Ministry of Public Health (MoPH). (2010). *AMS Afghanistan Mortality Survey*.

Ministry of Science & Technology (MoST). (2010). National Mission on Strategic Knowledge for Climate Change under National Action Plan on Climate Change. Mission Document. New Delhi: Government of India.

Ministry of the Environment (MoE). (2010). *Sector Vulnerability Profile: Health*. Supplementary Document to the NCCAS 2011–2016. Democratic Socialist Republic of Sri Lanka.

Ministry of Urban Development (MoUD). (2014). *National Mission on Sustainable Habitat: Adaptation and Mitigation Measures in the Field of Water Supply & Sanitation*. Central Public Health and Environmental Engineering Organisation. New Delhi: Government of India.

Ministry of Urban Development (MoUD). (2015). *National Urban Development Strategy (NUDS)*. Retrieved from <http://moud.gov.np/wp-content/uploads/2016/08/NUDS-2015-final-draft.pdf>.

Ministry of Water and Power (MoWP). (2010). *Annual Flood Report*. Government of Pakistan.

Ministry of Water Resources (MoWR). (2004). *Coastal Zone Policy*. Dhaka: Ministry of Water Resources (MoWR).

Ministry of Water Resources (MoWR). (2008). *National Water Mission under National Action Plan on Climate Change Comprehensive Mission Document*. Volume 2. New Delhi: Government of India, December.

Ministry of Water Resources (MoWR). (2009). *National Water Mission under the National Action Plan on Climate Change: Revised Comprehensive Mission Documents*. Volume 1 and 2. New Delhi: Government of India.

Mintzer, Irving M. and Leonard, J. Amber. (1994). "Visions of a Changing World." In Mintzer, Irving M. and Leonard, J. Amber (eds.) *Negotiating Climate Change: The Inside Story of the Rio Convention*. Cambridge: Cambridge University Press.

Miraynews. (2014). "Death Toll from Afghanistan Flooding Reaches 150." June 8.

Mirza, M. Monirul Qader. and Ahmad, Q.K. (eds.). (2005a). *Climate Change and Water Resources in South Asia*. London: Routledge.

Mirza, M. Monirul Qader. and Ahmad, Q.K. (2005b). "Climate Change and Water Resources in South Asia: An Introduction." In: Mirza, M. Monirul Qader. and Ahmad, Q.K. (eds.) *Climate Change and Water Resources in South Asia*. Leiden: A.A. Balkema Publishers.

Mirza, M. Monirul Qader., Warrick, R.A., Ericksen, N.J. and Kenny, G.J. (2001). "Are Floods Getting Worse in the Ganges, Brahmaputra and Meghna Basins?" *Environmental Hazards*, Vol. 3, No. 2.

Mirza, M.M. (2002). "Global Warming and Changes in the Probability of Occurrence of Floods in Bangladesh and Implications." *Global Environmental Change*, Vol. 12, No. 2.

Mishra, O.P. (2013). SAARC- Regional Progress Report on the Implementation of the Hyogo Framework for Action (2011–2013). A Regional HFA Monitor Update Published by Prevention Web. Available at <www.preventionweb.net/english/hyogo/progress/reports/>.

Mitchell, Jr. J.M., Dzerdzeevskii, B., Flohn, H., Hofmeyr, W.L., Lamb, H.H., Rao, K.N. and Wallen, C.C. (1966). *Climatic Change*. WMO Technical Note No. 79. Geneva: World Meteorological Organization.

Mitra, Devirupa. (2016). "What SAARC Has Done–and Failed to Do–since Its Last Summit." *WIRE*, March 15.

Mitrany, David. (1933). *The Progress of International Government*. London: Yale University Press.

Mitrany, David. (1943). *A Working Peace System: An Argument for the Functional Development of International Organization*. London: Royal Institute of International Affairs.

Mitrany, David. (1966). *A Working Peace System*. Chicago: Quadrangle.

Mitrany, David. (1975). "A Political Theory for a New Society." In: Groom, A.J.R. and Taylor, Paul. (eds.) *Functionalism: Theory and Practice in International Relations*. London: University of London Press.

Mittal, Surabhi. and Sethi, Deepti. (2009). *Food Security in South Asia: Issues and Opportunities, Indian Council for Research on International Economic Relations*. Working Paper No. 240. New Delhi: ICRIER.

MoAF. (2017). *National Forest Inventory Report: Stocktaking Nation's Forest Resources*. Volume 1. Thimphu, Bhutan: Ministry of Agriculture and Forestry.

MoE. (2009). *Climate Change Vulnerabilities in Agriculture in Pakistan*. Government of Pakistan Ministry of the Environment, Annual Report. Rawalpindi, Islamabad.

MoE/ADB. (2010). *Sector Vulnerability Profile: Agriculture and Fisheries*. Battaramulla, Sri Lanka: Ministry of Environment.

Mohan, Vishwa. (2014). "Climate Change May Lead India to War: UN Report." *The Times of India*, April 1.

Mohan, Vishwa. (2019). "Sea Levels Rising Faster, Indian Cities at High Flood Risk: IPCC." *The Times of India*, September 26.

MoMDE. (2016c). *Nationally Determined Contributions*. Ministry of Mahaweli Development and Environment Sri Lanka.

Mool, P.K., Bajracharya, S.R., Shrestha, B.R. (2005). *Glaciers, Glacial Lakes and Glacial Lake Outburst Floods in the Hindu Kush-Himalaya*. Proceedings of the International Karakorum Conference, Islamabad, Pakistan, 25–27 April, Abstract Volume.

Mool, P.K., Joshi, S.P. and Bajracharya, S.R. (2001). *An Inventory of Glaciers, Glacial Lakes and Glacial Lakes Outburst Floods, Monitoring an Early Warning Systems in the Hindu Kush-Himalayan Region*. Kathmandu: ICIMOD.

Mool, P.K., Wangda, D., Bajracharya, S.R., Joshi, S.P., Kunzang, K. and Gurung, D.R. (2001). *An Inventory of Glaciers, Glacial Lakes and Glacial Lakes Outburst Floods, Monitoring an Early Warning Systems in the Hindu Kush-Himalayan Region*. Kathmandu: International Centre for Integrated Mountain Development (ICIMOD).

Mooney, Christ. (2017). "Trump's Reasons for Leaving the Paris Climate Agreement Just Don't Add Up." *The Washington Post*, June 1.

Morgenthau, Hans J. (1978). *Politics among Nations: The Struggle for Power and Peace*. 5th revised edition. New York: Alfred A. Knopf.

Morner, Nils-Axel., Tooleyb, Michael. and Possnertc, Goran. (2004). "New Perspectives for the Future of the Maldives." *Global and Planetary Change*, Vol. 40.

Morris, James T. (2005). *World Food Programme–Annual Report 2005*.

MoSDWRD. (2018). *Sri Lanka Voluntary National Review on the Status of Implementing Sustainable Development Goals*. Ministry of Sustainable Development, Wildlife and Regional Development, Government of the Democratic Socialist Republic of Sri Lanka.

Mozaharul, Alam. and Tshering, Dago. (2004). *Adverse Impacts of Climate Change on Development of Bhutan: Integrating Adaptation into Policies and Activities, Capacity Strengthening in the Least Developed Countries (LDCS) for Adaptation to Climate Change (CLACC), CLACC.* Working Paper No. 2. Bangladesh Centre for advanced Studies (BCAS).

Muggah, Robert. (2019). "The World's Coastal Cities Are Going Under: Here's How Some Are Fighting Back." *World Economic Forum*, January 16.

Mukherji, Indra Nath. (2012). "SAFTA and Food Security in South Asia: An Overview." In: *Regional Trade Agreement and Food Security in Asia.* Bangkok: Food and Agriculture Organization.

Mukhopadhyay, Ranadhir., Karisiddaiah, S.M. and Mukhopadhyay, Julie. (2018). *Climate Change: Alternate Governance Policy for South Asia.* New York: Elsevier Science.

Müller, Benito. (2002). *The Global Climate Change Regime: Taking Stock and Looking Ahead.* Available at <www.wolfson.ox.ac.uk>. Accessed on June 18, 2013.

Munawar, Sana. (2016). *Bhutan Improves Economic Development as a Net Carbon Sink.* Washington, DC: Climate Institute.

Muniruzzaman, A.N.M. (2013). "Food Security in Bangladesh: A Comprehensive Analysis." *Peace and Security Review*, Vol. 5, No. 10, Second Quarter.

Muricken, Ajit. (2010). *Poverty and Vulnerability Cycles in South Asia: Narratives of Survival and Struggles.* Kathmandu: South Asia Alliance for Poverty Eradication.

Muthukumara, Mani., Chonabayashi, Shun., Bandyopadhyay, Sushenjit. and Mosier, Thomas. (2018). *South Asia's Hotspots: The Impacts of Temperature and Precipitation Changes on Living Standards.* Washington, DC: World Bank.

Myers, Norman. (1995). *Environmental Exodus: An Emergent Crisis in the Global Arena.* Washington, DC: Project of the Climate Institute.

Myers, Norman. (2002). "Environmental Refugees: A Growing Phenomenon of the 21st Century: Philosophical Transactions: Biological Sciences, Vol. 357, No. 1420." *Reviews and a Special Collection of Papers on Human Migration*, Royal Society.

Nabeel, Fazilda. (2019). "How India and Pakistan Are Competing over the Mighty Indus River." *Down to Earth*, February 23.

Naeem, Waqas. (2013). "Government Launches First-Ever National Policy on Climate Change." *Express Tribune*, February 27.

Najafizada, Said Ahmad Maisam. (2017). *Policy Research Institutions and Health Sustainable Development Goals: Building Momentum in South Asia.* Country Report, Afghanistan.

Najam, Adil. (2003). "The Human Dimensions of Environment Environmental Insecurity: Some Insights from South Asia." *ECSP Report*, No. 9.

Nambi, Aruvidai. (2014). *Adapting Climate Impacted Agriculture in South Asia.* Dhaka: Climate Action Network South Asia, March.

Nambudiri, Sudha. (2018). "DRDO Labs Send Ready-to-Eat Food for Kerala Flood Victims." *The Times of India*, August 20.

Namgyel, Thinley. (2003). *Bhutan: NAPA Process, National Environment Commission.* Thimphu: Royal Government of Bhutan, NAPA Regional Training Workshop.

Nan, Xu. (2012). "Beijing Floods: Not Enough Prevention." *The Guardian*, July 25.

Nanayakkara, Ruksgana. (2018). "If SDGs Are to be Truly Achieved in Afghanistan." *Voices for Transparency*, March 6.

Nandy, S.N., Dhyani, P.P. and Samal, P.K. (2006). "Resource Information Database of the Indian Himalaya." *ENVIS Monograph 3*, Centre on Himalayan Ecology, G.B. Pant Institute of Himalayan Environment and Development.

The Nansen Initiative. (2015). Climate Change, Disasters, and Human Mobility in South Asia and Indian Ocean. Background Paper Prepared by the Nansen Initiative Secretariat for the South Asia Regional Consultation Khulna, Bangladesh.

NAPA. (2010). *National Adaptation Plan of Action*. Kathmandu: Ministry of Environment.

National Aeronautics and Space Administration (NASA). (2018). *Long-Term Warming Trend Continued in 2017*. Available at <www.giss.nasa.gov/research/news/2018 0118/>. Accessed on January 20, 2018.

National Disaster Management Centre (NDMC). (2010). Strategic National Action Plan for Disaster Risk Reduction and Climate Change Adaptation 2010–2020, Provisional Draft by NDMC.

National Drinking Water Policy (NDWP). (2009). Ministry of Environment. Islamabad: Government of Pakistan.

National Environment Commission (NEC). (2011). *Second National Communication to the UNFCCC*. Thimphu, Bhutan: National Environment Commission.

National Environmental Protection Agency of Afghanistan (NEPA). (2012). *Afghanistan Initial National Communication to the United Nations Framework Convention on Climate Change*. Kabul, Afghanistan: NEPA.

National Institute of Population Research and Training (NIPORT), Mitra and Associates, and ICF International. (2013). *Bangladesh Demographic and Health Survey 2011*. Dhaka, Bangladesh and Calverton, MD: NIPORT, Mitra and Associates, and ICF International.

National Institute of Population Research and Training (NIPORT), Mitra and Associates, and ICF International. (2016). *Bangladesh Demographic and Health Survey 2014*. Dhaka, Bangladesh, and Rockville, MD, USA: NIPORT, Mitra and Associates, and ICF International.

National Mission on Sustainable Habitat (NMoSH). (2010). Ministry of Urban Development (MoUD), Government of India.

National Nutrition Council (NNC). (2013). Multi-Sector Action Plan for Nutrition, Vision 2016- Sri Lanka A Nourished Nation, Government of Sri Lanka.

National Nutrition Survey. (2011). Planning Commission, Government of Pakistan, Pakistan Institute of Development Economics. Islamabad, Pakistan.

National Oceanic and Atmospheric Administration (NOAA). (2007). *Climate Change*. NOAA National Weather Service, October.

National Oceanic and Atmospheric Administration (NOAA). (2012). Global Sea Level Rise Scenarios for the United States National Climate Assessment. NOAA Technical Report OAR CPO-1. Silver Spring, MD.

National Planning Commission (NPC). (2003). *Sustainable Development Agenda for Nepal*. Kathmandu, Nepal: National Planning Commission, His Majesty's Government of Nepal.

National Planning Commission (NPC). (2013). An Approach Paper to the Thirteenth Plan (FY 2013/14–2015/16). Nepal: Kathmandu: Government of Nepal.

National Planning Commission (NPC). (2015a). Nepal Earthquake 2015 Post Disaster Needs Assessment: Key Findings, Vol. A. Kathmandu: NPC, Government of Nepal.

National Planning Commission (NPC). (2015b). Nepal Earthquake 2015 Post Disaster Needs Assessment, Vol. B: Sector Reports. Kathmandu: NPC, Government of Nepal.

National Planning Commission (NPC) and Ministry of Population and Environment (MoPE). (2003). Sustainable Development Agenda for Nepal. Kathmandu, Nepal: Government of Nepal.

National Research Council (NRC). (2010). *Advancing the Science of Climate Change.* Washington, DC: The National Academies Press.

Nayak, Nihar R. (2015). "The Nepal Earthquake: Could SAARC Have Been Effective?" *IDSA Comment*, June 22.

NDMA. (2007). *National Disaster Management Guidelines: Preparation of State Disaster Management Plans.* National Disaster Management Authority, Government of India.

NEC. (2011). *Second National Communication to the UNFCCC.* Thimphu, Bhutan: National Environment Commission.

Neff, Stephen C. (1990). *Friends But No Allies: Economic Liberalism and the Law of Nations.* Columbia: Columbia University Press.

Nelson, Gerald C., Rosegrant, Mark W., Koo, Jawoo., Robertson, Richard., Sulser, Timothy., Zhu, Tingju., Ringler, Claudia., Msangi, Siwa., Palazzo, Amanda., Batka, Miroslav., Magalhaes, Marilia., Valmonte-Santos, Rowena., Ewing, Mandy. and Lee, David. (eds.). (2009). *Climate Change: Impact on Agriculture and Costs of Adaptation.* Food Policy Report. Washington, DC: International Food Policy Research Institute (IFPRI).

Nelson, Jazib. (2018). "Why Poverty Hasn't Reduced Significantly in Pakistan." *The Express Tribune*, January 22.

NEPA and UNEP. (2015). *Climate Change and Governance in Afghanistan.* Kabul: National Environmental Protection Agency and United Nations Environment Programme.

Nepal Climate Vulnerability Study Team (NCVST). (2009). "Mining Climate Change Lessons from Signature Events." In *Vulnerability through the Eyes of the Vulnerable: Climate Change Induced Uncertainties and Nepal's Development Predicaments.* Kathmandu, Nepal: Institute for Social and Environmental Transition.

Nepal Law Commission. (2004). *National Agricultural Policy.* Available at <www.lawcommission.gov.np/en/documents/2015/08/national-agricultural-policy-2004.pdf>.

Nepal Ministry of Environment (NMoE). (2010). *National Adaptation Programme of Action to Climate Change.* Nepal.

Neumann, J.E., Emanuel, K., Ravela, S., Ludwig, L., Kirshen, P., Bosma, K. and Martinich, J. (2015). "Joint Effects of Storm Surge and Sea-Level Rise on US Coasts: New Economic Estimates of Impacts, Adaptation, and Benefits of Mitigation Policy." *Climatic Change*, Vol. 129, No. 1.

Neumayer, E. and Pleumper, T. (2007). "The Gendered Nature of Natural Disasters: The Impact of Catastrophic Events on the Gender Gap in Life Expectancy, 1981–2002." *Annals of the Association of American Geographers*, Vol. 97, No. 3.

New Delhi Declaration of Environment Ministers on a Common SAARC Position before the UNGA Special Session on the Implementation of Agenda 21.

The New Indian Express. (2015). "Food Security Still a Worry among SAARC Members." April 21.

The New York Times. (2001). "U. S. Won't Follow Climate Treaty Provisions, Whitman Says." March 28. Available at <www.nytimes.com/2001/03/28/us/us-won-t-follow-climate-treaty-provisions-whitman-says.html>. Accessed on February 14, 2020.

The New York Times. (2014). "U.S. to Give $3 Billion to Climate Fund to Help Poor Nations, and Spur Rich Ones." November 14.

Niang, I., Ruppel, O.C., Abdrabo, M.A., Essel, A., Lennard, C., Padgham, J. and Urquhart, P. (2014). "Africa." In: Barros, V.R., Field, C.B., Dokken, D.J., Mastrandrea,

M.D., Mach, K.J., Bilir, T.E., Chatterjee, M., Ebi, K.L., Estrada, Y.O., Genova, R.C., Girma, B., Kissel, E.S., Levy, A.N., MacCracken, S., Mastrandrea, P.R. and White, L.L. (eds.), *Climate Change 2014: Impacts, Adaptation, and Vulnerability.* Part B: Regional Aspects, Contribution of Working Group II to the Fifth Assessment Report of the Intergovernmental Panel on Climate Change. Cambridge: Cambridge University Press.

Nicholls, R.J., Wong, P.P., Burkett, V.R., Codignotto, J.O., Hay, J.E., McLean, R.F., Ragoonaden, S. and Woodroffe, C.D. (2007). "Coastal Systems and Low-Lying Areas." In: Parry, M.L., Canziani, O.F., Palutikof, J.P., van der Linden, P.J. and Hanson, C.E. (eds.) *Climate Change 2007: Impacts, Adaptation and Vulnerability.* Contribution of Working Group II to the Fourth Assessment Report of the Intergovernmental Panel on Climate Change. Cambridge: Cambridge University Press.

Nightingale, Andrea J. (2017). "Power and Politics in Climate Change Adaptation Efforts: Struggles over Authority and Recognition in the Context of Political Instability." *Geoforum*, Vol. 84.

Nilsson, M., Griggs, D. and Visbeck, M. (2016). "Policy: Map the Interactions between Sustainable Development Goals." *Nature*, Vol. 534, June 16.

Nilsson, Sten. and Pitt, David. (1994). *Protecting the Atmosphere: The Climate Change and Its Context.* London: Earthscan.

Ninth Meeting of Ministers of Environment (Thimphu, Bhutan, September 29, 2011). In: *SAARC Ministerial Meetings 2000–2012.* (2012). Kathmandu: Secretariat of the South Asian Association for Regional Cooperation.

The Ninth Meeting of the SAARC Environment Ministers. (2011). Extended the Timeline for the SAARC Action Plan on Climate Change by Three Years to 2014, Thimphu, September 29.

The Ninth SAARC Summit. (1997). Maldives: Malé, May 12–14.

Nissanka, S.P., Punyawardena, B.V.R., Premalal, K.H.M.S. and Thattil, R.O. (2011). *Recent Trends in Annual and Growing Seasons' Rainfall of Sri Lanka.* Proceedings of the International Conference on the Impact of Climate Change in Agriculture. Faculty of Agriculture, University of Ruhuna, Mapalana, Sri Lanka, December 20, 2011.

Nishat, A., Mukherjee, N., Hasemann, A. and Roberts, E. (2013). *Loss and Damage from the Local Perspective in the Context of a Slow Onset Process: The Case of Sea Level Rise in Bangladesh.* Dhaka: ICCCAD.

Nizam, Ifham. (2013). "Lanka's Ambitious Economic Targets Could Be Derailed by Climate Change: Minister." *The Island*, February 19.

NOAA. (2018). Global Climate Report–Annual 2017: US National Oceanic and Atmospheric Administration's National Centers for Environmental Information (NCEI). Formerly the National Climatic Data Center (NCDC). Available at <www.ncdc.noaa.gov/sotc/global/201713>.

Nomman, A.M. and Schmitz, M. (2011). "Economic Assessment of the Impact of Climate Change on the Agriculture of Pakistan." *Business and Economic Horizon*, Vol. 4.

Noordwijk Declaration. (1989). The Noordwijk Declaration's Formulation for Limiting Overall Climate Change Had also Been Incorporated into Both the IPCC Report and the Second World Climate Conference (SWCC).

Nordas, Ragnhild. and Gleditsch, Nils Petter. (2007). "Climate Change and Conflict." *Political Geography*, Vol. 26, No. 6, cited in Elliott, Lorraine. (2011). *Policy Brief-Climate Change, Migration and Human Security in Southeast Asia.* Centre for Non-Traditional Security Studies, S. Rajaratnam School of International Studies, No., October 13.

Norgaard, Richard. B. (2001). "Dana Meadows and the Limits to Growth." *Ecological Economics*, Vol. 38.

NPC. (2010). *Poverty Measurement Practices in Nepal & Number of Poor*. Government of Nepal.

Obaidullah, A.N.M. (2010). *Integrated Energy Potential of South Asia: Vision 2020*. Islamabad: SAARC Energy Centre.

Oberthür, Sebastian. and Ott, Herman E. (1999). *The Kyoto Protocol: International Climate Policy for the 21st Century*. Berlin: Springer.

O'Brien, Karen., Asuncion, St. Clair. and Berit, Kristoffersen. (2010). "The Framing of Climate Change: Why It Matters." In: O'Brien, Karen., Asuncion Lera St., Clair. and Berit, Kristoffersen. (eds.) *Climate Change, Ethics and Human Security*. Cambridge, UK and New York: Cambridge University Press.

Office for the Coordination of Humanitarian Affairs (OCHA). (2014). *Humanitarian Bulletin Sri Lanka*, Colombo, No. 3, August.

Ohl, Cornelia. (2003). "Inducing Environmental Cooperation by the Design of Emissions Permits." In: Marsiliani, Laura. et al. (eds.) *Environmental Policy in International Perspective*. Dordrecht: Kluwer Academic Publishers.

Ojha, Hemant R., Ghimira, Sharad., Pain, Adam., Nightingale, Andrea J. and Khatri, Dil B. (2015). "Policy without Politics: Technocratic Control of Climate Change Adaptation Policy Making in Nepal." *Climate Policy*, Vol. 16, No. 4.

Ojha, Hemant., Persha, Lauren. and Chhatre, Ashwini. (2009). *Community Forestry in Nepal: A Policy Innovation for Local Livelihoods, 2020 Vision Initiative*. IFPRI Discussion Paper, 00913.

Okereke, C. (2010). "Climate Justice and the International Regime." *Wiley Interdisciplinary Reviews: Climate Change*, Vol. 1, No. 3.

Oldeman, L.R., Hakkeling, R.T.A. and Sombroek, W.G. (1991). *World Map of the Status of Human-Induced Soil Degradation: An Explanatory Note, Global Assessment of Soil Degradation GLASOD*. Wageningen: International Soil Reference and Information Centre (ISRIC).

Oli, Dhan Bahadur., Thapa, Y.B., Dubey, Pawan K. and Shrestha, Kumar. (2014). *Best Practices in Poverty Alleviation and SDGs in South Asia: A Compendium*. Kathmandu: SAARC Secretariat.

Olivier, J.G.J., Janssens-Maenhout, Greet., Muntean, M. and Peters, J.A.H.W. (2016). *Trends in Global CO2 Emissions: 2016 Report*. The Hague: PBL Netherlands Environmental Assessment Agency. Ispra: European Commission, Joint Research Centre.

Olivier, J.G.J., Janssens-Maenhout, Greet., Muntean, Marilena. and Peters, JA.H.W. (2013). *Trends in Global CO2 Emissions: 2013 Report*. The Hague: PBL Netherlands Environmental Assessment Agency.

Olivier, Jos G.J. and Peters, Jeroen A.H.W. (2018). *Trends in Global CO2 and Total Greenhouse Gas Emissions: 2018 Report*. The Hague: PBL Publication.

Olivier, J.G.J., Schure, K.M. and Peters, J.A.H.W. (2017). *Trends in Global CO2 and Total Greenhouse Gas Emissions: 2017 Report*. The Hague: PBL Netherlands Environmental Assessment Agency.

Omidi, Maryam. (2009). "Maldives Sends Climate SOS with Undersea Cabinet." *Reuters*, October 17.

O'Neill, K. (2009). *The Environment and International Relations*. Cambridge: Cambridge University Press.

Onuf, Nicholas Greenwood. (1989). *World of Our Making*. Columbia: University of South Carolina Press.

Oppenheimer, M., O'Neil, B.C., Webster, M. and Agrawala, S. (2007). "Climate Change: The Limits of Consensus." *Science*, Vol. 317.

Oreskes, Naomi. (2004). "Beyond the Ivory Tower: The Scientific Consensus on Climate Change." *Science*, Vol. 306, No. 5702, December.

Ortiz, R., Sayre, K.D., Govaerts, B., Gupta, R., Subbarao, G., Ban, T., Hodson, D.P., Dixon, J., Ortiz-Monasterio, I. and Reynolds, M.P. (2008). "Climate Change: Can Wheat Beat the Heat?" *Agriculture, Ecosystems and Environment*, Vol. 126.

Orville, Howard. (1958). "The Weather Weapon: New Race with the Reds." *Newsweek*, January 13.

Oskarsson, Katerina. (2012). "Afghanistan in Transition, Second International Tokyo Conference on Afghanistan." July. Available at <http://afghanistan-un.org/2012/07/tokyo-conference-on-afghanistan/>. Accessed on February 14, 2020.

Ostrom, Elinor. (2010). "Polycentric Systems for Coping with Collective Action and Global Environmental Change." *Global Environmental Change*, Vol. 20, No. 4.

Otto, F.E.L., Massey, N., van Oldenborgh, G.J., Jones, R.G. and Allen, M.R. (2012). "Reconciling Two Approaches to Attribution of the 2010 Russian Heat Wave." *Geophysical Research Letters*, Vol. 39, No. 4.

Oxfam. (2009). "Even the Himalayas Have Stopped Smiling." In: *Climate Change, Poverty and Adaptation in Nepal*. Summary Report. Nepal: Oxfam International, August.

Oxfam International. (2009). "Millions of Rural Poor in Nepal Could Face More Hunger as a Result of Climate Change: Situation Deeply Worrying." August 28. Available at <www.oxfam.org/en/tags/food-aid>. Accessed on February 14, 2020.

Oxley, Alan. (2005). *The Asia-Pacific Partnership on Clean Development and Climate New Prospects for Joint Strategies on Climate Change*. A report published by the Australian APEC Study Centre Monash University, Australia, November.

Pachauri, R.K. and Reisinger, A. (eds.). (2007a). *Climate Change 2007: Mitigation of Climate Change*. Synthesis Report. Contribution of Working Groups I, II and III to the Fourth Assessment Report of the Intergovernmental Panel on Climate Change. Geneva: Intergovernmental Panel on Climate Change.

Pachauri, R.K. and Reisinger, A. (eds.) (2007b). *Climate Change 2007: Synthesis Report*. Contribution of Working Groups I, II and III to the Fourth Assessment Report of the Intergovernmental Panel on Climate Change. Geneva.

Pachauri, R.K. and Meyer, L.A. (eds.). (2014). *Climate Change 2014: Synthesis Report*. Contribution of Working Groups I, II and III to the Fifth Assessment Report of the Intergovernmental Panel on Climate Change. Intergovernmental Panel on Climate Change. Geneva.

Padelford, Norman J. (1954). "Regional Organisation and the United Nations." *International Organization*, Vol. 8, No. 2.

Padma, T.V. (2007). "Bhutan's Balancing Act: Happiness vs. Development." August 16. Available at <www.scidev.net>. Accessed on February 14, 2020.

Pajhwok Afghan News (PAN). (2011). "Wheat Production Down By 15%." cited Ahmadzai, Saifullah. "Non-Traditional Security Issues in Afghanistan." In: Nayak, Nihar (ed.) (2013). *Cooperative Security Framework for South Asia*. New Delhi: IDSA Pentagon Press.

Pakistan Bureau of Statistics. (2015). *Compendium on Environment Statistics of Pakistan 2015*. Islamabad: Government of Pakistan.

Pakistan Environmental Protection Act (PEPA). (1997a). Act No. XXXIV of 1997, Pakistan.

Pakistan Environmental Protection Act (PEPA). (1997b). National Assembly of Pakistan Passed This Act on September 3, 1997, and by the Senate of Pakistan on November 7, 1997. The Act received the assent of the President of Pakistan on December 3, 1997.

Pallawala, Ranga., Muller, Scott A. and Woods, Ti. (2018). *FEATURE: Sri Lanka's Progress towards Multi-Level Climate Governance.* Climate & Development Knowledge Network, 23rd December.

Palmer, Norman D. (1975). "The Changing Scene in South Asia-Internal and External Dimensions." *Orbis*, Fall, cited by Tiwari, Chitra K. (1985). "South Asian Regionalism: Problems and Prospects." In: *Asian Affairs*, Taylor & Francis Ltd, Vol. 12, No. 2.

Palmujoki, Eero. (2001). *Regionalism and Globalism in South East Asia.* New York: Palgrave.

Pandey, Kiran. (2015). "South Asia Has Highest Number of Chronically Hungry People." *Down to Earth*, August 17.

Pandey, Navin. and Karki, Bipin. (2016). "Lessons from Nepal to the Rest of the World." *Culture of Peace New Network.* Available at <http://cpnn-world.org/new/?p=9301>. Accessed on February 14, 2020.

Pandey, V.C. (2004). *Environmental Security and Tourism in South Asia.* Volume 3. New Delhi: Isha Books.

Pant, Krishna Prasad. (2014). *The SAARC Food Bank for Food Security in South Asia.* Kathmandu: South Asia Watch on Trade, Economics and Environment (SAWTEE).

Parakoti, B. and Scott, D.M. (2002). Drought Index for Rarotonga (Cook Islands), Case Study Presented as Part of Theme 2, Island Vulnerability. Pacific Regional Consultation Meeting on Water in Small Island Countries, Sigatoka, Fiji Islands, July 29–August 3.

Paralkar, Rajashree. (2017). "Bangladesh Continues to Reduce Poverty But at Slower Pace." October 24. World Bank. Available at <www.worldbank.org/en/news/feature/2017/10/24/bangladesh-continues-to-reduce-poverty-but-at-slower-pace>. Accessed on February 14, 2020.

Parashar, Utpal. (2009). "Nepali Cabinet Meets at Everest Base Camp." *Hindustan Times*, December 4.

Pareira, Amantha. (2016). "Food Insecurity Follows Floods in Sri Lanka." *SciDevNet*, August 9.

Parik, Jyoti. (2009). "Climate Impact, Risk, Vulnerability and Adaptation." *The Adaptive Response to Climate Change.* Global Summit on Sustainable Development and Climate Change, Conference Organized by the Observer Research Foundation and The Rose-Luxemburg Stiftung, September 24.

Parikh, Kirit. (2011). Low Carbon Strategies for Inclusive Growth-Interim Report. The Planning Commission, Government of India, May.

Park, Jacob., Conca, Ken. and Finger, Matthias. (eds.). (2008). *The Crisis of Global Environmental Governance.* New York: Routledge.

Park, Young-Woo. (2013). The Environment and Climate Change: Outlook of Pakistan. United Nations Environment Programme (UNEP).

Parks, Bradley C. and Roberts, J. Timmons. (2008). "Inequality and the Global Climate Regime: Breaking the North-South Impasse." *Cambridge Review of International Affairs*, Vol. 21, No. 4, December.

Parry, Martin., Canziani, Osvaldo., Palutikof, Jean., Linden, Paul van der. and Hanson, Clair. (eds.). (2007). *Climate Change 2007: Impacts, Adaptation and Vulnerability*. Working Group II Contribution to the Fourth Assessment Report of the Intergovernmental Panel on Climate Change, Summary for Policymakers and Technical Summary. New York, US: Cambridge University Press.

Parry, Martin., Evans, Alex., Rosegrant, Mark W. and Wheeler, Tim. (2009). *Climate Change and Hunger-Responding to the Challenge*. Rome, Italy: World Food Programme.

Parsons, Rymn J. (2011). "Strengthening Sovereignty: Security and Sustainability in an Era of Climate Change." *Sustainability*, Vol. 3.

Parvaiz, Naim. and Neera, Pradham. (2000). First Annual South Asian Environment Assessment Conference. Proceedings of the First Annual South Asian Environment Assessment Conference, Regional Environment Assessment Programme IUCN Asia, Kathmandu, Nepal.

Parvin, Gulsan Ara. and Rajib, S. (2013). "Role of Microfinance Institutions (MFIs) in Coastal Community's Disaster Risk Reduction, Response and Recovery: A Case Study of Hatiya, Bangladesh." *Disasters*, Vol. 37.

Parvin, Gulsan A., Shimi, Annya Chanda., Shaw, Rajib. and Biswas, Chaitee. (2016). "Flood in a Changing Climate: The Impact on Livelihood and How the Rural Poor Cope in Bangladesh." *Climate*, Vol. 60, No. 4.

Paterson, Mathew. (2009). "Green Theory." In: Burchill, Scott., Linklater, Andrew., Devetak, Richard., Donnelly, Jack., Paterson, Matthew., Reus-Smit, Christian. and True, Jacqui. (eds.) *Theories of International Relations*. New York: Palgrave Macmillan.

Pathiraja, Kasun., Balaraman, Madhawan. and de Silva, Shanthi. (2014). *Study of Climate Change Adaptation Measures Lacking Funding in Sri Lanka*. Colombo: International Centre for Ethnic Studies.

Pathmarajah, Selvarajah. (2012). "Water Management Practices and Climate Change Adaptation: South Asian Experience." In: Anbumozhi, Venkatachalam., Breiling, Meinhard., Pathmarajah, Selvarajah. and Reddy, Vangimalla R. (eds.) *Climate Change in Asia and the Pacific-How Can Countries Adapt?* New Delhi: Sage Publications, India Pvt Ltd.

Paton, Douglas. and Johnston, David. (2017). *Disaster Resilience: An Integrated Approach*. Springfield, IL: Charles C. Thomas Publisher.

Patra, Jyotiraj. and Terton, Anika. (2017). *Review of Current and Planned Adaptation Action in Nepal*. CARIAA Working Paper #20. Ottawa: Collaborative Adaptation Research Initiative in Africa and Asia (CARIAA).

Pattanaik, Smruti S., Bisht, Medha. and Kartik, Bommakanti. (2015). *Fact Sheet-SAARC: A Journey through History*. Available at <https://idsa.in/system/files/SAARC-factsheet.pdf>. Accessed on August 30, 2015.

Paul, S.K. and Routray, J.K. (2010). "Flood Proneness and Coping Strategies: The Experiences of Two Villages in Bangladesh." *Disasters*, Vol. 34.

Pauwelyn, J., Wessel, R.A. and Wouters, J. (2014). "When Structures Become Shackles: Stagnation and Dynamics in International Lawmaking." *European Journal of International Law*, Vol. 25.

Pearce, Fred. (1994). "Greenhouse Target beyond 2000." *New Scientist*, September 3.

Peel, Jaqueline., Godden, Lee. and Keenan, Rodney J. (2012). "Climate Change Law in an Era of Multi-Level Governance." *Transnational Environmental Law*, Vol. 1, No. 2.

Peikar, Jawad. (2011). "Country Report: Afghanistan: Workshop on Climate Change and Its Impact on Agriculture." *Seoul, Republic of Korea*, December 13–16.

Pelden, Sonam. (2010). "Looking beyond Hydropower." *Bhutan Observer Online*, Thimphu, April 2.

Pelletier, D.L., Frongillo, E.A., Gervais, S., Hoey, L., Menon, P. and Ngo, T. (2012). "Nutrition Agenda Setting, Policy Formulation and Implementation: Lessons from the Mainstreaming Nutrition Initiative." *Health Policy Plan*, January 27(1).

Perera, Melani Manel. (2019). "Monsoon Rains, Drought and Landslides Are Some of the Environmental Disasters Afflicting the Lives of Sri Lankans." *Asia News*, July 19.

Pernetta, John C. (1992). "Impacts of Climate Change and Sea-Level Rise on Small Island States: National and International Responses." *Global Environmental Change*, Vol. 2, No 1, March.

Persson, Åsa. (2011). *Institutionalising Climate Adaptation Finance under the UNFCCC and beyond: Could an Adaptation 'Market' Emerge?* Working Paper No. 2011–03. Stockholm: Stockholm Environment Institute.

Pettengell, Catherine. (2010). *Climate Change Adaptation-Enabling People Living in Poverty to Adapt*. Oxfam International Research Report, April.

Pfeffer, W.T., Harper, J.T. and O'Neel, Shad. (2008). "Kinematic Constraints on Glacier Contributions to 21st-Century Sea Level Rise." *Science*, Vol. 321, No. 5894, pp. 1340–1343.

Phadnis, Urmila. (1989). *Ethnicity and Nation-Building in South Asia*. New Delhi: Sage.

Philander, S. George. (ed.). (2008). *Encyclopedia of Global Warming and Climate Change*. Volumes 1–3. Los Angeles: Sage Publications, Inc.

Phys.org. (2017). "Hottest Day Ever in Shanghai as Heat Wave Bakes China." July 21. Available at <https://phys.org/news/2017-07-hottest-day-shanghai-china.html>. Accessed on February 14, 2020.

Pickering, Kevin T. and Owen, Lewis A. (1997). *An Introduction to Global Environmental Issue*. New York: Routledge.

Pielke, Roger A., Jr. (2005). "Consensus about Climate Change?" *Science*, Vol. 308.

Podesta, John. (2019). The Climate Crisis, Migration, and Refugees. Prepared for the 2019 Brookings Blum Roundtable. Available at <www.brookings.edu/research/the-climate-crisis-migration-and-refugees/>. Accessed on February 14, 2020.

Pokharel, Jagadish Chandra. (2011). *Addressing Implications of Climate Change within a Regional Framework: Recent Initiatives*. Dhaka: Fourth South Asia Economic Summit (SAESIV).

Popovich, Nadja. and Fountain, Henry. (2017). "What Is the Green Climate Fund and How Much Does the U.S. Actually Pay?" *The New York Times*, June 2.

Popoviciu, Adrian-Claudiu. (2010). "David Mitrany and Functionalism: The Beginnings of Functionalism." In: *Revista Româna de Geografie Politica*. Year XII. Thessalcniki: Ianos.

Porter, G. and Brown, J. (1996). *Global Environmental Politics*. Boulder, CO: Westview Press.

Porter, J.R., Xie, L., Challinor, A.J., Cochrane, K., Howden, S.M., Iqbal, M.M., Lobell, D.B. and Travasso, M.I. (2014). "Food Security and Food Production Systems." In: Field, C.B., Barros, V.R., Dokken, D.J., Mach, K.J., Mastrandrea, M.D., Bilir, T.E., Chatterjee, M., Ebi, K.L., Estrada, Y.O., Genova, R.C., Girma, B., Kissel, E.S., Levy, A.N., MacCracken, S., Mastrandrea, P.R. and White, L.L. (eds.) *Climate*

Change 2014: Impacts, Adaptation, and Vulnerability. Part A: Global and Sectoral Aspects. Contribution of Working Group II to the Fifth Assessment Report of the Intergovernmental Panel on Climate Change: Cambridge, United Kingdom and New York, NY, USA: Cambridge University Press.

Poudel, Hemanta Raj. (2013). Developing Countries in the International Climate Negotiations with the Prospective of International Relation Theories: A Case of Nepal, Department of International Environment and Development Studies Master Thesis 60 Credits, Ås, Norwegian University of Life Science.

Prakash, Anjal., Saravanan, V.S. and Chourey, Jayati. (2011). *Interlacing Water and Human Health: Case Studies from South Asia (Water in South Asia).* New Delhi: Sage Publications Pvt. Ltd.

Prakash, Shravani. and Kalita, Pallavi. (2010). "SAARC Must Get Serious on Climate Pact." *The Economic Times*, May 4.

Prasad, H.A.C. and Kochher, J.S. (2009). *Climate Change and India-Some Major Issues and Policy Implications.* Working Paper No.2/2009-DEA, Department of Economic Affairs Ministry of Finance Government of India, March.

PRC. (2007). *China's National Assessment Report on Climate Change, People's Republic of China (PRC).* Beijing: Science Publishing House.

Preliminary Assessment of 2010 Flood Impact on Environment in Pakistan: Extent and Coverage Impacts and Adaptation Strategy. (2010). *Natural Resource Division.* Islamabad: Pakistan Agricultural Research Council.

President, George W. Bush. (2001). *President Bush's Speech on Global Climate Change,* June 11.

Price, Gareth et al. (2014). *Attitudes to Water in South Asia.* Chatham House Report. London: Royal Institute of International Affairs.

Prior, Ryan., Athas, Iqbal. and Mckirdy, Euan. (2017). "Sri Lanka Floods: Battle to Rescue Stranded as Death Toll Tops 180." *CNN*, May 30.

Proceedings of the 7th South Asia Economic Summit (SAES). (2014). *Towards South Asia Economic Union.* New Delhi: Research and Information System for Developing Countries, India, November 5–7.

Psaila, Stephanie. (2010). Small States at the United Nations. A dissertation presented to the Faculty of Arts, University of Malta, for the Degree of Master's in Contemporary Diplomacy, February.

Punyawardena, B.V.R. (2007). Impacts of Climate Change on Agriculture in Sri Lanka and Possible Response Strategies: Impacts, Adaptation and Mitigation, National Conference on Climate Change 2007, Centre for Climate Change Studies.

Punyawardena, B.V.R., Dissanaike, T. and Mallawatantri, A. (2013). Spatial Variation Climate Change Induced Vulnerability in Sri Lanka, An Analysis of the Components of Vulnerability at District Level. Published by the Department of Agriculture in collaboration with UNDP/SGP, Sri Lanka.

Punyawardena, B.V.R. and Premalal, K.H.M.S. (2013). Do Trends in Extreme Positive Rainfall Anomalies in the Central Highlands of Sri Lanka exist? Annals of the Sri Lanka Department of Agriculture, 15.

Puri, Raghav. (2017). India's National Food Security Act (NFSA): Early Experiences, LANSA Working Paper Series, No. 14. The National Food Security Act, 2013 No. 20 of 2013. Ministry of Law and Justice, No. 29, New Delhi, Tuesday, September 10, 2013/BHADRA 19, 1935 (SAKA).

Putnam, Robert D. (1998). "Diplomacy and Domestic Politics: The Logic of Two-Level Games." *International Organization*, Vol. 42, No. 3.

Quarterly Newsletter of SAARC Disaster Management Centre. (2008). Vol. 2, No. 3, India: New Delhi.

Rabbani, Golam., Rahman, A. Atiq. and Islam, Nazria. (2010). "Climate Change and Sea Level Rise: Issues and Challenges for Coastal Communities in the Indian Ocean Region." In: Michel, David. and Pandya, Amit. (eds.) *Coastal Zones and Climate Change*. Washington, DC: The Henry L. Stimson Center.

Rabbani, Golam., Shafeeqa, Fathimath. and Sharma, Sanjay. (2016). *Assessing the Climate Change Environmental Degradation and Migration Nexus in South Asia*. Dhaka: International Organization for Migration (IOM).

Radhakrishnan, R.K. (2011). "SAARC Drafts Disaster Response Agreement." *The Hindu*, May 30.

Rafferty, John P. (ed.). (2011). *The Living Earth-Climate and Climate Change*. New York: Britannica Educational Publishing.

Ragno, E., AghaKouchak, A., Love, C.A., Cheng, L., Vahedifard, F. and Lima, C.H.R. (2018). "Quantifying Changes in Future Intensity-Duration-Frequency Curves Using Multimodel Ensemble Simulations." *Water Resources Research*, Vol. 54, No. 3.

Rahman, A., Alam, M., Alam, S.S., Uzzaman, M.R., Rashid, M. and Rabbani, G. (2008). *Risks, Vulnerability and Adaptation in Bangladesh*. Human Development Report 2007/08, Human Development Report Office Occasional Paper, 2007/13.

Rahman, A., Rabbani, G., Muzammil, M., Alam, M., Thapa, S., Rakshit, R., and Iragaki, H. (2010). *Scoping Assessment of Climate Change Adaptation in Bangladesh*. Summary Report. Bangkok: Adaptation Knowledge Platform.

Rahman, M.A. (2011). *Study on the Changes of Coastal Zone: Chittagong to Cox's Bazar Along the Bay of Bengal*. Baltimore, MD, USA: Global Summit on Coastal Seas, EMECS 9.

Rahman, Mustafizur., Bari, Estiaque. and Farin, Sherajum Monira. (2017). *SAARC Food Bank (SFB): Institutional Architecture and Issues of Operationalisation*. Centre for Policy Dialogue (CPD) Working Paper 113. Dhaka, August.

Rahman, Rieta. (2007). "Kyoto and South Asian Environment." *South Asian Journal*, October–December.

Rahman, Sultan Hafeez. (2014). *Climate Change in South Asia-Strong Responses for Building a Sustainable Future*. Asian Development Bank, 6 ADB Avenue, Mandaluyong City, 1550 Metro Manila, Philippines, Publication Stock No, ARM102196.

Rahmstorf, S., Cazenave, A., Church, J.A., Hansen, J.E., Keeling, R.F., Parker, D.E. and Somerville, R.C.J. (2007). "Recent Climate Observations Compared to Projections." *Science*, Vol. 316, No. 5825.

Rahmstorf, S. and Coumou, D. (2011). "Increase of Extreme Events in a Warming World." *Proceedings of the National Academy of Sciences of the United States of America*, Vol. 108, No. 44.

Rai, Sandeep Chamling. and Gurung, Aarati. (2005). "Raising Awareness of the Impacts of Climate Change: Initial Steps in Shaping Policy in Nepal." *Mountain Research and Development*, Vol. 25, No. 4, November.

Raina, V.K. (2009). *Himalayan Glaciers: A State-of-Art Review of Glacial Studies, Glacial Retreat and Climate Change*. Ministry of Environment and Forest (MoEF), Discussion Paper, Government of India. Almora: G, B. Pant institute of Himalayan Environment and Development, Kosi-Katarmal.

Rajamani, Lavanya. (2012). "The Changing Fortunes of Differential Treatment in the Evolution of International Environmental Law." *International Affairs*, Vol. 88, No. 3.

Rajamani, Lavanya. (2016). "Differentiation and Equity in the Post-Paris Negotiations." In: Stavins, Robert N. and Stowe, Robert C. (eds.) *The Paris Agreement and Beyond: International Climate Change Policy Post-2020*. Cambridge: Harvard Project on Climate Agreements.

Rakib, M.A., Akter, M.S., Elahi, Mehjabin., Ali, M. and Hossain, Md Babul. (2015). "An Overview of Drought Hazards and Prospective Mitigation Approach in Bangladesh." *Advances in Research*, Vol. 5, No. 6. Article No. AIR.19233.

Raleigh, C., Jordan, L. and Salehyan, I. (2008). *Assessing the Impact of Climate Change on Migration and Conflict*. Washington, DC, USA: The Social Development Department, the World Bank Group.

Ramamasy, Selvaraju. (2007). "Climate Variability and Change: Adaptation to Drought in Bangladesh." *FAO, Asian Disaster Preparedness Center, A Resource Book and Training Guide*.

Ramamasy, Selvaraju. and Baas, Stephan. (2007). *Climate Variability and Change: Adaptation to Drought in Bangladesh a Resource Book and Training Guide*. Rome: Asian Disaster Preparedness Center Food and Agriculture Organization of The United Nations.

Ramanathan, Veerabhadran. and Feng, Y. (2008). "On Avoiding Dangerous Anthropogenic Interference with the Climate System: Formidable Challenges Ahead." *Proceedings of the National Academy of Sciences*, Vol. 105, No. 38.

Ramanathan, Veerabhadran. and Xu, Yangyang. (2010). "The Copenhagen Accord for Limiting Global Warming: Criteria, Constraints, and Available Avenues." *The Scripps Institution of Oceanography*, Vol. 107, No. 18, May 4.

Ramesh, Randeep. (2008). "Paradise Almost Lost: Maldives Seek to Buy a New Homeland." *The Guardian*, November 10.

Rana, Imran. (2013). "Pakistan Extremely Vulnerable Due to Climate Change." *The Express Tribune with the International Herald Tribune*, June 7.

Rankin, Jennifer. (2019). "Central European Countries Block EU Moves towards 2050 Zero Carbon Goal." *The Guardian*, 20 June.

Rashid, H.E. (1991). *Geography of Bangladesh*. Dhaka, Bangladesh: The University Press Ltd.

Rashid, Harun ur. (2012). Doha Climate Conference Bangladesh. *PriyoAustralia. com.au*, December 9.

Rasul, Ghulam. (2012). Building Capacity on Climate Change Adaptation in Coastal Areas of Pakistan (CCAP): An European Union Funded WWF Pakistan Project, Pakistan Meteorological Department (PMD), Climate Data and Modelling Analysis of the Indus Ecoregion, Karachi.

Rasul, Ghulam., Chaudary, Q.Z., Mehmood, A., Hydar, K.W. and Dahe, Q. (2011). "Glaciers and Glacial Lakes under Changing Climate in Pakistan." *Pakistan Journal of Meteorology*, Vol. 8, No. 15.

Ratna, Minu Sinha. (2015). "Climate Change Can Trigger Security Concerns in South Asia." *Down to Earth*, November 12.

Ratnayake, U. and Herath, G. (2005). "Changes in Water Cycle: Effect on Natural Disasters and Ecosystems: Sri Lanka National Water Development Report." In: Wijesekera, N.T.S., Imbulana, K.A.U.S. and Paris, N.B. (eds.) *World Water Assessment Program*. Paris.

Rattani, Vijeta. (2018). *Coping with Climate Change: An Analysis of India's National Action Plan on Climate Change*. New Delhi: Centre for Science and Environment.

Ravell, Andrea. (2008). "Valuing the Environment: Environmental Economics and the Limits to Growth Debate." In: Buckingham, Susan. and Turner, Mike. (eds.) *Understanding Environment Issues*. London: Sage.

Redclift, Michael. and Sage, Colin. (1999). "Resources, Environmental Degradation and Inequality." In Hurrell, Andrew. and Woods, Ngaire. (eds.) *Inequality, Globalization, and World Politics*. New York: Oxford University Press.

Redclift, M.R. and Grasso, M. (2015). *Handbook on Climate Change and Human Security*. Cheltenham, United Kingdom: Edward Elgar.

Regional Environment Assessment Programme (REAP)-IUCN Asia. (1999). First Annual South Asian Environment Assessment Conference. Proceedings of the First Annual South Asian Environment Assessment Conference, Kathmandu, Nepal.

Regmi, Bimal Raj., Pandit, Anil., Pradhan, Bandana., Kovats Sari. and Lama, Pooja. (2008). *Capacity Strengthening in the Least Developed Countries-(LDCS) for Adaptation to Climate Change (CLACC)*. Climate Change and Health in Nepal, CLACC Working Paper 3.

Regmi, Bimal Raj. and Paudyal, A. (2009). *Climate Change and Agrobiodiversity in Nepal: Opportunities to Include Agrobiodiversity Maintenance to Support Nepal's National Adaptation Programme of Action (NAPA)*, Pokhara, Nepal: LI-BIRD for Platform for Agrobiodiversity Research.

Reid, Colin T. (2000). "Environmental Citizenship and the Courts." *Environmental Law Review*, Vol. 2.

Reig, P., Maddocks, A. and Gassert, F. (2013). *World's 36 Most Water Stressed Countries*. World Resources Institute.

Reisinger, A., Kitching, R.L., Chiew, F., Hughes, L., Newton, P.C.D., Schuster, S.S., Tait, A. and Whetton, P. (2014). "Australia." In: *Climate Change 2014: Impacts, Adaptation and Vulnerability. Part B: Regional Aspects, Contribution of Working Group II to the Fifth Assessment Report of the Intergovernmental Panel on Climate Change*. Cambridge: Cambridge University Press.

Ren, G. (2007). *Climate Change and Water Resources in China*. Beijing: China Meteorological Press.

Renner, M. (2011). *Water and Energy Dynamics in the Greater Himalayan Region: Opportunities for Environmental Peace-Building*. Oslo: Norwegian Peacebuilding Centre.

Report of the United Nations Conference on Environment and Development. (June 3–14, 1992). Rio de Janeiro.

Report of the United Nations Conference on the Human Environment (UNCHE). (1973). *The United Nations Conference on the Human Environment, Having Met at Stockholm from 5 to 16 June 1972*. New York: United Nations.

Republic of the Maldives (RoM). (2010). *Strategic National Action Plan for Disaster Risk Reduction and Climate Change Adaptation 2010–2020*. Provisional Draft, Malé.

Resnick, Brian. (2017). "Trump Wants a Better Deal Than Paris on Climate: What's Better Than 'Nonbinding'?" *Vox*, June 1.

Reusswig, F. and Lass, W. (2010). "Post-Carbon Ambivalences-the New Climates Change Discourse and the Risks of Climate Science." *Science, Technology and Innovation Studies*, Vol. 6, No. 2.

Reuveny, R. (2007). "Climate Change Induced Migration and Violent Conflict." *Political Geography*, Vol. 26, No. 6.

RGB. (2011). *Bhutan National Human Development Report 2011-Sustaining Progress: Rising to the Climate Challenge.* Thimphu: Royal Government of Bhutan.

Richards, Katherine. (2015). *Malnutrition in Bangladesh-Harnessing Social Protection for the Most Vulnerable.* London: Save the Children Fund.

Richtel, Matt. and Santos, Fernanda. (2016). "Wildfires, Once Confined to a Season, Burn Earlier and Longer." *New York Times*, April 12.

Riley, Charles. (2017). "Natural Disasters Caused $175 Billion in Damage in 2016." *CNN Money*, January 4.

Rinchen, Sonam. (2008). "Monitoring Climate Change." *Bhutan Observer Online*, Thimphu, September 16.

Ringler, Claudia. and Anwar, Arif. (2013). "Water for Food Security: Challenges for Pakistan." *Water International*, Vol. 38, No. 5. Special issue on "Water for food security: challenges for Pakistan" with contributions by IWMI authors.

The Rio Declaration on Environment and Development. (1992).

Roberts, David. (2017). "The 5 Biggest Deceptions in Trump's Paris Climate Speech." *Vox*, June 2.

Romero-Lankao, P., Smith, J.B., Davidson, D.J., Diffenbaugh, N.S., Kinney, P.L., Kirshen, P., Kovacs, P. and Villers-Ruiz, L. (2014). "North America." In: *Climate Change 2014: Impacts, Adaptation and Vulnerability. Part B: Regional Aspects, Contribution of Working Group II to the Fifth Assessment Report of the Intergovernmental Panel on Climate Change.* Cambridge: Cambridge University Press.

Rootes, Chris. (1999). "Environmental Movements: From Local to Global." *Environmental Politics*, Vol. 8, No. 1.

Rosamond, B. (2000). *Theories of European Integration.* Basingstoke/New York: St. Martin's Press.

Rothchild, Donald. (1967). "Functionalism and World Politics: A Study Based on United Nations Programs Financing Economic Development by James Patrick Sewell." *The Journal of Modern African Studies*, Vol. 5, No. 1.

Roul, Avilash. (2014). "SAARC to Rescue India on Climate Change Before Lima?" *Society for the Study of Peace and Conflict Studies*, New Delhi, November 25.

Rourke, John T. (2005). *International Politics on the World Stage.* New York: McGraw Hill.

Rourke, John T. and Boyer, Mark A. (2004). *International Politics on the World Stage-Brief.* New York: McGraw Hill.

Roy, Jayashree. (2007). Sundarbans-Can They Be Saved?" In: *The Hindu Survey of the Environment.* New Delhi: The Hindu, Special Issue.

Roy, Jayashree., Ghosh, Anupa. and Baruah, Gopa. (2006). *The Economics of Climate Change: A Review of Studies in the Context of South Asia with a Special Focus on India.* A Report Submitted to the Stern Review on the Economic of Climate Change.

Royal Government of Bhutan (RGB). (1998). *The Middle Path: National Environment Strategy for Bhutan.* Thimphu, Bhutan: National Environment Commission.

Royal Government of Bhutan (RGB). (1999). *Bhutan 2020: A Vision for Peace, Prosperity and Happiness, Planning Commission.* Thimphu: Planning Commission Royal Government of Bhutan, Bhutan.

Royal Government of Bhutan (RGB). (2000a). *Initial National Communication, National Environment Commission.* Thimphu: Planning Commission Royal Government of Bhutan, Bhutan.

Royal Government of Bhutan (RGB). (2000b). *Environmental Assessment Act, 2000.* Thimphu, Bhutan: National Environment Commission Tashichho Dzong.

Royal Government of Bhutan (RGB). (2002). *Initial National Communication, National Environment Commission.* Thimphu: Planning Commission Royal Government of Nepal, Bhutan.

Royal Government of Bhutan (RGB). (2004). *Brief Report on State of the Environment, the National Environment Commission.* Thimphu: Planning Commission Royal Government of Bhutan, Bhutan.

Royal Government of Bhutan (RGB). (2006a). *National Adaptation Programme of Action.* Thimphu: National Environment Commission Royal Government of Bhutan.

Royal Government of Bhutan (RGB). (2006b). *National Communication to the UNFCCC.* Thimphu: Planning Commission Royal Government of Bhutan, Bhutan.

Royal Government of Bhutan (RGB). (2008). *National Adaptation Programme of Action.* Thimphu: Planning Commission Royal Government of Nepal, Bhutan.

Royal Government of Bhutan (RGB). (2009a). *Strategizing Climate Change for Bhutan, National Environment Commission and United Nations Environment Programme.* Thimphu: Bhutan, National Environment Commission and United Nations Environment Programme.

Royal Government of Bhutan (RGB). (2009b). *Tenth Five Year Plan-2008–2013- Vol. 1.* Main Document, Gross National Happiness Commission. Thimphu: National Environment Commission & Department of Agriculture, Royal Government of Bhutan.

Royal Government of Bhutan (RGB). (2009c). *Strategizing Climate Change for Bhutan.* Thimphu, Bhutan: National Environment Commission and United Nations Environment Programme.

Royal Government of Bhutan (RGB). (2011). *Turning Vision into Reality: The Development Challenges Confronting Bhutan, Eleventh Round Table Meeting.* Thimphu: Gross National Happiness Commission.

Royal Government of Bhutan (RGB). (2013a). *Environmental Impact Assessment and Management Plan.* Thimphu, Bhutan: Ministry of Works & Human Settlement.

Royal Government of Bhutan (RGB). (2013b). *National Environment Commission Secretariat.* Available at <www.nec.gov.bt/nec1/index.php/about-nec/>. Accessed on May 22, 2013.

Royal Government of Bhutan (RGB). (2015a). *Communication of the INDC of the Kingdom of Bhutan.* Thimphu, Bhutan: National Environment Commission.

Royal Government of Bhutan (RGB). (2015b). Press Release, Bhutan Deposits Its Instrument of Acceptance to the Doha Amendment to the Kyoto Protocol under the United Nations Framework Convention on Climate Change 2015 Treaty Event, United Nations Headquarters, New York, 28 September 2015, MFA/MD/GA-6/2015/1178, Thimphu, Bhutan.

The Royal Society. (2014). "Global Response to Climate Change." Available at <https://royalsociety.org/policy/publications/2005/global-response-climate-change/>. Accessed on February 14, 2020.

Ruggie, J.G. (1998). *Constructing the World Polity: Essays on International Institutionalization.* New York: Routledge.

Russett, B. and O'Neal, J. (2001). *Triangulating Peace: Democracy, Interdependence, and International Organizations.* New York: WW Norton & Company.

Ryan, Joe. (2017). "Rich Nations Fail to Help Developing World Fight Climate Change." *Bloomberg*, November 6.

Saad, L. (2012). "In U.S., Global Warming Views Steady Despite Warm Winter." *Gallup Poll*. Available at <www.gallup.com/poll/153608/Global-Warming-Views-Steady-Despite-Warm-Winter.aspx>. Accessed on 26th October, 2014.

SAARC. (1992a). *Regional Study on the Causes and Consequences of Natural Disasters and the Protection and Preservation of Environment*. Kathmandu: SAARC Secretariat.

SAARC. (1992b). *Regional Study on Greenhouse Effect and Its Impact on the Region*. Kathmandu: SAARC Secretariat.

SAARC. (1999). *SAARC Ministerial Meetings April 1986–August 1999*. Kathmandu: SAARC Secretariat.

SAARC. (2005). *Comprehensive Framework on Disaster Management*. New Delhi, India.

SAARC. (2011). *SAARC Agreement on Rapid Response to Natural Disasters*. Maldives: Malé.

SAARC. (2012a). *SAARC Ministerial Meetings 2000–2012*. Kathmandu: Secretariat of the South Asian Association for Regional Cooperation.

SAARC. (2012b). *Second Meeting of the Inter-Governmental Expert Group on Climate Change*. SAARC Secretariat, Kathmandu, April 16–17.

SAARC. (2014a). *Eighteenth SAARC Summit: Kathmandu Declaration* Kathmandu: SAARC Secretariat, November, 26–27.

SAARC. (2014b). *Post-2015 Drr Framework for SAARC Region (Hfa2) SDMC*. New Delhi: SAARC Disaster Management Centre, March.

SAARC. (2015). Brief on Regional Cooperation under SAARC in the Areas of Environment, Climate Change and Natural Disasters. SAARC Secretariat (Environment Division). [This unpublished document of SAARC was given by UgyenSamdrup, the Personal Assistant at SAARC Secretariat (Environment Division) during Library Work at SAARC Secretariat: Kathmandu].

SAARC. (2016a). *A Guide to the SAARC Food Bank*. Kathmandu: A Publication of the SAARC Food Bank Board, South Asian Association for Regional Cooperation, SAARC Secretariat.

SAARC. (2016b). SAARC *Plan of Action on Labour Migration*. Consultative Workshop on SAARC Plan of Action for Cooperation on Matters Related to Migration, Kathmandu.

SAARC. (2019). "Press Release-the Fourth Meeting of the SAARC Agriculture Ministers held in Thimphu." *Bhutan*, June 28.

SAARC Action Plan on Climate Change. (2008). *(Adopted by the SAARC Ministerial Meeting on Climate Change) Dhaka, 3 July 2008*, it was later endorsed by 15th SAARC Summit, Colombo, August 3.

SAARC Agreement on Rapid Response to Natural Disasters, (Signed at the Seventeenth SAARC Summit) Addu City, Maldives, 10–11 November 2011).

SAARC Agriculture Ministers' Meeting. (1996). Pakistan: Islamabad, October 8–9, Cited in SAARC. (1999). *SAARC Ministerial Meetings- April 1996–August 1999*. Kathmandu: SAARC Secretariat.

SAARC Coastal Zone Management Centre. (2016). *Republic of Maldives: Malé*. Available at <http://saarc-sec.org/saarc-regional-centres> Accessed on August 20, 2018.

SAARC Convention on Cooperation on Environment (Signed during the Sixteenth SAARC Summit) Thimphu, 28–29 April 2010.

SAARC Declaration on Food Security, Adopted by the Extraordinary Session of the Agriculture Ministers. (2008). India: New Delhi, November 5.

SAARC Disaster Management Centre (SDMC). (2008). *SDMC Newsletter*. Vol. II, No. 3, New Delhi, July.

SAARC Disaster Management Framework. (2007). *SAARC Disaster Management Centre*. Available at <www.preventionweb.net/english/professional/policies/v. php?id=60968> Accessed on January 10, 2016.

SAARC Environment Action Plan (Adopted by the Third Meeting of the SAARC Environment Ministers) Malé, 15–16 October 1997.

SAARC Environment Ministers Conference. (1997). Malé, Maldives, October 15–16, cited in SAARC. (1999). *SAARC Ministerial Meetings April 1986–August 1999*. Kathmandu: SAARC Secretariat.

SAARC Environment Ministers' Conference. (1997). Republic of Maldives: Malé, [Annex XIII of the conference document], October 15–16.

SAARC Forestry Centre (SFC). (2008). *Bhutan Broadcasting Service: Thimphu*, June 13. Available at < www.sfc.org.bt/index.html>. Accessed on June 20, 2015.

SAARC Ministerial Statement on Cooperation on Environment: Delhi Statement, Adopted by the Eighth Meeting of the SAARC Environment Ministers. (2009). New Delhi, October 20–21.

SAARC Social Charter. (2004). April 1. Available at <www.jus.uio.no/english/services/library/treaties/02/2-03/saarc-social-charter.xml>. Accessed on February 14, 2020.

SAARC Workshop on Climate Change and Disasters: Emerging Trends and Future Strategies. (2008). Nepal: Kathmandu, August 21–22, Available at <http://saarc-sdmc.nic.in>. Accessed on December 12, 2014.

Saba, D.S. (2001). "Afghanistan: Environmental Degradation in a Fragile Ecological Setting." *International Journal of Sustainable Development and World Ecology*, Vol. 8, No. 4.

Saddiqui, Rehana. (2010). "Environmental Issues and Policy Response in Pakistan." *South Asian Survey*, April–June.

Safi, Michael. (2018). "Death Toll Climbs in Karachi Heatwave." *The Guardian*, May 22.

Saha, Autri. and Talwar, Karan. (2010). "India's Response to Climate Change: The 2009 Copenhagen Summit and Beyond." *NUJS Law Review*, Vol. 3; *NUJS Law Review*, Vol. 159.

Saidi, Saideh. (2017). "Climate Change and Human Mobility: Displacement in Afghanistan." *Iran Review*, January 24.

Saini, Shweta. and Gulati, Ashok. (2015). The National Food Security Act (NFSA) 2013-Challenges, Buffer Stocking and the Way Forward. Indian Council for Research on International Economic Relations (ICRIER).

Saleem, Rubab. (2008). "Food Crisis in Pakistan." *Pakistan Times*, Islamabad, May 18.

Saleh, Faisal. (2007). "Climate Change Affecting Pakistan's Environment." *Daily Times*, September 26.

Salma, S., Rehman, S. and Shah, M.A. (2012). "Rainfall Trends in Different Climate Zones of Pakistan." *Pakistan Journal of Meteorology*, Vol. 9, No. 17.

Salman, Aneel. (2014). Mainstreaming Community-Based Climate Change Adaptation in Pakistan. Series on Vulnerability and Resilience, Occasional Paper 30, Communication Unit, LEAD Pakistan, April.

Salman, Ayesha. (ed.). (2011). Institutional Arrangements for Climate Change in Pakistan. Project Report Series # 19th, Sustainable Development Policy Institute, Pakistan.

Salter, Richard E. (2009). United States Agency for International Development (USAID) Technical Report. Development of a National Biodiversity Strategy for Afghanistan- Second Mission Report. Afghanistan: Kabul, July.

Salve, Harish. (2004). "Justice between Generations: Environment and Social Justice." In: Kirpal, B.K., Desai, Ashok, H., Subramanium, Gopal., Dhavan, Rajeev. and Ramchandran, Raju. (eds.) *Supreme But Not Infallible*. New Delhi: Oxford University Press.

Sand, Peter. (1990). *Lessons Learned in Global Environmental Governance*. Washington: World Resources Institute, Cited in Laferrière, Eric. and Stoett, Peter J. (1999). *International Relations Theory and Ecological Thought-Towards a Synthesis*. London: Routledge: London.

Sandalow, David. (2018). *Guide to Chinese Climate Policy 2018*. Columbia: SIPA/ The Center on Global Energy Policy.

Sandbrook, R. (1993). "Live and Learn." *New Statesman and Society*, January.

Sang, Lucia I. Suarez. (2019). "Cyclone Fani Lashes Eastern India, Killing at Least 3 and Displacing Millions." *Fox News*, May 3.

Sangal, P.P. (2008). "India's Climate Change Action Plan." *The Economic Times*, New Delhi, July 29.

Sangomla, Akshit. (2018). "Kerala Battles Worst Floods since 1924, Says CM." *Down to Earth*, August 13.

Sarkar, Md Sujahangir Kabir., Sadeka, Sumaiya., Al Feardous, Md Ashiq. and Ahmed, Siddique. (2013). "Human Development Performance of Bangladesh: South Asian Perspective." *Journal of Applied Sciences Research*, Vol. 9, No. 4.

Sarofim, M.C., Saha, S., Hawkins, M.D., Mills, D.M., Hess, J., Horton, R., Kinney, P., Schwartz, J. and St. Juliana, A. (2016). *Chapter 2: Temperature-Related Death and Illness: The Impacts of Climate Change on Human Health in the United States: A Scientific Assessment*. Washington, DC: U.S. Global Change Research Program.

Satpathy, Subhranshu Kumar. (2016). Implementation of National Food Security Act (NFSA) in Odisha. *Odisha Review*.

Savage, Matthew., Dougherty, Bill., Hamza, Mohammed., Butterfiled, Ruth. and Bharwani, Sukaina. (2009a). Socio-Economic Impacts of Climate Change in Afghanistan. Report DFID CNTR 08 85, Stockholm Environment Institute, Oxford, UK.

Savage, Matthew., Dougherty, Bill., Hamza, Mohammed., Butterfield, Ruth. and Bharwani, Sukaina. (2009b). *Socio-Economic Impacts of Climate Change in Afghanistan: A Report to the Department of International Development*. Stockholm: Stockholm Environment Institute.

Sawe, Benjamin Elisha. (2018). *South Asia: Constituent Countries and Their Populations and Economies, World Atlas*, August 15. Available at <www.worldatlas. com/articles/the-population-and-economy-of-the-south-asian-countries.html>. Accessed on 15 August 2019.

Saxena, N.C. (2011). Hunger, Under-Nutrition and Food Security in India. Working Paper 44, Chronic Poverty Research Centre, Indian Institute of Public Administration, New Delhi.

Sayeed, Saad. (2018). "Pakistan Heatwave Kills 65 People in Karachi-Welfare Organization." *Reuters*, May 22.

Schenck, Lisa M. (2008). "Climate Change 'Crisis'-Struggling for Worldwide Collective Action." *Colorado Journal of International Environmental Law and Policy*, Vol. 19, No. 3.

Scheyvens, R. (2011). "The Challenge of Sustainable Tourism Development in the Maldives: Understanding the Social and Political Dimensions of Sustainability." *Asia Pacific Viewpoint*, Vol. 52.

Scholte, Jan Aart. (2005). *Globalization: A Critical Evaluation*. New York: Palgrave Macmillan.

Schubert, R., et al. (2008). *Climate Change as a Security Risk*. Members of the German Advisory Council on Global Change (WBGU). London: Earthscan.

Schultz, Kai. (2017). "Maldives Faces Bleak Future as Shores Recede." *Gulf News*, April 7.

SDGF. (2017). *Case Study-Healthy Children, Healthy Afghanistan: Best Practices and Lessons Learned*. Sustainable Development Goals Fund.

Sediqi, Qadir. (2018). "Flash Floods in Afghanistan Kill at Least 34." *Reuters*, May 15.

Senaratne, A., Perera, N. and Wickramasinghe, K. (2009). Mainstreaming Climate Change for Sustainable Development in Sri Lanka: Towards a National Agenda for Action. *Research Studies: Working Paper Series No. 14*. Institute of Policy Studies of Sri Lanka. Colombo, Sri Lanka.

Seraj, Shykh. (2013). "In the Quagmire of Festivity and Reality." *The Daily Star*, April 14.

Seventeenth SAARC Summit. (2011). Maldives: Addu City, Maldives, November 10–11.

The Seventeenth SAARC Summit Addu Declaration "Building Bridges." Issued on 11.11.11, in Addu, Maldives.

Seventh Meeting of the SAARC Environment Ministers. (2006). Bangladesh: Dhaka, May 24–24.

Shahan, Asif Mohammad. and Jahan, Ferdous. (2017). *Opening the Policy Space: The Dynamics of Nutrition Policy Making in Bangladesh*. Agropolis International, Global Support Facility for the National Information Platforms for Nutrition initiative. France: Montpellier.

Shahid, S. and Behrawan, H. (2008). "Drought Risk Assessment in the Western Part of Bangladesh." *Natural Hazards*, Vol. 46, No. 3.

Shaig, Ahmed. (2006). Climate Change Vulnerability and Adaptation Assessment of the Land and Beaches of Maldives. Technical Papers to Maldives National Adaptation Plan of Action for Climate Change, Malé, Ministry of Environment, Energy and Water.

Shakun, Jeremy D. and Carlson, Anders E. (2010). "A Global Perspective on Last Glacial Maximum to Holocene Climate Change." *Quaternary Science Reviews*, Vol. 29, No. 15–16.

Shaljan, A.M. (2004). "Population, Gender and Development in Maldives." *Economic and Political Weekly*, May 1.

Shamsuddoha, Md., Khan, Hannan S.M. Munjurul., Tanjir, Hossain. and Sajid, Raihan. (2011). *Displacement and Migration from the Climate Hot-Spots: Causes and Consequences*. Dhaka: Center for Participatory Research and Development and Action Aid Bangladesh.

Shanahan, David. (2017). "South Asia." In: Ear, Jessica., Cook, Alistair D.B. and Canyon, Deon V. (eds.) *Disaster Response Regional Architectures Assessing Future Possibilities*. Honolulu: Daniel K. Inouye Asia-Pacific Center for Security Studies.

Shankar, K.R., Nagasree, K., Nirmala, G., Prasad, M., Venkateswarlu, B. and Rao, C.S. (2015). "Climate Change and Agricultural Adaptation in South Asia." In: Filho, L. (ed.) *Handbook of Climate Change Adaptation*. Berlin: Springer.

Shannon, Sarah., Smith, Robin., Wiltshire, Andy., Payne, Tony., Huss, Matthias., Betts, Richard., Caesar, John., Koutroulis, Aris., Jones, Darren. and Harrison, Stephan. (2019). "Global Glacier Volume Projections under High-End Climate Change Scenarios." *The Cryosphere*, Vol. 13.

Shareef, A. Waheed, Ahmed., Ali, Ahmed., Laila, Aishath., Mohamed, Miruza., Asif, Mohamed., Khaleel, Zammath. and Riyaza, Fathimath. (2015). *Baseline Analysis of Adaptation Capacity and Climate Change Vulnerability Impacts in the Tourism Sector*. Malé: Ministry of Tourism, Malé, Republic of Maldives.

Sharif, Ibrat., Nasir, Naznin., Khanum, Roufa. and Khan, AS Moniruzzaman. (2016). "Climate Change Adaptation Policies in Bangladesh: Gap Analysis through a Gender Lens." *BRAC University Journal*, Vol. 11, No. 2.

Sharma, Ashok. (2008). "Climate Change to Impact Indian Agriculture: IARI." *Financial Express*, January 28.

Sharma, Bharat R., Rao, V.K., Vittal, K.P.R., Ramakrishna, Y.S., Amarasinghe, Upali. (2010). "Estimating the Potential of Rainfed Agriculture in India: Prospects for Water Productivity Improvements." *Agricultural Water Management*, Vol. 97, No. 1.

Sharma, E., Chettri, N., Tsering, K., Shrestha, AB., Fang, Jing, Mool, P. and Eriksson, M. (2009). *Climate Change Impacts and Vulnerability in the Eastern Himalayas*. Kathmandu: ICIMOD.

Sharma, Krishnavatar. (2017). "India Has 139 Million Internal Migrants: They Must Not Be Forgotten." *World Economic Forum*, October 1.

Sharma, Sanchita. (2018). "Working Toilets Will Guarantee a Swachh Bharat." *Hindustan Times*, March 17.

Sharma, Sanchita. and Thapa, D. (2013). *Taken for Granted: Nepali Migration to India*. Kathmandu: Research Paper III, Centre for the Study of Labour and Mobility (CESLAM).

Sharma, Sheel Kant. (2011). "South Asian Regionalism: Prospects and Challenges." *Indian Foreign Affairs Journal*, Vol. 6, No. 3, July–September.

Sharma, Suman. (2011a). Existential threat to Human Security in South Asia and Regional Response: A Case Study of Climate Change and SAARC Initiatives. Paper Presented at the Third Global International Studies Conference: Porto, Organized by the World International Studies Committee.

Sharma, Sudhir., Kishan, Ram. and Doig, Alison. (2014). *Low-Carbon Development in South Asia-Leapfrogging to a Green Future*. London: Christian Aid.

Shaw, Rajib. (2010). "Climate Change Adaptation Research in South Asia: An Overview." *Asian Journal of Environment and Disaster Management (AJEDM)*, Vol. 2, No. 4.

Shaw, Rajib., Mallick, Fuad. and Islam, Aminul (eds.). (2013). *Urban Poverty, Climate Change and Health Risks for Slum Dwellers in Bangladesh, Climate Change Adaptation Actions in Bangladesh*. London: Springer.

Shearman, D. and Smith, J.W. (2007). *The Climate Change Challenge and the Failure of Democracy*. Westport: Praeger Publishers.

Shepherd, A., Mitchell, T., Lewis, K., Lenhardt, A., Jones, L., Scott, L. and Muir-Wood, R. (2013). *The Geography of Poverty, Disasters and Climate Extremes in 2030*. London, UK: Overseas Development Institute.

Sheppard, Cameron. (2019). "Researcher Studies Monsoons in South Asia-Periods of Drought, Too Much Rainfall Can Impact Agriculture." *The Daily Evergreen*, April 29.

Shiekh, Anis uddin. (2013). "Poorest Countries of the World: Projections upto 2018." *Dunya News*. Available at <http://dunyanews.tv/world_poorest_countries.pdf>. Accessed on February 14, 2020.

Shivakoti, B.R., Lopez-Casero, F., Kataoka, Y. and Shrestha, S. (2015). Climate Change, Changing Rainfall and Increasing Water Scarcity: An Integrated Approach for Planning Adaptation and Building Resilience of Smallholder Subsistence Livelihoods in Nepal IGES Research Report 2014–07. Japan: Hayama, Institute for Global Environmental Strategies.

Shivakumar, Mannava, V.K. and Stefanski, Robert. (2011). "Climate Change in South Asia." In: Lal, R., Sivakumar, M.V.K., Faiz, M.A., Mustafizur Rahman, A.H.M. and Islam, K.R. (eds.) *Climate Change and Food Security in South Asia*. Netherlands: Springer.

Shivay, Y.S. and Rahal, Anshu. (2008). *Effect of Global Warming on Crop Productivity*. Kurukshetra.

Shrestha, Arun B. and Bajracharya, Sagar R. (eds.). (2013). *Case Studies on Flash Flood Risk Management in the Himalayas in Support of Specific Flash Flood Policies*. Kathmandu: ICIMOD.

Shrestha, Arun B., Wake, C.P., Dibb, J.E. and Mayewski, P.A. (2000). "Precipitation Fluctuations in the Nepal Himalaya and Its Vicinity and Relationship with Some Large Scale Climatological Parameters." *International Journal of Climatology*, Vol. 20, No. 3.

Shrestha, Arun B., Wake, C.P., Mayewski, P.A. and Dibb, J.E. (1999). "Maximum Temperature Trends in the Himalaya and Its Vicinity: Analysis Based on Temperature Records from Nepal for the Period 1971–94." *Journal of Climate*, Vol. 12.

Shrestha, B., Mool, P.K. and Bajracharya, S.R. (2007). *Impact of Climate Change on Himalayan Glaciers and Glacial Lakes: Case Studies on GLOF and Associated Hazards in Nepal and Bhutan*. Kathmandu: ICIMOD.

Shrestha, B.D., Dhakal, B. and Rai, M.R. (2007). *Disaster Preparedness and Integrated Watershed Management Plan of Jugedi Stream, Kabilas VDC, Chitwan District, Nepal (2007–2012)*. Kathmandu: Practical Action.

Shrestha, Uttam Babu., Gautam, Shiva. and Bawa, Kamaljit S. (2012). "Widespread Climate Change in the Himalayas and Associated Changes in Local Ecosystems." *PLoS One*, Vol. 7, No. 5. Available at <https://doi.org/10.1371/journal.pone.0036741>. Accessed on February 14, 2020.

Shrivastava, N.K., Srivastava, Sanjay K., Soma, P., Chiranjeevi, O., Hegde, V.S., Shivakumar, S.K. and Dave, Darshana R. (2008). "International Charter on 'Space and Major Disasters': An Assessment of Outreach in South Asia." Annual Report, Vol. 1, No. 1.

Sidhu, W.P.S. and Sandhu, Rohan. (2014). "Reinvigorating SAARC: India's Opportunities and Challenges." In: Mehta, Vikram S. and Sidhu, W.P.S. (eds.) *Reinvigorating SAARC: India's Opportunities and Challenges*. New Delhi: Brookings India.

Sills, David L. (ed.). (1972). *International Encyclopaedia of the Social Sciences*. Volume 13. London: The MacMillan Company & The Free Press.

Simmons, Beth A. (1998). "Compliance with International Agreements." *Annual Review of Political Science*, Vol. 1, pp. 75, 93.

Singh, Surender P., Bassignana-Khadka, Isabella., Bhaskar Singh, Karky. and Sharma, Eklabya. (2011). *Climate Change in the Hindu Kush-Himalayas: The State of Current Knowledge*. Kathmandu: ICIMOD.

Singh, Timon. (2012). "Pakistan Rejects Foreign Aid for Combating Climate Change." *The Express Tribune*, July 23.

Singha, Komol. (2010). "Environment and Development in Bhutan." *South Asian Journal*, April–June.

Sinha, Anushree., Bauer, Armin. and Bullen, Paul. (2015). *The Environments of the Poor in South Asia: Simultaneously Reducing Poverty, Protecting the Environment and Adapting to Climate Change*. London: Oxford University Press.

Sitarz, Daniel. (ed.). (1993). *Agenda 21: The Earth Summit Strategy to Save our Planet*. Boulder, CO: Earthpress.

Sixteenth SAARC Summit. (2010). Bhutan: Thimphu, April 28–29.

The Sixteenth Summit. (2010). Bhutan: Thimphu, April 28–29.

The Sixth Meeting of Ministers of Environment. (2004). Bhutan: Thimphu, June 12–13.

The Sixth Summit in Colombo, Sri Lanka on 21 December 1991.

Sklias, Pantelis. (2010). "India's Position at the Copenhagen Climate Change Conference: Towards a New Era in the Political Economy of International Relations?" *Research Journal of International Studies*, August 15.

Skymet Weather Team. (2017). Dependency of Indian Agriculture on Monsoon, March 27.

Slaughter, Anne-Marie. (2015). The Paris Approach to Global Governance, Project Syndicate, December 28.

Smelser, Neil J. and Baltes, Paul B. (eds.). (2001). *International Encyclopaedia of Social and Behavioural Sciences*. Volume. 4. New York: Elsevier.

Smil, Vaclav. (1993). *Global Ecology: Environmental Change and Social Flexibility*. London: Routledge.

Smit, D. and Vevekananda, J. (2007). "A Climate of Conflict: The Links between Climate Change, Peace and War." *International Alert*, November.

Smith, J.B., Huq, S., Lenhart, S., Mata, L.J., Nemesova, I. and Toure, S. (eds.). (1996). *Vulnerability and Adaptation to Climate Change, Interim Results from the U.S. Country Studies Program*. Dordrecht: Kluwer Academic Publishers.

Smith, J.B., Klein, R.J.T. and Huq, S. (2003). *Climate Change, Adaptive Capacity, and Development*. London, UK: Imperial College Press.

Smith, K.R., Woodward, A., Campbell-Lendrum, D., Chadee, D.D., Honda, Y., Liu, Q., Olwoch, J.M., Revich, B. and Sauerborn, R. (2014). "Human Health: Impacts, Adaptation, and Co-Benefits." In: Field, C.B., Barros, V.R., Dokken, D.J., Mach, K.J., Mastrandrea, M.D., Bilir, T.E., Chatterjee, M., Ebi, K.L., Estrada, Y.O., Genova, R.C., Girma, B., Kissel, E.S., Levy, A.N., Maccracken, S., Mastrandrea, P.R., White, L.L. (eds.) *Climate Change 2014: Impacts, Adaptation, and Vulnerability, Part A: Global and Sectoral Aspects*. Contribution of Working Group II to the Fifth Assessment Report of the Intergovernmental Panel on Climate Change. Cambridge, UK and New York, NY, USA: Cambridge University Press.

Solomon, S., Qin, D., Manning, M., Chen, Z., Marquis, M., Averyt, K.B., Tignor, M. and Miller, H.L. (eds.) (2007). *Climate Change 2007: The Physical Science Basis*. Contribution of Working Group I to the Fourth Assessment Report of the Intergovernmental Panel on Climate Change. Cambridge: Cambridge University Press.

Sood, Shefali. (2017). Discourse and Responsibility: Climate Change Refugees in South Asia. Thesis submitted to the Faculty of the College of Literature, Science and the Arts at the University of Michigan in Partial Fulfillment for the Requirements for the Degree of Bachelor of Arts.

South Asia Disaster Report: 2007–2011. (2011). New Delhi: SAARC Disaster Management Centre (SDMC).

South Asia Disaster Report: 2009. (2010). New Delhi: SAARC Disaster Management Centre (SDMC).

South Asia Disaster Report: 2011. (2013). New Delhi: SAARC Disaster Management Centre (SDMC).

South Asia Environment Outlook (SAEO). (2009). Nairobi: United Nations Environment Programme (UNEP).

South Asian Association for Regional Cooperation (SAARC). (1992). *Regional Study on the Causes and Consequences of Natural Disasters and the Protection and Preservation of the Environment.* Kathmandu: Published by the SAARC Secretariat.

Sovacool, B.J. (2012). "Expert Views of Climate Change Adaptation in the Maldives." *Climatic Change,* Vol. 114, No. 2.

Special Meeting of Ministers of Environment. (2005). Maldives: Malé, June 25.

Special Operation Afghanistan So 200635. (2014–2018). Capacity Development in Support of the Strategic Grain Reserve in Afghanistan.

Special Session of the Environment Ministers in the aftermath of the Indian Ocean Tsunami (2005). Malé, July 25.

Springate-Baginski, Oliver. and Blaikie, Piers M. (eds.). (2007). *Forests, People and Power: The Political Ecology of Reform in South Asia.* London: Earthscan Publications Ltd.

Sridharan, Kripa. (2008). *Regional Organisations and Conflict Management: Comparing ASEAN and SAARC, Crisis States.* Working Papers Series No. 2.

Sri Lanka Disaster Management (SLDM) Act. (2005). No. 13 of 2005. Colombo: Sri Lanka.

Sri Lanka Ministry of Environment (SLMOE). (2010). *National Climate Change Adaptation Strategy for Sri Lanka: 2011–2016.* Colombo: Ministry of Environment.

Sri Lanka Ministry of Forestry and Environment (SLMFE). (2000). *Initial National Communication under the United Nations Framework Convention on Climate Change: Sri Lanka.* Colombo: Sri Lanka Government.

Srinivasan, Sharmadha. (2014). "SAARC Summit Debriefing: The 18th South Asian Association for Regional Cooperation Was, in a Word, Disappointing." *The Diplomat,* December 2.

Sriskantharajah, Karthika. (2014). *FAO Sri Lanka News Update, 2014: Technical Report.* FAO Sri Lanka News Update-Issue 19. Sri Lanka: Food and Agriculture Organization of the United Nations.

Srivastava, R. (2012). "Internal Migration in India: An Overview of Its Features, Trends and Policy Challenges." In: Zerah, M.H., Dupont, V., Tawa Lama-Rewal, S. and Faetanini, M. (eds.) *Urban Policies and the Right to the City in India: Rights, Responsibilities and Citizenship.* New Delhi: UNESCO and Centre de Sciences Humaines.

Srivastava, R.S., Sutradhar, R., Abrar, C.R., Reza, Md.S., Adhikari, J. and Gurung, G. (2014). Impact of Internal Labour Migration to the Construction Sector in South Asia on Poverty and Wellbeing. Paper Presented at KNOMAD Conference on Internal Migration and Urbanization, Dhaka.

Srivastava, Ravi. and Pandey, Arvind Kumar. (2017). *Internal and International Migration in South Asia: Drivers, Interlinkage and Policy Issues: Discussion Paper.* New Delhi: United Nations Educational, Scientific and Cultural Organization (UNESCO).

Stafford-Smith, M., Griggs, D., Gaffney, O., Ullah, F., Reyers, B., Kanie, N. and O'Connell, D. (2016). "Integration: The Key to Implementing the Sustainable Development Goals." *Sustainability Science*, Vol. 12, No. 6.

Stanton, E.A. and Ackerman, F. (2007). *Florida and Climate Change: The Costs of Inaction.* Medford, MA: Global Development and Environment Institute, Tufts University.

Starke, J.G. (1989). *Introduction to International Law.* 10th Edition. London: Butterworths.

The Statement by H.E. Maumoon Abdul Gayoom. (1997). Kyoto, Japan (COP 3), Maldives, December 4.

Statement by Mr. Abdullahi Majeed (Maldives). Milan, Italy (COP 9 Round-Table). (December 11, 2003).

Statements and Declarations of SAARC Summits of the Heads of State or Government (1985–2010). (2010). *South Asian Association for Regional Cooperation, Institute of Foreign Affairs.* Kathmandu: IFA Tripureshwor.

The Statesman. (2008). "Mosquito Menace Global Warming to Blame." November 9.

Stavins, R. (2009). *What Hath Copenhagen Wrought? A Preliminary Assessment of the Copenhagen Accord.* Available at <http://belfercenter.ksg.harvard.edu/analysis/stavins/?p=464>. Accessed on April 14, 2010.

Stern, Nicholas. (2006). *The Economics of Climate Change: The Stern Review.* London: HM Treasury.

Sternberg, T. (2012). "Chinese Drought, Bread and the Arab Spring." *Applied Geography*, Vol. 34.

Sterrett, Charlotte. (2011). *Oxfam Research Reports: Review of Climate Change Adaptation Practices in South Asia, Climate Concern.* Melbourne, Australia, November 16.

Stojanov, Robert., Kelman, Ilan., Procházka, David., Němec, Daniel. and Duží, Barbora. (2017a). "Climate Change and Migration in Maldives." *Georgetown Journal of International Affairs* (requested paper), Vol. 183, No. 4.

Stojanov, Robert., Duží, Barbora., Němec, Daniel. and Procházka, David. (2017b). *Slow Onset Climate Change Impacts in Maldives and Population Movement from Islanders' Perspective.* KNOMAD Working Paper 20.

Stojanov, Robert., Kelman, Ilan., Ullah, Akm Ahsan., Duží, Barbora., Procházka, David., Kavanová, Klára., and Utová, Bla. (2016). "Local Expert Perceptions of Migration as a Climate Change Adaptation in Bangladesh." *Sustainability*, Vol. 8, No. 12, p. 1223. doi: 10.3390/su8121223

Stubbs, Richard. (2000). "Singing on to Liberalization: AFTA and the Politics of Regional Economic Cooperation." *The Pacific Review*, Vol. 13, No. 2.

Sturcke, James. (2005). "India Floods Death Toll Nears 700." *The Guardian*, July 29.

Suberi, Bhagat., Tiwari, Krishna R., Gurung, D.B., Bajracharya, Roshan M. and Sitaula, Bishal K. (2018). "People's Perception of Climate Change Impacts and Their Adaptation Practices in Khotokha Valley, Wangdue, Bhutan." *Indian Journal of Traditional Knowledge*, Vol. 17, No. 1.

Suja, Maria. (2017). "Fish and Farming." *The Maldives Journal*, December 14.

The Sun. (2018). "Stifling Temperatures in Pakistan Have Left People Collapsing in the Streets and Begging for Water as a 45C Heatwave Is Said to Have Killed at Least 65 People." May 22.

Sunday Observer. (2013). "Climate Change Adaptation, a Must-Minister Susil Premajayantha." February 22.

SUN Movement. (2014). Annual Progress Report SUN Movement Compendium of Country Profiles, The SUN Movement Secretariat, September.

Susskind, L. (1994). *Environmental Diplomacy: Negotiating More Effective Global Agreements.* Oxford: Oxford University Press.

Suter, K. (1999). "The Club of Rome: The Global Conscience." *Contemporary Review*, Vol. 275, No. 1602.

Swain, Ashok. (2002). "Environmental Cooperation in South Asia." In: Conca, K. and Dabelko, G.D. (eds.) *Environmental Peacekeeping.* London: Woodrow Wilson Center Press and Johns Hopkins University Press.

Swaminathan, M.S. (2008). "For an Action Plan for Bihar." *The Hindu*, September 5.

Swanstrom, Niklas. (2002). Regional Cooperation and Conflict Management: Lesson from the Pacific Rim. Uppsala, Sweden: PhD Thesis of Uppsala University.

Sweet, W.V., Kopp, R.E., Weaver, C.P., Obeysekera, J., Horton, R.M., Thieler, E.R. and Zervas, C. (2017). *Global and Regional Sea Level Rise Scenarios for the United States.* NOAA Tech. Rep. NOS CO-OPS 083. National Oceanic and Atmospheric Administration, National Ocean Service, Silver Spring, MD.

Syed, Fasahat H. (1999). "The Concept of Regional Cooperation among Indian Ocean Countries." In: Syed, Fasahat H. (ed.) *Regional Cooperation Among Indian Ocean Countries.* Islamabad: Asia Printers.

Synthesis Report from Climate Change, Global Risks, Challenges & Decisions, Copenhagen (March 10–12, 2009).

Syvitksi, James P.M., Kettner, Albert J., Overeem, Irina., Hutton, Eric W.H., Hannon, Mark T., Brakenridge, G. Robert., Day, John., Vörösmarty, Charles., Saito, Yoshiki., Giosan, Liviu. and Nicholls, Robert J. (2009). "Sinking Delta Due to Human Activities." In: *Nature Geoscience.* New York: Macmillan Publishers, p. 2.

Talley, W.K. and Ng, M. (2017). "Hinterland Transport Chains: Determinant Effects on Chain Choice." *International Journal of Production Economics*, Vol. 185.

Tallmeister, Julia. (2013). "Is Immigration a Threat to Security?" *E-International Relations Students*, August 24.

Tanter, Raymond. (1969). "Review: A Working Functionalism?" *The Journal of Conflict Resolution*, Vol. 13, No. 3, September.

Taylor, L. (2012). *The Nutrition Agenda in Bangladesh: Too Massive to Handle.* Sussex, UK: Institute of Development Studies.

The Telegraph. (2015). "India Heatwave: Death Toll Passes 2,500 as Victim Families Fight for Compensation." June 3.

Tesfaye, K., Zaidi, P., Gbegbelegbe, S., Boeber, C., Getaneh, F., Seetharam, K., Erenstein, Olaf. and Stirling, Clare. (2017). "Climate Change Impacts and Potential Benefits of Heat-Tolerant Maize in South Asia." *Theoretical and Applied Climatology*, Vol. 130.

Text of Agenda 21 Was Revealed at the United Nations Conference on Environment and Development (UNCED) or Earth Summit, held in Rio de Janeiro in 1992, 20 Years after the 1st World Environmental Summit. 178 Governments Voted to Adopt the Programme.

Thapa, Bishal. (2013). *Thimphu Statement on Climate Change: A Mere Rhetoric.* Kathmandu: South Asia Watch on Trade, Economics and Environment (SAWTEE).

Thimphu Statement on Climate Change, (Adopted by Heads of State or Government, Sixteenth SAARC Summit) Thimphu, 28–29 April 2010.

Thinley, Karma. (2009). *A Country Report: Higher Education in the Kingdom of Bhutan-Cherishing Dreams and Confronting Challenges.* Thimphu: Tertiary Education Division Department of Adult and Higher Education Ministry of Education Royal Government of Bhutan.

Thinley, Sangay. and Chopel, Dendup. (2012). "The Natural Order: People's Faith and Environment Management in Bhutan." In: Delinic, Tomislav. and Pandey, Nishchal N. (eds.) *Regional Environmental Issues: Water and Disaster Management.* Kathmandu: Nepal: Centre for South Asian Studies.

Thompson, Alexander. (2016). "The Future of the Financial Mechanism: Analysis and Proposals." In: Stavins, Robert N. and Stowe, Robert C. (eds.) *The Paris Agreement and Beyond: International Climate Change Policy Post-2020.* Cambridge: Harvard Project on Climate Agreements.

Thompson, Paul M. and Sultana, Parvin. (1996). "Distribution and Social Impacts of Flood Control in Bangladesh." *The Geographical Journal*, Vol. 67, Part 1.

Tian, H., Ren, W., Tao, B., Sun, G., Chappelka, A., Wang, X., Pan, Shufen., Yang, Jia., Liu, Jiyuan., Felzer, Ben S., Melillo, Jerry M. and Reilly, John. (2016). "Climate Extremes and Ozone Pollution: A Growing Threat to China's Food Security." *Ecosystem Health and Sustainability*, Vol. 2, No. 1.

The Times of India. (2010). "Pakistan Floods: Death Toll Crosses 1, 300." August 1.

The Times of India. (2014). "India to Fence over Water to Stop Immigration from Bangladesh." August 24.

The Times of India. (2015a). "2.8 Million Nepalese in Need of Humanitarian Aid: United Nations." June 2.

The Times of India. (2015b). "Disaster Monitoring System for SAARC Soon." April 27.

The Times of India. (2015c). "Nepal Earthquake: 2,500 Dead, People Shell-Shocked." April 26.

Timilsina-Parajuli, Laxmi., Timilsina, Yajna. and Parajuli, Rajan. (2014). "Climate Change and Community Forestry in Nepal: Local People's Perception." *American Journal of Environmental Protection*, Vol. 2, No. 1.

Tirwa, Badan. (2008). "Managing Health Disaster." *Bhutan Observer Online: Thimphu*, January 19.

Tiwari, K.R., Balla, M.K., Pokharel, R.K. and Rayamajhi, S. (2012). Climate Change Impact, Adaptation Practices and Policy in Nepal Himalaya. This Is a Paper Submitted for Presentation at UNU-WIDER Conference on 'Climate Change and Development Policy', held in Helsinki on September 28–29, 2012. This is not a formal publication of UNU-WIDER and may reflect work-in-progress.

Tol, R.S.J. (2007). "The Double Trade-Off between Adaptation and Mitigation for Sea Level Rise: An Application of FUND." *Mitigation and Adaptation Strategies for Global Change*, Vol. 12.

Tolba, Mostafa. (ed.). (1988). *Evolving Environmental Perceptions: From Stockholm to Nairobi.* London: Butterworth.

Tornikoski, Sami. (2015). *LHI Special Report 2015.* World Wide Fund For Nature (WWF), WWF Bhutan Programme, Thimphu.

Trabacchi, Chiara. and Stadelmann, Martin. (2013). *Making Adaptation a Private Sector Business: Insights from the Pilot Program for Climate Resilience in Nepal.* South Asia: Climate Policy Initiative, December.

Tshering, Doley. (2007). "Emerging Environmental Issues in Bhutan." *South Asian Journal*, October–December.

Tussie, Diana. (2006). "Introduction." In: Tussie, Diana. (ed.) *The Environment and International Trade Negotiations Developing Country Stakes.* London: Macmillan Press Ltd.

The Twelfth Summit Meeting of the South Asian Association for Regional Cooperation (SAARC). (January 2004). Pakistan: Islamabad, January 4–6.

Twilley, R.R., Barron, E.J., Gholz, H.L., Harwell, M.A., Miller, R.L., Reed, D.J., Rose, J.B., Siemann, E.H., Wetzel, R.G. and Zimmerman, R.J. (2001). *Confronting Climate Change in the Gulf Coast Region: Prospects for Sustaining Our Ecological Heritage.* Cambridge, MA: Union of Concerned Scientists.

Ullah, Wahid., Nihei, Takaaki., Nafees, Muhammad., Zaman, Rahman. and Ali, Muhammad. (2018). "Understanding Climate Change Vulnerability, Adaptation and Risk Perceptions at Household Level in Khyber Pakhtunkhwa." *Pakistan, International Journal of Climate Change Strategies and Management*, Vol. 10, No. 3.

Umar, Ghulam. (1992). *SAARC: An Analytical Survey.* New Delhi: Renaissance Publishing House.

UN. (2005). Available at <www.UN.org>. Accessed on February 14, 2020, cited in Karthikeyan, T.C. (2007). *Environmental Security of Maldives.* M. Phil Thesis, Jawaharlal Nehru University: New Delhi.

UN. (2007). *The Millennium Development Goals Report.* New York: United Nations Department of Economic and Social Affairs, UN.

UNAMA. (2016). *Water Rights: An Assessment of Afghanistan's Legal Framework Governing Water for Agriculture.* United Nations Assistance Mission in Afghanistan, Rule of Law Unit.

UN Department for Economic and Social Affairs (UNDESA). (2015). *World Population Prospects: 2015 Revision.* New York: United Nations Department of Economic and Social Affairs, Population Division.

Underdal, Arild. (2001). "One Question, Two Answers." In: Miles, Edward L. and Underdal, Arild. (eds.) *Explaining Environmental Regime Effectiveness: Confronting Theory with Evidence.* Cambridge MA: The MIT Press.

UNDP. (2004). *A Global Report Reducing Disaster Risk: A Challenge for Development.* New York: United Nations Development Programme.

UNDP (United Nations Development Programme). (2012). Technical Review and Social Impact Assessment. Reducing Climate Change-Induced Risks and Vulnerabilities from Glacial Lake Outburst Floods in the Punakha, Wangdue and Chamkhar Valleys.

UNDP. (2015). *The Millennium Development Goals Report 2015.* United Nations, New York. Available at <www.un.org/millenniumgoals/2015_MDG_Report/pdf/MDG%202015%20rev%20%28July%201%29.pdf>. Accessed on July 6, 2014.

UNDP. (2016). *Pakistan's New Poverty Index Reveals That 4 Out of 10 Pakistanis Live in Multidimensional Poverty*, Jun 20. Available at <www.pk.undp.org/content/pakistan/en/home/presscenter/pressreleases/2016/06/20/pakistan-s-new-poverty-index-reveals-that-4-out-of-10-pakistanis-live-in-multidimensional-poverty/>. Accessed on February 14, 2020.

UNDP. (2017a). "Afghanistan Launches US $ 71 Million Initiative to Prepare Rural Communities for Climate Change." *Climate Change Adaptation*, December 13.

UNDP. (2017b). "Protect Landscapes to Protect Everything': Bhutan Announces National Push for Climate Resiliency and Conservation." *Climate Change Adaptation*, November 11.

UNDP. (2019). "Green Climate Fund Approves $25.3 Million for Climate-Resilient Agriculture in Bhutan." *Climate Change Adaptation*, July 9.

UNEP. (2004). *Afghanistan Post-Conflict Environmental Assessment*. Kabul, Afghanistan: UNEP.

UNEP. (2007). Is Climate Change relevant for Afghanistan? Press Release, May 6.

UNEP. (2008). *Biodiversity and Wetlands Working Group Final Thematic Report*. Nairobi, February 11.

UNEP. (2009). *Recent Trends in Melting Glaciers, Tropospheric Temperatures over the Himalayas and Summer Monsoon Rainfall over India*. Nairobi: United Nations Environment Programme (UNEP).

UNEP. (2016). *The Adaptation Finance Gap Report 2016*. Nairobi, Kenya: United Nations Environment Programme (UNEP).

UNFCCC. (1992). *United Nations Framework Convention on Climate Change, United Nations*. Available at <https://unfccc.int/files/essential_background/background_publications_htmlpdf/application/pdf/conveng.pdf>. Accessed on February 14, 2020.

UNFCCC. (2002). *A Guide to the Climate Change Convention and its Kyoto Protocol*. United Nations Framework Convention on Climate Change. Bonn: Climate Change Secretariat.

UNFCCC. (2005). *Climate Change, Small Island Developing States*. Bonn, Germany: Issued by the Climate Change Secretariat (UNFCCC).

UNFCCC. (2007a). *Climate Change: Impacts, Vulnerabilities and Adaptation in Developing Countries*. Bonn: Climate Change Secretariat.

UNFCCC. (2007b). *The Nairobi Work Programme on Impacts, Vulnerability and Adaptation to Climate Change*. Bonn: Climate Change Secretariat.

UNFCCC. (2008). *Kyoto Protocol Reference Manual on Accounting of Emissions and Assigned Amount*. Bonn: Germany, Climate Change Secretariat (UNFCCC).

UNFCCC. (2010). Report of the Conference of the Parties on Its Sixteenth Session, held in Cancun from 29 November to 10 December 2010, FCCC/CP/2010/7/Add.1.

UNFCCC. (2011). The Least Developed Countries Reducing Vulnerability to Climate Change, Climate Variability and Extremes, Land Degradation and Loss of Biodiversity: Environmental and Developmental Challenges and Opportunities. Contribution to the Fourth UN Conference on Least Developed Countries, UNFCCC Secretariat, Bonn, May.

UNFCCC. (2016). Decision 1/CP.21, Adoption of the Paris Agreement (UN Doc. FCCC/CP/2015/10, Add.1, 29 January 2016).

U.N. Framework Convention on Climate Change (UNFCCC). (1992). S. Treaty Doc No. 102-38, 1771, U.N.T.S. 107, May 9.

UNICEF. (2015). Progress Report 2013–2015, Results for Children in Pakistan, Stop Stunting, UNICEF Pakistan.

UNICEF. (2017). "The State of The World's Children 2017." Available at <www.unicef.org/publications/files/SOWC_2017_ENG_WEB.pdf>. Accessed on February 14, 2020.

UNICEF Nutrition Program in Afghanistan. (October 2008–September 2009). Available at <www.usaid.gov/node/50831>. Accessed on February 14, 2020.

UNICEF, WHO (World Health Organization), and World Bank. (2018). *Levels and Trends in Child Malnutrition: UNICEF/WHO/World Bank Group Joint Child Malnutrition Estimates: Key Findings of the 2018 Edition.* Available at <www.who.int/nutgrowthdb/2018-jme-brochure.pdf?ua=1&ua=1>. Accessed on February 14, 2020.

United Nations (UN). (1992). *Framework Convention on Climate Change.* New York: United Nations.

United Nations (UN). (2009). *Risk and Poverty in Changing Climate: Global Assessment Report on Disaster Risk Reduction.* Geneva, Switzerland: United Nations.

United Nations (UN). (2013). *World Economic and Social Survey 2013: Sustainable Development Challenges.* New York: Department of Economic and Social Affairs, E/2013/50/Rev. 1S T/ESA /34 4.

United Nations Children's Fund (UNICEF). (2012). *Climate Change Adaptation and Disaster Risk Reduction in the Education Sector: Resource Manual.* New York: UNICEF Division of Communication, November.

United Nations Development Assistance Framework (UNDAF). (2012). Bhutan One Programme 2014–2018, Revised December 3.

United Nations Development Group. (2014). *Thematic Paper on MDG1 Eradicates Extreme Poverty and Hunger.* Washington, DC: United Nations Development Group.

United Nations Development Programme (UNDP). (2011). *Human Development Report 2011: Sustainability and Equity: A Better Future for All.* Basingstoke, UK: Palgrave Macmillan.

United Nations Development Programme Maldives (UNDP). (2006). *Developing a Disaster Risk Profile for Maldives.* Malé, Maldives: United Nations Development Programme.

United Nations Economic and Social Commission for Asia and the Pacific (ESCAP). (2012). *Regional Cooperation for Inclusive and Sustainable Development: South and South-West Asia Development Report 2012–2013.* New York: Routledge.

United Nations Environment Programme (UNEP). (1997). *Global Environment Outlook.* New York: Oxford University Press.

United Nations Environment Programme (UNEP). (2005). *Maldives Post-Tsunami Environmental Assessment.* Nairobi, Kenya: UNEP.

United Nations Environment Programme (UNEP). (2007). *Multilateral Environmental Agreement: Negotiator's Handbook.* Nairobi: University of Joensuu.

United Nations Environment Programme (UNEP). (2009). *South Asia Environment Outlook 2009.* Nairobi and New Delhi: United Nations Environment Programme.

United Nations Framework Convention on Climate Change (UNFCCC). (1992). *United Nations.* Available at <http://unfccc.int/files/essential_background/background_publications_htmlpdf/application/pdf/conveng.pdf>. Accessed on February 14, 2020.

United Nations Office for the Coordination of Humanitarian Affairs (OCHA). (2007). *Maldives: Coastal Flooding OCHA Situation Report No. 3.* New York.

United Nations Research Institute for Social Development (UNRISD). (2012). *Social Dimensions of Green Economy.* UNRISD Research and Policy Brief 12. Geneva.

United Nations Research Institute for Social Development (UNRISD). (2016). *Policy Innovations for Transformative Change: Implementing the 2030 Agenda for Sustainable Development.* Geneva.

UN News. (30 March 2010). "Human Rights Abuses Exacerbating Poverty in Afghani-stan." UN Report Available at <https://news.un.org/en/story/2010/03/334042-human-rights-abuses-exacerbating-poverty-afghanistan-un-report-finds> accessed on April 4, 2020, cited in Ahmadzai, Saifullah. "Non-Traditional Security Issues in Afghanistan." In: Nayak, Nihar (ed.) (2013). *Cooperative Security Framework for South Asia*. New Delhi: IDSA Pentagon Press.

United States Agency for International Development (USAID). (2010). *Asia-Pacific Regional Climate Change Adaptation Assessment*. Final Report: Findings and Recommendation.

United States Agency for International Development (USAID). (2018). *Bangladesh: Nutrition Profile*. Dhaka: United States Agency for International Development.

UO Oxfam. (2008). *Climate Change: As If Development Mattered*. Dhaka: Unnayan Onneshan (UO) and Oxfam GB Bangladesh (Mimeo).

Uprety, Batu Krishna. (2013). "It Has Taken Some Steps, But Nepal Has a Long Way to Go to Address Climate Change." *Republica*. Available at <www.myrepublica.com>. Accessed on February 14, 2020.

Ura, Karma. (2015). *The Experience of Gross National Happiness as Development Framework*. Manila, Philippines: Asian Development Bank.

Ura, Karma., Alkire, Sabina., Zangmo, Tshoki. and Wangdi, Karma. (2012). *An Extensive Analysis of GNH Index*. Thimphu: The Centre for Bhutan Studies.

USAID. (2007). *Review of Policies and Institutional Capacity for Early Warning and Disaster Management in Sri Lanka (August 10–16, 2006)*, Prepared for U.S. Agency for International Development, January.

USAID. (2017). *Food Assistance Fact Sheet Bangladesh*. Available at <www.usaid.gov/bangladesh/food-assistance>. Accessed on February 14, 2020.

US Department of State Archive. (2007). Asia-Pacific Partnership on Clean Develop-ment and Climate Second Ministerial Meeting Communique, New Delhi, October 15. Available at <https://2001-2009.state.gov/g/oes/rls/or/93803.htm>. Accessed on February 14, 2020.

U.S. Environmental Protection Agency. (2017). *Final Report on Review of Agency Actions that Potentially Burden the Safe*. Efficient Development of Domestic Energy Resources Under Executive Order 13783, October 25.

USGCRP. (2016). *The Impacts of Climate Change on Human Health in the United States: A Scientific Assessment*. Crimmins, A., Balbus, J., Gamble, J.L., Beard, C.B., Bell, J.E., Dodgen, D., Eisen, R.J., Fann, N., Hawkins, M.D., Herring, S.C., Jan-tarasami, L., Mills, D.M., Saha, S., Sarofim, M.C., Trtanj, J. and Ziska, L. (eds.). Washington, DC: U.S. Global Change Research Program.

USGCRP. (2018). *Impacts, Risks, and Adaptation in the United States: Fourth National Climate Assessment*. "Volume II." In: Reidmiller, D.R., Avery, C.W., East-erling, D.R., Kunkel, K.E., Lewis, K.L.M., Maycock, T.K. and Stewart B.C. (eds.) Washington, DC: U.S. Global Change Research Program.

US Treaty Document 102-38. (October 7, 1992). Resolution of Advice and Consent to Ratification Agreed to in the Senate by Division Vote.

Vahedifard, F., AghaKouchak, A. and Robinson, J.D. (2015). "Drought Threatens California's Levees." *Science*, Vol. 349, No. 6250.

Vahedifard, F., Robinson, J.D. and AghaKouchak, A. (2016). "Can Protracted Drought Undermine the Structural Integrity of California's Earthen Levees?" *Jour-nal of Geotechnical and Geoenvironmental Engineering*, Vol. 142, No. 6.

Van Epp, Timothy., Berwick, Stephen., Van Epp, Marissa. and Bashari, Mujtaba. (2017). *FAA 119 Biodiversity Assessment with Summary Assessment of Climate Vulnerability and Other Environmental Threats and Opportunities to Inform USAID/ Afghanistan Program Design (Biodiversity-Plus Assessment)*. Kathmandu: USAID.

Vasilyan, Syuzanna. (2006). *The Policy of Regional Cooperation in the South Caucasus*. Genk: Centro Argentino de Estudios Internacionales Área CEI y Países Bálticos.

Venkatesh, Shreeshan. (2018a). "By 2100, Heat Waves in South Asia Could Be Too Deadly to Survive: Study." *Down to Earth*, March 29.

Venkatesh, Shreeshan. (2018b). "Kerala Floods Reveal the Horror That Is Climate Change." *Down to Earth*, August 23.

Vespa, Matthew. (2002). "Note." In: *Climate Change 2001: Kyoto at Bonn and Marrakech*, 29 Ecology L.Q. 395.

Vhurumuku, Elliot., Nanayakkara, Laksiri., Petersson, Anders., Kumarasiri, R.H.W.A. and Rupasena, L.P. (2012). *Food Security in the Northern and Eastern Provinces of Sri Lanka: A Comprehensive Food Security Assessment Report Sri Lanka 2012*. United Nations World Food Programme, Colombo: Ministry of Economic Development and Hector Kobbekaduwa Agrarian Research and Training Institute.

Victor, David G. (2001). *The Collapse of the Kyoto Protocol and the Struggle to Slow Global Warming*. New York: Princeton University Press.

Vidal, John. (2006). "Nepal's Farmers on the Front Line of Global Climate Change." *The Guardian*, December 2.

Vidal, John. (2012). *Guardian Professional Networks*. Rio+20: Earth Summit Dawns with Stormier Clouds Than in 1992, June 19.

Vidal, John. (2013). "Climate Change: How a Warming World Is a Threat to Our Food Supplies." *The Guardian*, April 13.

Vidal, John. and Kelly, Annie. (2013). "Bhutan Set to Plough Lone Furrow as World's First Wholly Organic Country." *The Guardian*, February 11.

Vince, Gaia. (2019). "The Heat Is on over the Climate Crisis: Only Radical Measures Will Work." *The Guardian*, May 18.

Vinke, K., Martin, M.A., Adams, S., Baarsch, F., Bondeau, A., Coumou, D., Donner, R.V., Menon, A., Perette, M., Rehfeld, K., Robinson, A., Rocha, M., Schaeffer, M., Schwan, S., Serdeczny, O. and Svirejeva-Hopkins, A. (2017). "Climatic Risks and Impacts in South Asia: Extremes of Water Scarcity and Excess." *Regional Environmental Change*, Vol. 17, No. 6.

Vitousek, P.M., Lubchenco, J., Mooney, H.A. and Melillo, J. (1997). "Human Domination of Earth's Ecosystems." *Science*, Vol. 277.

Vivekananda, Janani. (2010). *Climate Change, Governance and Fragility: Rethinking Adaptation-Lessons from Nepal*. IFP Democratisation and Transitional Justice Cluster. Rue Belliard: International Alert, December.

Vivekananda, Janani. (2017). *Action on Climate and Security Risks: Review of Progress 2017*. Hague, Netherlands: Adelphi, the Center for Climate and Security.

von Grebmer, Klaus., Bernstein, Jill., Hammond, Laura., Patterson, Fraser., Sonntag, Andrea., Klaus, Lisa Maria., Fahlbusch, Jan., Towey, Olive., Foley, Connell., Gitter, Seth., Ekstrom, Kierstin. and Fritschel, Heidi. (2018). *2018 Global Hunger Index: Forced Migration and Hunger*. Bonn and Dublin: Welthungerhilfe and Concern Worldwide.

von Stechow, Christoph., Minx, Jan C., Riahi, Keywan., Jewell, Jessica., McCollum, David L., Callaghan, Max W., Bertram, Christoph., Luderer, Gunnar. and Baiocchi,

Giovanni. (2016). "2°C and SDGs: United they stand, divided they fall?" *Environment Research Letters*, Vol. 11, No. 3.

Waheed, Mirza. (2016). "India's Crackdown in Kashmir: Is This the World's First Mass Blinding?" *Guardian*, November 8.

Wahlen, Catherine Benson. (2018). 2019 Climate Index Finds No Country Performed Well Enough to Receive Top Ranking, USAID, December 13.

Wallerstein, Immanuel. (1974). *The Modern World System: Capitalist Agriculture and the Origins of the European World-Economy in the Sixteenth Century*. New York: Academic Press.

Wallerstein, Immanuel. (1979). "The Rise and Future Demise of the World Capitalist System." In: Wallerstein, Immanuel. (ed.) *The Capitalist World System*. Cambridge: Cambridge University Press.

Waltz, Kenneth N. (2001). *Man, the State, and War: A Theoretical Analysis*. New York: Columbia University Press.

Wang, J., Huang, J. and Rozelle, S. (2010). *Climate Change and China's Agricultural Sector: An Overview of Impacts, Adaptation and Mitigation*. Issue Brief No. 5. Geneva: International Centre for Trade and Sustainable Development (ICTSD).

Wang, J., Huang, J., Rozelle, S., Huang, Q. and Zhang, L. (2009). "Understanding the Water Crisis in Northern China: What Government and Farmers Are Doing." *International Journal of Water Resources Development*, Vol. 25, No. 1.

Wang, Sonam Wangyel., Lee, Woo-Kyun. and Son, YoWhan. (2017). "An Assessment of Climate Change Impacts and Adaptation in South Asian Agriculture." *International Journal of Climate Change Strategies and Management*, Vol. 9, No. 1.

Wang, Y.J., Ren, F.M. and Zhang, X.B. (2014). "Spatial and Temporal Variations of Regional High Temperature Events in China." *International Journal of Climatology*, Vol. 34, No. 10.

Ward, Barbara. and Dubos, Rene. (1972). *Only One Earth: The Care and Maintenance of a Small Planet*. London: Deutsch.

Warraich, Faizan Ali. (2017). "Pakistan Has World's 6th Largest Population." *The Nation*, January 6.

Water and Energy Commission Secretariat. (2002). *Executive Summary: Water Resources Strategy Nepal*. Kathmandu: Water and Energy Commission Secretariat Singha Durbar, Nepal.

Water Law of Afghanistan Article 1, Official Gazette No. 980. (2009). Kabul: Government of the Islamic Republic of Afghanistan (GIRA), April 26.

Water Resources Planning Organization (WARPO). (2001). *National Water Management Plan, Vol. 2: Main Report*. Ministry of Water Resources, Government of the People's Republic of Bangladesh.

Water Sector Strategy (WSS). (2008). *Afghanistan National Development Strategy*. Kabul: Government of Islamic Republic of Afghanistan.

Water Security Risk Index (WSRI). (2010). *Maplecroft*. Available at <http://maplecroft.com/about/news/water-security.html>. Accessed on February 14, 2020.

Watson, Robert T. and the Core Writing Team. (eds.). (2001). *Climate Change 2001: Synthesis Report*. Contribution of Working Groups I, II and III to the Third Assessment Report of the Intergovernmental Panel on Climate Change, University Press, Cambridge.

Watson, Robert T., Zinyowera, Marufu C. and Moss, Richard H. (eds.). (1998). *The Regional Impacts of Climate Change: An Assessment of Vulnerability*. A Special

Report of Working Group II of the Inter-Governmental Panel on Climate Change, Cambridge and New York: Cambridge University Press.

Webb, Edward L. and Gautam, Ambika Prasad. (2001). "Effects of Community Forest Management on the Structure and Diversity of a Successional Broadleaf Forest in Nepal." *The International Forestry Review*, Vol. 3, No. 2.

Weber, Cynthia. (2001). *International Relations Theory: A Critical Introduction*. New York: Routledge.

Weekly Compilation of Presidential Documents. (1997). "Remarks at the National Geographic Society." *Weekly Compilation of Presidential Documents*, Vol. 33, No. 43, October 22 and 27, pp. 1629–1634.

Weerakoon, W.M.W. and De Costa, W.A.J.M. (2009). Impacts of Climate Change on Rice Production in Sri Lanka. Climate Change & Its Impacts on Agriculture, Forestry and Water. Paper present at National Conference, Kandy.

Wehner, M.F., Arnold, J.R., Knutson, T., Kunkel, K.E. and LeGrande, A.N. (2017). "Droughts, Floods, and Wildfires." In: Wuebbles, D.J., Fahey, D.W., Hibbard, K.A., Dokken, D.J., Stewart, B.C. and Maycock, T.K. (eds.). *Climate Science Special Report: Fourth National Climate Assessment*. Volume 1. Washington, DC: U.S. Global Change Research Program.

Wei, K. and Chen, W. (2009). "Climatology and Trends of High Temperature Extremes across China in Summer." *Atmospheric and Oceanic Science Letters*, Vol. 2.

Weigall, David. (2002). *International Relations: A Concise Companion*. Great Britain: Arnold Publishers.

Weiner, Myron. (1978). *Sons of the Soil*. Princeton, NJ: Princeton University Press.

Welsh, H.A. and Willerton, J.P. (1997). "Regional Cooperation and the CIS: West European Lessons and Post-Soviet Experience." *International Politics*, Vol. 34, No. 1.

Wendt, Alexander. (1992a). "Levels of Analysis vs. Agents and Structures: Part III." *Review of International Studies*, Vol. 18, No. 2.

Wendt, Alexander. (1992b). "Anarchy Is What States Make of It: The Social Construction of Power Politics." *International Organization*, Vol. 46, No. 2.

Wendt, Alexander. (1994). "Collective Identity Formation and the International State." *American Political Science Review*, Vol. 88, No. 2.

Wendt, Alexander. (1998). "Constitution and Causation in International Relations." *Review of International Studies*, Vol. 24.

Wendt, Alexander. (1999). *Social Theory of International Politics*. Cambridge: Cambridge University Press.

Werz, Michael., and Hoffman, Max. (2015). "Climate Change, Migration, and the Demand for Greater Resources: Challenges and Responses." *SAIS Review of International Affairs*, Vol. 35, No. 1.

Westing, Arthur, H. (1980) *Warfare in a Fragile World: Military Impact on the Human Environment*. London: Taylor and Francis.

WFP. (2015). *Bhutan SPR 2015*. Rome: World Food Programme.

WFP. (2018). *Pakistan Country Strategic Plan (2018–2022)*, Rome: World Food Programme, February 2.

White, Stacey. (2015). *A Critical Disconnect: The Role of SAARC in Building the DRM Capacities of South Asian Countries*. Washington, DC: Brookings Institution.

WHO. (2015). *Climate and Health Country Profile-2015*. Bangladesh: World Health Organization, Geneva.

Whyte, Ian. (2008). *World without End? Environmental Disaster and the Collapse of Empire*. London: I B Travis and Co. Pvt Ltd.

Wickramasekara, Piyasiri. (2011). *Labour Migration in South Asia: A Review of Issues, Policies and Practices*. International Migration Paper No. 108. Geneva: International Labour Office.

Wiegandt, Ellen. (2001). "Climate Change, Equity, and International Negotiations." In: Luterbacher, Urs. and Sprinz, Detlef F. (eds.) *International Relations and Global Climate Change*. Cambridge: MIT Press.

Wijaya, A.S. (2014). "Climate Change, Global Warming and Global Inequity in Developed and Developing Countries (Analytical perspective, Issue, Problem and Solution)." *IOP Conference Series: Earth and Environmental Science*, January 19.

Wijenayake, Vositha. (2013). *South Asian Policymakers and Experts to Gear Up for New Climate Challenges*. Dhaka: Climate Action Network South Asia, December 13.

Wijenayake, Vositha. (2014). *Regional Collaboration to Address Climate Change*. Dhaka: Climate Action Network South Asia, November 13.

Wissenburg, M. (1998). *Green Liberalism: The Free and the Green Society*. London: UCL Press.

Wolfe, W., Ziska, L., Petzoldt, C., Seaman, A., Chase, L. and Hayhoe, K. (2007). "Projected Change in Climate Thresholds in the Northeastern U.S.: Implications for Crops, Pests, Livestock, and Farmers." *Mitigation and Adaptation Strategies for Global Change*, Vol. 13, Nos. 5–6.

Wolinsky, Yael. (1994). International Bargaining Under the Shadow of the Electorate. PhD Thesis of Department of Political Science. Chicago: University of Chicago.

Wong, P.P., Losada, I.J., Gattuso, J.P., Hinkel, J., Khattabi, A., McInnes, K.L., Saito, Y. and Sallenger, A. (2014). "Coastal Systems and Low-Lying Areas." In: *Climate Change 2014: Impacts, Adaptation, and Vulnerability*. Part A: *Global and Sectoral Aspects*, Contribution of Working Group II to the Fifth Assessment Report of the Intergovernmental Panel on Climate Change, Field, C.B., Barros, V.R., Dokken, D.J., Mach, K.J., Mastrandrea, M.D., Bilir, T.E., Chatterjee, M., Ebi, K.L., Estrada, Y.O., Genova, R.C., Girma, B., Kissel, E.S., Levy, A.N., MacCracken, S., Mastrandrea, P.R. and White. L.L. (eds.). Cambridge and New York: Cambridge University Press.

Wong, T.E., Bakker, A.M.R. and Keller, K. (2017). "Impacts of Antarctic Fast Dynamics on Sea-Level Projections and Coastal Flood Defense." *Climatic Change*, Vol. 144, No. 2.

Woodbridge, Michael. (2015). From MDGs to SDGs: What Are the Sustainable Development Goals? CLEI BRIEFING SHEET: Urban Issues, No. 01, Bonn, November.

Woodworth, P.L. (2005). "Have There Been Large Recent Sea Level Changes in the Maldives Islands?" *Global and Planetary Change*, Vol. 49.

Workshop Report. (2010). Government of Nepal Ministry of Environment National Adaptation Programme of Action (NAPA) to Climate Change. A Brainstorming Workshop on Establishing a National Climate Change Knowledge Management Platform in Nepal Monday, January 18.

Worland, Justin. (2018). "The U.S. Isn't the Only Major Country Not Meeting Its Climate Goals." *Time*, November 27.

World Bank. (2005). Afghanistan-Poverty, Vulnerability and Social Protection: An Initial Assessment. Washington, DC, Document of the World Bank, March 5.

World Bank. (2009a). *Why Is South Asia Vulnerable to Climate Change?* Available at <https://reliefweb.int/report/india/why-south-asia-vulnerable-climate-change>. Accessed on February 15, 2020.

World Bank. (2009b). *South Asia Climate Change Strategy*. A Draft Report, World Bank, South Asian Region.

World Bank. (2012a). *Bangladesh and Maldives Respond to Climate Change Impacts*. The World Bank, Washington DC, Press Release, December 7.

World Bank. (2012b). *Food Security in South Asia*, October 22. Available at <www.worldbank.org/en/news/feature/2012/10/22/food-security-south-asia>. Accessed on May 28, 2016.

World Bank. (2012c). *Turn Down the Heat: Why a 4 °C Warmer World Must be Avoided*, November 19. Available at <http://documents.worldbank.org/curated/en/865571468149107611/Turn-down-the-heat-why-a-4-C-warmer-world-must-be-avoided>. Accessed on February 15, 2020.

World Bank. (2013a). *Goal 1: Reducing Poverty and Hunger*. Available at <http://econ.worldbank.org/WBSITE/EXTERNAL/EXTDEC/EXTDECPROSPECTS/0,contentMDK:23112529~pagePK:64165401~piPK:64165026~theSitePK:476883,00.html>. Accessed on October 16, 2014.

The World Bank. (2014). Dhaka Government of India and World Bank Sign $153 Million Agreement for Odisha Disaster Recovery Project, July 14.

World Bank. (2014). *Turn Down the Heat: Confronting the New Climate Normal*. Washington, DC: World Bank.

World Bank. (2015). *South Asia Food and Nutrition Security Initiative (SAFANSI)*. Available at <www.worldbank.org/en/region/sar/brief/foodnutrition-security-initiative-safansi>. Accessed on January 1, 2016.

World Bank. (2018). *Building a Climate-Resilient South Asia*, April 20. Available at <www.worldbank.org/en/news/feature/2018/04/20/building-a-climate-resilient-south-asia>. Accessed on July 5, 2018.

World Bank Group. (2009b). *World Development Report 2009: Reshaping Economic Geography*. Washington: International Bank for Reconstruction and Development and World Bank.

World Commission for Environment and Development (WCED). (1987). *Our Common Future*. Oxford: Oxford University Press.

World Development Indicators. (2015). *International Bank for Reconstruction and Development*. Washington: The World Bank.

World Food Program (WFP). (2009). *Nepal Food Security Bulletin*. Kathmandu, Issue 25, July–October.

World Food Programme (WFP). (2005). *Tsunami: WFP Operation Overview*, December 1. Available at <www.wfp.org/stories/tsunami-wfp-operation-overview>. Accessed on January 10, 2013.

The World Food Programme (WFP). (2010). *Nepal-Current Issues and What the World Food Programme Is Doing*, June 17. Available at <www.wfp.org/countries/nepal>. Accessed on February 14, 2020.

World Food Programme (WFP). (2015). India Country, Strategic Plan: 2015–2018. New Delhi.

World Food Programme (WFP). (2016a). *Afghanistan-Current Issues and what the World Food Programme Is Doing*. Available at <www.wfp.org/countries/afghanistan>. Accessed on February 14, 2020.

The World Food Programme (WFP). (2016b). *Bhutan-Current Issues and What the World Food Programme Is Doing*. Available at <www.wfp.org/countries/bhutan>. Accessed on February 14, 2020.

World Food Programme (WFP). (2017). Food and Nutrition Assistance to Vulnerable Returnees and Refugees in Eastern Afghanistan and People Displaced by Conflict-Standard Project Report 2016, World Food Programme in Afghanistan, Islamic Republic of (AF), July 14.

World Food Programme (WFP). (2018). Pakistan Country Strategic Plan (2018–2022), February 2.

World Health Organization (WHO). (2008). *Protecting Health from Climate Change: World Health Day 2008.* WHO Library Cataloguing-in-Publication Data. Geneva: WHO Press.

World Health Organization (WHO). (2015). *Review of Climate Change and Health Activities in Sri Lanka.* New Delhi: Regional Office for South Asia.

World Health Organization (WHO). (2016). *Climate and Health Country Profile-2015, Pakistan.* Geneva, Switzerland: World Health Organization.

World Meteorological Organization (WMO). (2005). *World Climate Programme.* Geneva, Switzerland: World Health Organization.

World Meteorological Organization (WMO). (2016). *Provisional WMO Statement on the Status of the Global Climate in 2016.* Geneva, Switzerland: World Health Organization.

Worldwatch Institute. (2006). *Vital Signs 2006–2007.* New York: W.W. Norton.

Wright, Quincy. (1970). *The Study of International Relations.* Bombay: The Times of India Press.

WWF. (2005). *An Overview of Glaciers, Glacier Retreat, and Subsequent Impacts in Nepal, India and China.* Kathmandu: WWF Nepal.

WWF Pakistan. (2012). *Delta Wide Hazard Mapping: A Case Study of KetiBundar, KharoChann, and Jiwani, Draft Report.* Karachi: GIS Laboratory.

Wyett, Kelly. (2013). "Escaping a Rising Tide: Sea Level Rise and Migration in Kiribati." *Asia & the Pacific Policy Studies*, Vol. 1, No. 1, October.

Xu, J. (2009). *The Trend of Land Use and Land Cover Change and Its Impacts on Biodiversity in the Himalayas.* Lalitpur, Kathmandu: International Mountain Biodiversity Conference.

Yamane, Akiko. (2003). Rethinking Vulnerability to Climate Change in Sri Lanka. Paper Submitted for the 9th International conference on Sri Lanka Studies. Matara, Sri Lanka, Address for Correspondence Dept of Geography and Environmental Studies, Monash University, Melbourne, Australia, November 28–30.

Yangka, Dorji., Newman, Peter., Rauland, Vanessa. and Devereux, Peter. (2018). "Sustainability in an Emerging Nation: The Bhutan Case Study." *Sustainability*, Vol. 10, p. 1622.

Yasmin, Shakila. (2018). "Climate Financing Mechanism in Bangladesh: Does the Climate Change Trust Fund Play Its Role Properly?" *European Journal of Social Sciences Studies*, Vol. 3, No. 3.

Yassi, Annalee., Kjellström, Tord., Kok, Theo de. and Guidotti, Tee. (2001). *Basic Environmental Health.* Oxford: Oxford University Press.

Ye, L., Tang, H., Wu, W., Yang, P., Nelson, G., Mason-D'Croz, Daniel. and Palazzo, Amanda. (2013). *Chinese Food Security and Climate Change: Agriculture Futures.* Economics Discussion Papers, No, 2013–2. Kiel Institute for the World Economy.

Yeung, Jessie., Pokharel, Sugam. and Gupta, Swati. (2019). "More Than 100 Dead and 6 Million Affected by Flooding across South Asia." *CNN*, July 17.

Yohe, Gary. (2012). "Economics and Environmental Studies." *Review of Development Economics*, Vo. 16, No. 3.

Yohe, Gary., Jacobsen, Mark. and Gapotchenko, Taras. (1999). "Spanning Not-Implausible Futures to Assess Relative Vulnerability to Climate Change and Climate Variability." *Global Environmental Change*, Vol. 9, No. 3.

You, Qing-Long., Ren, Guo-Yu., Zhang, Yu-Qing., Ren, Yu-Yu., Sun, Xiu-Bao., Zhan, Yun-Jian., Shrestha, Arun Bhakta. and Krishnan, Raghavan. (2017). "An Overview of Studies of Observed Climate Change in the Hindu Kush Himalayan (HKH) Region." *Advances in Climate Change Research*, Vol. 8, No. 3.

Young, Oran R. (1989). "The Politics of International Regime Formation: Managing Natural Resources and the Environment." *International Organization*, Vol. 43, No. 3.

Younus, M. and Harvey, N. (2013). "Community-Based Flood Vulnerability and Adaptation Assessment: A Case Study from Bangladesh." *Journal of Environmental Assessment Policy and Management*, Vol. 15.

Yousaf, Kamran. (2016). "New Delhi Torpedoes 19th SAARC Summit." *Express Tribune*, September 28.

Yu, J. and Wang, Y. (2009). "Response of Tropical Cyclone Potential Intensity over the North Indian Ocean to Global Warming." *Geophysical Research Letters*, Vol. 36, p. L03709. doi:10.1029/2008GL036742.

Yu, Jia-Yuh. and Chiu, Ping-Gin. (2011). "Contrasting Various Metrics for Measuring Tropical Cyclone Activity." *Terrestrial, Atmospheric and Oceanic Sciences*, Vol. 23, No. 3.

Yusuf, Suhail. (2011) "Pakistan's First Climate Change Policy Ready", *Dawn*, May 26.

Zabarenko, Deborah. (2012). "Ocean's Acidic Shift May Be Fastest in 300 Million Years." *Reuters*, March 1.

Zafarullah, Habib. and Huque, Ahmed Shafiqul. (2015). Regional Synchrony in Sustainable Development: Harmonising Regulatory Policies in Environmental Conservation in South Asia. Paper Presented to the Panel on 'Regulation and Governance in Developing and Emerging Economies' at the International Research Society for Public Management Conference 2015: 'Shaping the Future – Re-Invention or Revolution?' University of Birmingham, March 30–April 1.

Zafarullah, Habib. and Huque, Ahmed Shafiqul. (2018). "Climate Change, Regulatory Policies and Regional Cooperation in South Asia." *Public Administration and Policy*, Vol. 21, No. 1.

Zaher, Mostapha. (2016). *Climate Change in Afghanistan What Does It Mean for Rural Livelihoods and Food Security?* Sweden: World Food Programme.

Zahid, Maida. and Rasul, Ghulam. (2010). "Rise in Summer Heat Index over Pakistan." *Pakistan Journal of Meteorology*, Vol. 6, No. 12.

Zahid, Maida. and Rasul, Ghulam. (2012). "Changing Trends of Thermal Extremes in Pakistan." *Climatic Change*, Vol. 113, No. 3.

Zahir, Hussein., Asis, Mohamed., Rasheed, Ahmed., Musthafa, Zahid Mohamed., Latheef, Aishath Thimna. and Mohamed, Ibrahim. (2016). *Second National Communication of Maldives to the United Nations Framework Convention on Climate Change*. Ministry of Environment and Energy. Republic of Maldives: Malé.

Zhai, J. (2010). "Spatial Variation and Trends in PDSI and SPI Indices and Their Relation to Streamflow in Ten Large Regions of China." *Journal of Climate*, Vol. 23.

Zhai, P.M. and Zou, X.K. (2005). "Changes of Temperature and Precipitation and Their Effects on Drought in China during 1951–2003." *Advances in Climate Change Research*, Vol. 1.

Zhang, J. and Wang, G. (2007). *Impact Research of Climate Change on Hydrology and Water Resources*. Beijing: Science Publishing House.

Zhang, Z., Duan, Z., Chen, Z., Xu, P. and Li, G. (2010). "Food Security in China: The Past, Present and Future." *Plant Omics Journal*, Vol. 3, No. 6.

Zhao, Y., Sultan, B., Vautard, R., Braconnot, P., Wang, H.J. and Ducharne, A. (2016). "Potential Escalation of Heat-Related Working Costs with Climate and Socioeconomic Changes in China." *Proceedings of the National Academy of Sciences*, Vol. 113, No. 17, USA, April 26.

Zhidong, Li. (2016). "Climate Change Measures in China after Paris Agreement." *IEEJ Energy Journal Special*, Issue, June.

Zia, M.S., Mahmood, T., Baig, M.B. and Aslam, M. (2004). "Land and Environmental Degradation and Its Amelioration for Sustainable Agriculture in Pakistan." *Quarterly Science Vision*, Vol. 9, Nos. 1–2.

Zillman, John W. (1997). *The IPCC: A View from the Inside*. Melbourne: Australian APEC Study Centre.

Zubair, L., Hansen, J., Chandimala, J., Siraj, M.R.A., Siriwardhana, M., Ariyaratne, K., Bandara, I., Bulathsinghala, H., Abeyratne, T. and Samuel, T.D.M.A. (2005). *Current Climate and Climate Change Assessments for Coconut and Tea Plantations in Sri Lanka*. IRI, FECT, NRMS and UoP Contribution to AS12 Project Report to be Submitted to START, Washington, DC, USA.

Index

Page numbers in bold indicate a table on the corresponding page.